Accelerator Radiation Physics for Personnel and Environmental Protection

T0175713

J. Donald Cossairt
Matthew Quinn

Fermi National Accelerator Laboratory

CRC Press
Taylor & Francis Group
Boca Raton London New York

CRC Press is an imprint of the
Taylor & Francis Group, an **informa** business

CRC Press
Taylor & Francis Group
6000 Broken Sound Parkway NW, Suite 300
Boca Raton, FL 33487-2742

First issued in paperback 2020

© 2019 by Taylor & Francis Group, LLC
CRC Press is an imprint of Taylor & Francis Group, an Informa business

No claim to original U.S. Government works

ISBN-13: 978-1-138-58901-8 (hbk)
ISBN-13: 978-0-367-77984-9 (pbk)

This book contains information obtained from authentic and highly regarded sources. Reasonable efforts have been made to publish reliable data and information, but the author and publisher cannot assume responsibility for the validity of all materials or the consequences of their use. The authors and publishers have attempted to trace the copyright holders of all material reproduced in this publication and apologize to copyright holders if permission to publish in this form has not been obtained. If any copyright material has not been acknowledged please write and let us know so we may rectify in any future reprint.

Except as permitted under U.S. Copyright Law, no part of this book may be reprinted, reproduced, transmitted, or utilized in any form by any electronic, mechanical, or other means, now known or hereafter invented, including photocopying, microfilming, and recording, or in any information storage or retrieval system, without written permission from the publishers.

For permission to photocopy or use material electronically from this work, please access www.copyright.com (http://www.copyright.com/) or contact the Copyright Clearance Center, Inc. (CCC), 222 Rosewood Drive, Danvers, MA 01923, 978-750-8400. CCC is a not-for-profit organization that provides licenses and registration for a variety of users. For organizations that have been granted a photocopy license by the CCC, a separate system of payment has been arranged.

Trademark Notice: Product or corporate names may be trademarks or registered trademarks, and are used only for identification and explanation without intent to infringe.

Library of Congress Cataloging-in-Publication Data

Names: Cossairt, J. Donald, 1948-author. | Quinn, Matthew (Radiation and laser safety officer), author.
Title: Accelerator radiation physics for personnel and environmental
protection / J. Donald Cossairt and Matthew Quinn.
Description: Boca Raton, FL : CRC Press, Taylor & Francis Group, [2019]
Identifiers: LCCN 2018060150| ISBN 9781138589018 (hbk : alk. paper) | ISBN 1138589012
(hbk : alk. paper) | ISBN 9780429491634 (ebook) | ISBN 0429491638 (ebook)
Subjects: LCSH: Radiation--Safety measures. | Particle accelerators--Shielding (Radiation)
Classification: LCC QC795.32.S3 C67 2019 | DDC 539.7/20289--dc23
LC record available at https://lccn.loc.gov/2018060150

Visit the Taylor & Francis Web site at
http://www.taylorandfrancis.com

and the CRC Press Web site at
http://www.crcpress.com

Accelerator Radiation Physics for Personnel and Environmental Protection

Contents

Preface

The advancement of particle accelerators is now well into its second century considering Röntgen's x-ray tube to be a particle accelerator. This scientific field, one of ongoing development, has achieved maturity, not stagnation. Accelerators now pervade nearly every facet of both modern scientific research and everyday life. They are utilized in virtually all branches of science ranging from the frontiers of physics (particle, nuclear, atomic, and condensed matter) to engineering, chemistry, biology, geology, and the environmental sciences. Important practical applications of accelerators are now found in industrial applications and even in agriculture. Accelerators are now ubiquitous in medicine. Among "radiological installations," particle accelerators are now the type most commonly encountered by members of the public.

This book is developed from a graduate course taught at the U.S. Particle Accelerator School (USPAS) beginning in 1993. This all began in 1992 when the founding director of USPAS, Melvin Month of Brookhaven National Laboratory, identified the need for the USPAS to offer a course on this topic. Mel then conspired with A. Lincoln Read, a prominent Fermilab physicist, to motivate Fermilab Director John Peoples, Jr., to "conscript" one of the authors (JDC), at that time serving as the head of Fermilab's Environment, Safety, and Health Section, to set time aside from considerable administrative duties to develop a course on this subject and deliver it at the upcoming USPAS Session at Florida State University in January 1993. Success was apparently achieved because the course has been conducted at 13 subsequent USPAS sessions. This success reflects the contributions of following people in addition to the authors of this book who have served as instructors in the USPAS course: Vernon Cupps, Nancy Grossman, Lincoln Read, Diane Reitzner, Sayed Rokni, Reginald Ronningen, Scott Schwahn, Kamran Vaziri, and Vaclav Vylet.

Consolidating this material into a book has been proposed by all of the directors of USPAS and pushed toward realization by one of them; William Barletta. The development of the material that comprises this book has been an ongoing effort attempting to keep up with developments in the field. The content has been selected to address the major elements of radiation physics issues that are encountered at accelerators of all particle types and energies. Some topics not commonly thought to be within the domain of the discipline of health physics are included at an introductory level in support of the theme of the book: charged particle optics, synchrotron radiation "light" sources, meteorology, and hydrogeology. The goal of this work is to provide a solid general background in the subject. The hope is that this book provides a basis for further knowledge development. Some readers may be disappointed to find that the operation of modern Monte Carlo codes is not covered in specific detail. Due to the ever-changing improvements being made to these codes, it is believed that instruction in the details is best left to the sponsors of these important and highly effective tools.

Acknowledgments

As authors, in addition to the ongoing support of Fermilab directors, we are grateful for the enthusiastic support of the successive leaders of Fermilab's environment, safety, and health organization, now the Environment, Safety, Health, and Quality Section—William Griffing, Nancy Grossman, and Martha Michels. We are also appreciative of the outstanding support, encouragement, assistance with content, and even some coverage of our regular assignments by our professional colleagues in Fermilab's Radiation Physics organization— John Anderson, Jr., Nino Chelidze, Kathy Graden, William Higgins, Susan McGimpsey, Diane Reitzner, Wayne Schmitt, Madelyn Schoell, Kamran Vaziri, and Michael Vincent.

We acknowledge the efforts of all who came before us in the development of radiation physics at Fermilab. Our special scientific mentors have been Robert D. Bent (Indiana University Bloomington), Ralph H. Thomas (Lawrence Berkeley National Laboratory), and Alexander J. Elwyn (Fermi National Accelerator Laboratory) for JDC and Ani Aprahamian (University of Notre Dame) and John C. Roeske (Loyola University Medical Center) for MQ. We furthermore acknowledge the work of our colleagues worldwide in this field.

We specifically dedicate this work to our spouses, Claudia Cossairt and Sara Quinn, and to our children.

J. Donald Cossairt and Matthew Quinn
Fermi National Accelerator Laboratory
Batavia, Illinois
December 2018

Authors

J. Donald Cossairt, PhD, is a Distinguished Scientist at the Fermi National Accelerator Laboratory (Fermilab) in Batavia, Illinois. He earned a bachelor of arts degree in physics and mathematics at Indiana Central College (now the University of Indianapolis) (1970) and master of science and doctor of philosophy degrees in nuclear physics at Indiana University Bloomington (1972, 1975). His career began with a postdoctoral appointment in nuclear physics research at the Texas A&M University Cyclotron Institute, and then he transitioned to radiation physics with his move to Fermilab in 1978. He is a member of the American Physical Society, a Fellow Member of the Health Physics Society, a Distinguished Emeritus Member of the National Council on Radiation Protection and Measurements, and a Certified Health Physicist. Cossairt has numerous publications in health physics, nuclear physics, and particle physics. He received a G. William Morgan Lectureship Award from HPS in 2011. He has been an instructor of the radiation physics, regulation, and management course for 14 sessions of the U.S. Particle Accelerator School and was co-academic dean of the Professional Development School of the Health Physics Society held at Oakland, California, in 2008.

Matthew Quinn, PhD, is Senior Radiation Safety Officer and Laser Safety Officer at the Fermi National Accelerator Laboratory (Fermilab) in Batavia, Illinois. He has worked on shielding assessments, operational radiation safety, radioanalytical measurements, and laser safety. Quinn is a three-time instructor of the radiation physics, regulation, and management course at the U.S. Particle Accelerator School, serves as the vice chair of the Department of Energy Energy Facility Contractors Group (EFCOG) Laser Safety Task Group, and is the president-elect of the Accelerator Section of the Health Physics Society. He earned a bachelor of science degree in physics at Loyola University Chicago (2000) and master of science and doctor of philosophy degrees in nuclear physics at the University of Notre Dame (2005, 2009), and was a postdoctoral researcher in the Department of Radiation Oncology at Loyola University Medical Center before joining Fermilab in 2010.

1

Basic Radiation Physics Concepts
and Units of Measurement

1.1 Introduction

Our study begins with a review of the standard terminology of radiation physics. The most important physical and radiological quantities and the system of units used to measure them are introduced. Due to its importance at most accelerators, the results of the special theory of relativity are reviewed. The energy loss by ionization along with the multiple Coulomb scattering of charged particles is also summarized.

1.2 Units of Measure and Physical Quantities

To develop an understanding of accelerator radiation physics, it is necessary to introduce the quantities of importance and the units by which they are measured. Over the years various systems of units have been employed. Presently in the United States there is a slow migration toward the use of the *Système Internationale* (SI) nearly universally employed elsewhere. This requires the practitioner to understand the interconnections of the units, both those "customary" in the United States and SI, due to the diversity found in both the scientific literature and government regulations. As always in technical work the wise practitioner conducts a careful unit analysis of all calculations to assure meaningful results.

Several quantities commonly used in physics are relevant to the subdiscipline of radiation physics. The unit of measure of the *energy* of particles is nearly always the *electron volt* (eV). This choice is usually much more convenient than the use of the Joule, the SI unit of energy: 1.0 eV is the kinetic energy of a particle carrying one electron's worth of electric charge (positive or negative) after acceleration through an electric potential of one volt. It is equal to 1.602×10^{-19} Joules. Multiples in common use at accelerators are the keV (10^3 eV), MeV (10^6 eV), GeV (10^9 eV), and TeV (10^{12} eV). Nearly always in discussions of particles at accelerators, the "energy" of an accelerated particle refers to its *kinetic energy*, a term further defined in Section 1.6.

The number of particles that transverse a unit area per unit time is the *flux density* ϕ, where

$$\phi = \frac{d^2n}{dAdt} \tag{1.1}$$

d^2n is the differential number of particles crossing surface area element dA during time dt. For radiation fields where the particles move in a multitude of directions rather than in a parallel or nearly parallel beam, ϕ is more generally understood to be the number crossing a sphere of cross-sectional area dA per unit time. The units of flux density are $\mathrm{cm^{-2}\,s^{-1}}$ or $\mathrm{m^{-2}\,s^{-1}}$ (SI) with other units of time (e.g., hours, minutes, days, and years) often encountered in published literature. The *fluence* Φ, the number of particles that cross such a surface area element during some time interval, $t_i < t < t_f$, is simply the time integral of the flux density,

$$\Phi = \int_{t_i}^{t_f} \phi(t)dt \tag{1.2}$$

Obviously, the units of fluence Φ are inverse area (e.g., $\mathrm{cm^{-2}}$, $\mathrm{m^{-2}}$).

When particles interact with matter, the *cross section* σ is an extremely important quantity. It represents the effective "size" of the atom or nucleus for the occurrence of some particular process such as particle interactions, nuclear transformations, etc. Consider a beam of particles of fluence Φ (particles $\mathrm{cm^{-2}}$) incident on a thin slab of absorber of thickness dx. The absorbing medium has N atoms $\mathrm{cm^{-3}}$. The number of incident particles that interact and are "lost" from the original fluence $-d\Phi$ is given by

$$-d\Phi = \sigma N \Phi dx \tag{1.3}$$

where σ is most commonly measured in units of $\mathrm{cm^2}$. N is determined from

$$N = \frac{\rho N_A}{A} \tag{1.4}$$

where ρ is the *density* of the material ($\mathrm{g\,cm^{-3}}$), N_A is *Avogadro's number* (6.022141×10^{23} atoms gram-mole^{-1}), and A is the atomic weight (i.e., the mass number). Cross sections are often tabulated in units of *barns*; 1.0 barn is 10^{-24} $\mathrm{cm^2}$ (10^{-28} $\mathrm{m^2}$). Submultiples such as the mb (10^{-3} barn, 10^{-27} $\mathrm{cm^2}$), μb (10^{-6} barn, 10^{-30} $\mathrm{cm^2}$), etc., are in common use. If only one physical process is present and if one starts with an initial fluence Φ_o, after some distance x (cm) of material a simple integration reveals the fluence $\Phi(x)$ to be

$$\Phi(x) = \Phi_o e^{-N\sigma x} \tag{1.5}$$

Exponential attenuation, a phenomenon ubiquitous in radiation physics, can be characterized by the *linear absorption coefficient* μ;

$$\mu = N\sigma (\mathrm{cm^{-1}}) \tag{1.6}$$

and it follows that the *attenuation length* $\lambda = 1/N\sigma$ (cm) is thus the reciprocal of μ.

Alternatively, the *mass attenuation length* $\lambda_m = \rho/N\sigma$ ($\mathrm{g\,cm^{-2}}$) is used where ρ is the density ($\mathrm{g\,cm^{-3}}$). Some authors use the symbol λ for λ_m; thus care with units is always needed. The term *interaction length* for this quantity is commonly used for high-energy hadrons (see Chapter 4).

1.3 Radiological Standards

Standards or limits on occupational and environmental exposure to ionizing radiation are now instituted worldwide. In general, individual nations, or subnational entities, incorporate guidance provided by international or national bodies into their laws and regulations. The main international body that develops radiological standards designed to promote occupational and environmental radiological health is the International Commission on Radiological Protection (ICRP). The International Atomic Energy Agency (IAEA) also has involvement with this topic. The separate, but somewhat similarly named, International Commission on Radiation Units and Measurements (ICRU) is more focused on units of measurements utilized in radiation protection. In the United States the National Council on Radiation Protection and Measurements (NCRP) carries out a scope of work analogous to that of both the ICRP and the ICRU. In the United States the Environmental Protection Agency (USEPA) is the primary federal agency responsible for establishing basic radiation protection requirements that are implemented in other governmental agencies. For particle accelerators, radiation protection requirements are set forth by other U.S. federal agencies, notably the U.S. Department of Energy (DOE) for accelerators within its national laboratory system, and by individual states. Currently the U.S. Nuclear Regulatory Commission (NRC) does not regulate the operation of particle accelerators but does under some circumstances regulate the use of radioactive materials at non-DOE facilities including some particle accelerators. Certain aspects of state regulations pertaining to accelerators are reflective of general NRC requirements. The regulation of accelerator facilities varies considerably among individual states and some local jurisdictions; the authority having jurisdiction should be consulted to obtain an accurate understanding of applicable regulatory requirements.

1.4 Units of Measure for Radiological Quantities

Absorbed dose, usually denoted D, is the energy absorbed per unit mass of material. It is a purely physical quantity, one that is directly measurable, at least in principle. The customary unit of absorbed dose is the *rad*, while the SI unit is the *Gray* (Gy). One rad is defined to be 100 ergs g^{-1}; thus 1.0 rad = 6.24151×10^{13} eV g^{-1}. One Gray (Gy) is defined as 1.0 J kg^{-1} and is equal to 100 rads. Thus 1.0 Gy = 6.24151×10^{15} eV g^{-1}. The concept of absorbed dose is applicable to any material and is also used for some forms of nonionizing radiation. It is commonly used to quantify both radiation exposures to human beings and potential radiation damage in the delivery of energy to materials and equipment components.

Radiation protection would be very simple if the deleterious effects of ionizing radiation were correlated in a very simple functional dependence, ideally a linear one, with absorbed dose. However, the results of a large body of science support the conclusion that these effects are also correlated with the types of particles and their energies. This correlation, a topic of ongoing scientific research, is known to be complex, and the current picture remains incomplete. To meet the pressing needs of assuring workplace and environmental radiation safety, approximations and special dosimetric quantities and units have been devised. The definition and usage of these quantities has evolved with time along with the subfield of accelerator radiation protection as the experts in the field represented by bodies such as the ICRP, the ICRU, and the NCRP have worked to improve the system of radiological protection. Thus, in reading archival scientific literature specifically on accelerator radiation protection, one finds that older papers sometimes utilize slightly different concepts and units than do more modern ones.

The ongoing development of this topic is rather complicated, and only summary results are discussed here. Using the publications of ICRP as approximate time markers, the older system of radiation protection was established by about 1973 by the ICRP (1973) and in this book is called the *1973 Radiation Protection System*, referenced from here on as the *1973 System*. The more current system was initially announced in 1991 by the ICRP (ICRP 1991) and in this book is called the *1990 Radiation Protection System*, referenced from here on as the *1990 System*. The refinement of the 1990 System by the ICRP has continued (ICRP 1996; ICRP 2007). In addition to external radiation fields of major interest here, these systems address internal radiation exposures, here germane only to a small part of the content of Chapters 7 and 8. While the 1990 System is fully adopted outside the United States, its use within the United States can be said in the year 2018 to be "piecemeal." The 1990 System now applies to radiation protection at U.S. DOE facilities (CFR 2007; DOE 2011a,b), while the 1973 System remains in use by the U.S. Nuclear Regulatory Commission, by the U.S. Occupational Health and Safety Administration, and in some instances by individual states. The USEPA uses a mixture of the two systems as of 2018. Familiarity with both systems is clearly needed to understand scientific publications and regulatory requirements.

1.4.1 Synopsis of the 1973 Radiation Protection System

In the 1973 System *dose equivalent* H_{equiv} is used to account for the fact that different particle types have biological effects that are enhanced, per given absorbed dose, over those due to the standard reference radiation taken to be 250 keV photons (Cember and Johnson 2009). This quantity has the same physical dimensions as absorbed dose. The customary unit of measure is the *rem*, while the SI unit is the *Sievert* (Sv) (1.0 Sv = 100 rem). The concept of dose equivalent is applied only to radiation exposures received by human beings. The dimensionless *quality factor*, usually denoted by Q, is used to reflect this enhancement by connecting H_{equiv} with D through

$$H_{equiv} = QD \tag{1.7}$$

Thus, H_{equiv} (rem) = QD (rads) or H_{equiv} (Sv) = QD (Gy). Q is dependent on both particle type and energy, and for any radiation field its value is an average over all components. It is formally defined to have a value of unity for the referenced 250 keV photons. In the 1973 System, the value of Q ranges from unity for photons, electrons, and muons up to a value as large as 20 for α-particles (^{4}He nuclei) of a few MeV in kinetic energy. For neutrons, in the 1973 System the value of Q ranges from 2 to approximately 11.

Q is defined to be a function of linear energy transfer (LET) L. LET is approximately equivalent to the stopping power, or rate of energy loss for charged particles (see Section 1.7) and is conventionally expressed in units of keV μm^{-1}. All ionizing radiation ultimately manifests itself through charged particles. Thus the parameter LET correlates reasonably well with localized radiation damage to the biological structures of interest. For the common situation where a spectrum of energies and a mixture of particle types are present, the value of Q for the complete radiation field is an average over the spectrum of LET present weighted by the absorbed dose as a function of LET, $D(L)$:

$$\langle Q \rangle = \frac{\int_0^\infty dL\, Q(L) D(L)}{\int_0^\infty dL\, D(L)} \tag{1.8}$$

The *dose equivalent per fluence conversion factor* for various particles P_{equiv}, as a function of energy, is a very important quantity quite useful in practical work. It provides the connection between the fluence and the dose equivalent:

$$H_{equiv} = P_{equiv}\Phi \tag{1.9}$$

Values of P_{equiv} as well as the relationship between Q and L consistent with the 1973 System for neutron radiation fields were published by the NCRP (1971).

1.4.2 Synopsis of the 1990 Radiation Protection System

The 1990 System revises the details, but not the character, of this system with motivation to more accurately incorporate results from research in radiobiology. For example, a more complicated model is used to better reflect the role of exposures to individual organs in human radiation risk. Thus, several quantities in addition to H_{equiv} have been defined as the metric of radiation detriment. These quantities are also applicable only to exposures to humans.

Since these quantities are encountered in the scientific literature, it is relevant to define them here. *Ambient dose equivalent* $H_{amb}(d)$ (J kg^{-1}, i.e., Sv) is the dose equivalent, measured at each point in a radiation field that would be produced in the corresponding expanded and aligned field in the *ICRU sphere* at depth d on the radius opposing the direction of the aligned field. The ICRU sphere, a mathematical construct developed as a surrogate for the human body by the ICRU, has a diameter of 0.3 m, a density of 1.0 g cm^{-3}, and a "tissue equivalent" elemental composition of 76.2% oxygen, 11.1% carbon, 10.1% hydrogen, and 2.6% nitrogen. "Expanded" means the radiation field encompasses the sphere, and "aligned" means that the measurement is independent of the angular distribution of the radiation field (Kaye and Laby 2008). *Personal dose equivalent* $H_{pers}(d)$ (J kg^{-1}, i.e., Sv) is the equivalent dose in soft tissue defined at depth d below a specified point in the body. For the primary concern of whole-body exposures, d is taken here to be 10 mm for these two quantities. *Equivalent dose* E_{equiv} (J kg^{-1}, i.e., Sv) is the absorbed dose in an organ or tissue multiplied by the relevant *radiation weighting factor* w_R that is qualitatively analogous to Q and discussed in more detail later. In the 1990 System the ICRP has retained the use of the term Q as a function of LET. In 2007 the ICRP chose to define the value of w_R to be unity for photons of all energies and has not chosen photons of specific reference energy such as 250 keV as was done in the 1973 System (ICRP 2007). The *effective dose* H_{eff} (J kg^{-1}, i.e., Sv) is the sum of all absorbed doses weighted by radiation weighting factors and by the correct organ weighting factors of the entire body. Where applicable it includes internal dose from uptakes of radioactive materials.

These quantities are of two types. *Operational quantities* such as $H_{amb}(d)$ and $H_{pers}(d)$ are, at least in principle, measurable and can be used to determine the properties of radiation fields to estimate and demonstrate compliance with specified standards. *Protection quantities* such as H_{eff} and E_{equiv} are used to determine conformance with numerical limits and action levels published in radiation protection standards. They are theoretical in nature and not directly measurable.

Given our primary interest here in external radiation fields generated by particle accelerators, H_{eff} is selected as the metric of radiation detriment in this book under the 1990 System. Thus, one is interested in the *effective dose per fluence* P_{eff} as a function of particle energy as used in

$$H_{eff} = P_{eff}\Phi \qquad (1.10)$$

This choice is confounded by an additional complexity. The use of the dosimetry quantities requires a selection of exposure geometry from standardized models. These models specify the orientation of the exposed person relative to the radiation source, a condition likely undefined in a typical workplace or environmental setting. Two of the models, called "ROT" and "ISO" in the scientific literature, appear to best match workplace conditions at accelerators. ROT (rotational) geometry is defined to be that where the human body is irradiated by a parallel beam of ionizing radiation from a direction orthogonal to the long axis of the body rotating at a uniform rate about its long axis (ICRP 1996). While for practical applications in radiation fields at accelerators, this "rotisserie" picture is preferable to the alternatives that involve a static orientation, it is clearly imperfect. In ISO (isotropic) geometry, the fluence per unit solid angle is independent of direction (ICRP 1996).

The importance of neutron radiation fields at accelerators will be justified in subsequent chapters. Using results available from the literature based on ROT (preferred) and ISO (when values for ROT were unavailable), Cossairt and Vaziri (2009) have provided values of P_{eff} for neutrons from "thermal" energies to the energies of the accelerated beams.

The 1990 System includes *radiation weighting factors* w_R used to connect absorbed dose to the protection quantity H_{eff} by replacing Q with w_R and H_{equiv} with H_{eff} in Equation 1.7. As with Q, a value of unity for w_R is tied to the radiobiological effects of low energy photons (≈ 200 keV) (ICRP 2007). In general, the values of w_R in the 1990 System are larger than those of Q for the same radiation field. However, the most recent ICRP guidance (ICRP 2007) has reduced some of this increase in w_R for some energy domains of neutrons. In contrast, for most neutron energies, the values of P_{eff} are less than those for P_{equiv}. Readers should note that people continue to refer to "quality factors" and "effective quality factors" and use the symbol Q. This apparent ambiguity includes contents of the ICRP publications referenced here. In this system the effective quality factor of a given radiation field is still determined with Equation 1.8. Furthermore, the use of terminology is not always precise as practitioners sometimes even refer to H_{eff} as "dose equivalent." It is also common to see the symbol H without subscripts, forcing the reader to precisely identify the system or quantity being used. Fortunately, the differences are small in many practical circumstances where averages over a large domain of particle energies are involved. Thus, it is often not important to distinguish between the two systems.

It should be noted that other references including the ICRP reports cited commonly use more esoteric, less intuitive symbols for the dosimetric quantities. These are as follows: H for dose equivalent, $H^*(d)$ for ambient dose equivalent, $H_p(d)$ for personal dose equivalent, E for effective dose, and $H_{T,R}$ for equivalent dose. In particular, the use of the letter E for effective dose has been avoided here because of the unacceptable potential for confusion of effective dose with standard usage of this symbol for the quantity *energy* in the physical sciences.

1.4.3 Values of Radiation Protection Quantities

Figure 1.1 gives the relationship between Q and LET in the 1973 and 1990 Systems.

Figure 1.2 provides values of Q in the 1973 System as a function of particle energy for a variety of particles and energies.

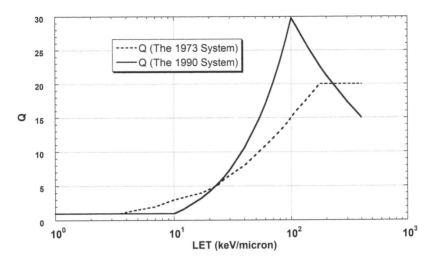

FIGURE 1.1
Quality factor Q of charged particles as specified in the 1973 System and in the 1990 System as a function of collision stopping power (LET). (From ICRP. 1996. *Conversion coefficients for use in radiological protection against external radiation.* ICRP Publication 74. Oxford, UK: Pergamon Press. http://www.icrp.org/. Used with permission.)

The results shown in Figure 1.2 are based on ionization due to the *primary particles* only. For particles subject to the strong (or nuclear) interaction, the inclusion of *secondary particles* produced in numbers that increase as a function of energy results in larger values of Q at higher energy. The distinction between primary and secondary particles is clarified in subsequent chapters. For example, under the 1973 System with secondary

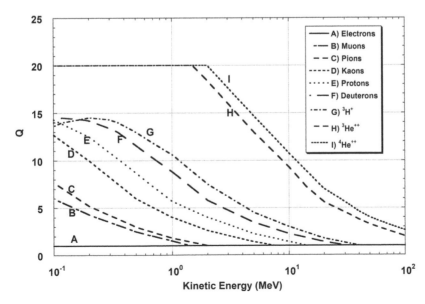

FIGURE 1.2
Quality factor Q of several types of charged particles and ions as a function of energy, as recommended by the ICRP according to the 1973 System. (From ICRP. 1973. *Data for protection against ionizing radiation from external sources: Supplement to ICRP Publication 15. ICRP Publication 21: Data for protection against ionizing radiation from external sources.* Washington, DC: Pergamon Press. http://www.icrp.org/. Used with permission.)

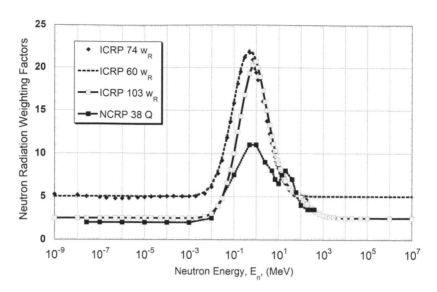

FIGURE 1.3

Radiation weighting factors for neutrons as a function of neutron kinetic energy. "NCRP 38 Q" values are those of the 1973 System and were values published by the NCRP (NCRP 1971). Those labeled "ICRP 60 w_R," "ICRP 74 w_R," and "ICRP 103 w_R" are found in (ICRP 1991, 1996, 2007), respectively, and are representative of the 1990 System. (Reprinted from Cossairt, J. D., and K. Vaziri. 2009. *Health Physics* 96:617–628. Rights to this figure are reserved by Fermi Research Alliance, LLC as manager and operator of the Fermi National Accelerator Laboratory.)

particles included, the value of Q for protons rises with energy to a value of 1.6 at 400 MeV and 2.2 at 2000 MeV (Patterson and Thomas 1973). Figure 1.3 gives the values of quality factor Q and radiation weighting factors w_R for neutrons as a function of energy in the two systems.

Dose per fluence factors P are often more useful than are radiation weighting or quality factors in practical work. Figure 1.4 provides dose equivalent per fluence values for a representative sample of charged particles over a wide range of energies according to the 1990 System.

Muons, as will be seen later, can be of prominent importance at high-energy accelerators. For muons, the dose equivalent per fluence P has been found to be about 40 fSv m² (4.0 × 10⁻⁴ μSv cm²) for 100 MeV < E_μ < 200 GeV (Stevenson 1983). This is equivalent to 2.5 × 10⁴ muons cm⁻² mrem⁻¹. At lower energies range-out of muons in the human body with consequential higher *energy deposition* (see Section 1.7) gives a dose per fluence factor of 260 fSv m² (equivalent to 3850 muons cm⁻² mrem⁻¹). Figure 1.5 provides dose per fluence factors for neutrons and photons.

In principle, values of dose equivalent per fluence factors can be calculated for any particle. For the higher energies, behavior for other particles that are subject to *strong interactions* (i.e., *nuclear interactions*) is similar to that of neutrons. Representative results have been documented by the group led by M. Pelliccioni (Ferrari et al. 1996; Ferrari et al. 1997a,b,c; Pelliccioni 2000). Especially for high-energy phenomena, many of these quantities have been tabulated by Fassò et al. (1990). As an example of this quantity for more "exotic" particles possibly of importance at future accelerators, Figure 1.6 gives values of P for muon neutrinos, ν_μ (Cossairt et al. 1997), that are in agreement with the calculations of Mokhov and Van Ginneken (1999).

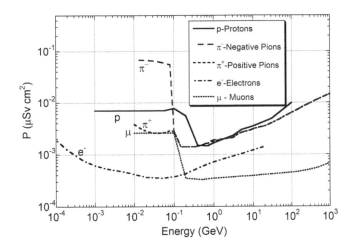

FIGURE 1.4
Dose equivalent per fluence for various charged particles *P* as a function of energy. The curve for muons is valid for both negative and positively charged muons. (Adapted from the tabulations of Fassò, A. et al. 1990. *Landolt-Börnstein numerical data and functional relationships in science and technology new series; Group I: Nuclear and particle physics. Volume II: Shielding against high energy radiation*, ed. H. Schopper. Berlin, Germany: Springer-Verlag.)

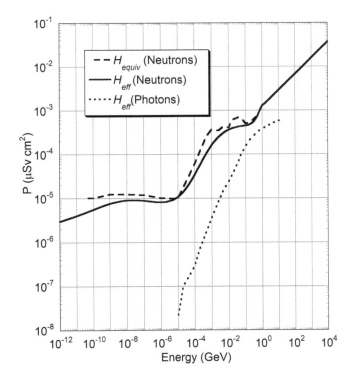

FIGURE 1.5
Dose per fluence factors for photons and neutrons *P* as a function of energy. (Photon results adapted from the tabulations of Fassò 1990. Neutron results adapted from Cossairt, J. D., and K. Vaziri. 2009. *Health Physics* 96:617–628. Rights to this figure are reserved by Fermi Research Alliance, LLC as manager and operator of the Fermi National Accelerator Laboratory.)

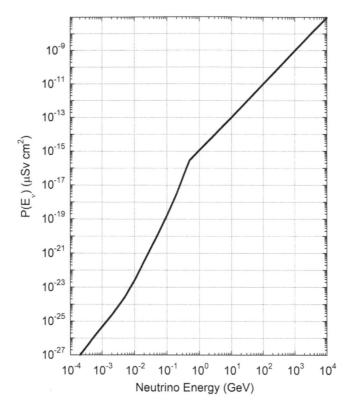

FIGURE 1.6
Dose equivalent per fluence factors for muon neutrinos P as a function of energy. (Reprinted from Cossairt, J. D. et al. 1997. *Health Physics* 73:894–898. Rights to this figure are reserved by Fermi Research Alliance, LLC as manager and operator of the Fermi National Accelerator Laboratory.)

For a radiation field containing a mixture of n different components (e.g., different particle types), one determines the dose equivalent H (or H_{eff} in the 1990 System) by

$$H = \sum_{i=1}^{n} \int_{E_{min}}^{E_{max}} dE P_i(E)\Phi_i(E) \tag{1.11}$$

where $\Phi_i(E)$ is the fluence of particles of type i with energy between E and dE, and $P_i(E)$ is the dose per fluence.

1.5 Physical Constants and Atomic and Nuclear Properties

Tables 1.1 and 1.2 give physical constants and atomic and nuclear properties as tabulated and regularly updated by the Particle Data Group based at the Lawrence Berkeley National Laboratory (Beringer et al. Particle Data Group 2012). The content of these tables is used as reference material throughout the rest of this text and in the solutions to the problems. Most of these quantities are discussed subsequently in more detail.

TABLE 1.1

Physical Constants

Quantity	Symbol, Equation	Value[a,b]
Speed of light	c	2.99792458×10^8 m s^{-1}
Planck constant	h	$6.62606957(29) \times 10^{-34}$ J s
Planck constant, reduced	$\hbar = h/2\pi$	$1.054571726(47) \times 10^{-34}$ J s
		$= 6.58211928(15) \times 10^{-22}$ MeV s
Electron charge	e	$1.602176565(35) \times 10^{-19}$ C
		$= 4.80320450(11) \times 10^{-10}$ esu
Electron mass	m_e	$0.510998928(15)$ MeV/c^2
		$= 9.10938291(40) \times 10^{-31}$ kg
Proton mass	m_p	$938.272046(21)$ MeV/c^2
		$= 1.672621777(74) \times 10^{-27}$ kg
		$= 1.007276466812(90)$ u
		$= 1836.15267245(75)$ m_e
Neutron mass	m_n	$939.565379(21)$ MeV/c^2
		$= 1.0086649160(04)$ u
Neutron mean-life	τ_n	$880.1(1.1)$ s
Deuteron mass	m_d	$1875.612859(41)$ MeV/c^2
Unified atomic mass unit (u)	(mass ^{12}C atom)/12	$931.494061(21)$ MeV/c^2
	$= (1$ g$)/N_A$	$= 1.660538921(73) \times 10^{-27}$ kg
Avogadro number	N_A	$6.02214129(27) \times 10^{23}$ mol^{-1}
(Electric) permittivity of free space	$\varepsilon_o = 1/\mu_o c^2$	$8.854187817 \times 10^{-12}$ Farad m^{-1}
(Magnetic) permittivity of free space	μ_o	$4\pi \times 10^{-7}$ Newton Ampere^{-2}
Classical radius of the electron	$r_e = e^2/4\pi\varepsilon_o m_e c^2$	$2.8179403267(27) \times 10^{-15}$ m
Fine structure constant	$\alpha = e^2/4\pi \varepsilon_o \hbar c$	$1/137.035999074(44)$
Boltzmann constant	k	$1.3806488(13) \times 10^{-23}$ J K^{-1}
		$= 8.6173324(78) \times 10^{-5}$ eV K^{-1}
	1 barn	10^{-28} m$^2 = 10^{-24}$ cm^2
	1 eV	$1.602176565(35) \times 10^{-19}$ J
	1 Gauss	10^{-4} Tesla
	1 erg	10^{-7} J
	1 fm	10^{-15} m
	1 atmosphere	760 torr $= 1.01325 \times 10^5$ N m^{-2} (Pa)
	0°C	273.15°K

Source: Adapted from Beringer, J. et al. (Particle Data Group). 2012. *Physical Review* D86:010001. This report is updated and periodically republished. Current tabulations are available online at http://pdg.lbl.gov/.

[a] The one-standard deviation uncertainties in the last digits are given in parentheses.

[b] N = Newton, A = Ampere, C = Coulomb, J = Joule, esu = electrostatic unit, u = atomic mass unit.

1.6 Summary of Relativistic Relationships

The results of the *special theory of relativity* are evident at nearly all accelerators. The *rest energy* W_o of a particle of rest mass m_o is given by

$$W_o = m_o c^2 \tag{1.12}$$

where c is the velocity of light. The *total energy* in free space W is given by

$$W = mc^2 = \frac{m_o c^2}{\sqrt{1-\beta^2}} = \gamma m_o c^2, \text{ with } \gamma = \frac{W}{W_o} = \frac{1}{\sqrt{1-\beta^2}} \tag{1.13}$$

TABLE 1.2

Atomic and Nuclear Properties of Materials

Material	Z	A[b]	Nuclear Total Cross Section[c] σ_T (barn)	Nuclear Inelastic Cross Section[c] σ_{in} (barn)	Nuclear Collision Length[d] λ_T (g cm^{-2})	Nuclear Interaction Length[d] λ_{in} (g cm^{-2})	Minimum Stopping Power[e] dE/dx (MeV cm^2 g^{-1})	Radiation Length X_o (g cm^{-2})	Density ρ (g cm^{-3}) or (g l^{-1}) for Gas
H_2[a]	1	1.00794	0.0387	0.033	42.8	52.0	4.103	63.04	0.071 (0.084)
He[a]	2	4.00260202	0.133	0.102	51.8	71.0	1.937	94.32	0.125 (0.166)
Li	3	6.941	0.211	0.157	52.2	71.3	1.639	82.78	0.534
Be	4	9.012182	0.268	0.199	55.3	77.8	1.595	65.19	1.848
C	6	12.0107	0.331	0.231	59.2	85.8	1.742	42.70	2.210[f]
N_2[a]	7	14.0067	0.379	0.265	61.1	89.7	1.825	37.99	0.807 (1.250)
O_2[a]	8	15.9994	0.420	0.292	61.3	90.2	1.801	34.24	1.141 (1.332)
Al	13	26.9815386	0.634	0.421	69.7	107.2	1.615	24.01	2.699
Si	14	28.0855	0.660	0.440	70.2	108.4	1.664	21.82	2.329
Ar[a]	18	39.948	0.868	0.566	78.8	126.2	1.519	19.55	1.396 (1.662)
Fe	26	55.845	1.120	0.703	81.7	132.1	1.451	13.84	7.874
Cu	29	63.546	1.232	0.782	81.2	137.3	1.403	12.86	8.960
W	74	183.84	2.767	1.65	110.4	191.9	1.145	6.76	19.30
Pb	82	207.2	2.960	1.77	114.1	199.6	1.122	6.37	11.350
U	92	238.02891	3.378	1.98	118.6	209.0	1.081	6.00	18.95
Air[a]					61.3	90.1	1.815	36.62	(1.205)
H_2O					58.5	83.3	1.992	36.08	1.00
Shielding concrete[g]					65.1	97.5	1.711	26.57	2.4
SiO_2 (fused quartz)					65.2	97.8	1.699	27.05	2.2
NaI					93.1	154.6	1.305	9.49	3.667
Polystyrene, scintillator (CH)					58.5	81.7	1.936	43.79	1.06

(Continued)

TABLE 1.2 (Continued)

Atomic and Nuclear Properties of Materials

Material	Z	A[b]	Nuclear Total Cross Section[c] σ_T (barn)	Nuclear Inelastic Cross Section[c] σ_{in} (barn)	Nuclear Collision Length[d] λ_T (g cm^{-2})	Nuclear Interaction Length[d] λ_{in} (g cm^{-2})	Minimum Stopping Power[e] dE/dx (MeV cm^2 g^{-1})	Radiation Length X_o (g cm^{-2})	Density ρ (g cm^{-3}) or (g l^{-1}) for Gas
Polyethylene (CH$_2$)					56.7	78.5	2.079	44.77	0.89–0.95[h]
CO$_2$[a]					60.7	88.9	1.819	36.20	(1.842)
Methane[a] (CH$_4$)					54.0	73.8	2.417	46.47	(0.667)
Ethane[a] (C$_2$H$_6$)					55.0	75.9	2.304	45.66	(1.263)
LiF					61.0	88.7	1.614	39.26	2.635

Source: Adapted from Beringer, J. et al. (Particle Data Group). 2012. *Physical Review* D86:010001. This report is updated and periodically republished. Current tabulations are available online at http://pdg.lbl.gov/ and from previous publications of the Particle Data Group.

a Parameters for materials that are gases at NTP (20°C and 1 atm) with values in parentheses or as cryogenic liquids at their 1.0 atmosphere boiling point if the *value* is given without parentheses.

b Averaged over naturally occurring isotopes.

c These are energy dependent. The values tabulated are approximate for the high-energy limit extracted from previous editions of the *Particle Data Group* publications. The inelastic cross section is obtained by subtracting the elastic cross section from the total cross section.

d These quantities are the mean free path between all collisions (λ_T) or inelastic interactions (λ_{in}) and are also energy dependent. The values quoted are for the high-energy limit as discussed in Chapter 4.

e This is the minimum value of the ionization stopping power for heavy particles. It is calculated specifically for pions, and the results are slightly different for other particles.

f The tabulated values are for pure graphite; industrial graphite may vary between 2.1 and 2.3 g cm^{-3}.

g This is for "standard" shielding blocks, typical composition of O$_2$ (52%), Si (32.5%), Ca (6%), Na (1.5%), Fe (2%), Al (4%), inclusive of reinforcing iron bars. The original reference cites a value of density of 2.3 g cm^{-3}, 2.4 g cm^{-3} is more typically found at particle accelerators.

h A wide range of densities is found for this material with density of 0.89 g cm^{-3} given in the cited reference.

where $\beta = v/c$, and v is the velocity of the particle in a given frame of reference. The relationship between the quantities β and γ is obvious. Similarly, the *relativistic mass m* of a particle moving at velocity β is given by

$$m = \frac{m_o}{\sqrt{1-\beta^2}} = \gamma m_o \tag{1.14}$$

The kinetic energy E is

$$E = W - W_o = (m - m_o)c^2 \tag{1.15}$$

and

$$\beta = \sqrt{1 - \left(\frac{W_0}{W}\right)^2} \tag{1.16}$$

The *momentum p* of a particle in terms of its relativistic mass m and velocity v is

$$p = mv = \gamma m_o \beta c = \frac{Wm_o}{m_o c^2}\left[\sqrt{1-\left(\frac{W_o}{W}\right)^2}\right]c = \frac{1}{c}\left[\sqrt{W^2 - W_o^2}\right] = \frac{1}{c}\sqrt{E(E+2W_o)} \tag{1.17}$$

so that at high energies ($E \gg W_o$), $p \approx E/c \approx W/c$, while at low energies ($E \ll W_o$) one has the familiar nonrelativistic $p^2 \approx 2(W_o/c^2)E = 2m_o E$.

It is usually most convenient to work in a system of units where energy is in units of eV, MeV, etc. Velocities are then expressed in units of the speed of light (β), momenta are expressed as energy divided by c (e.g., MeV/c, etc.), and masses are expressed as energy divided by c^2 (e.g., MeV/c², etc.). In these *energy units*, W is *numerically* equal to m. One thus avoids the explicit inclusion of numerical values for c or c^2, a major simplification in both performing calculations and readability.

The *decay length* at a given velocity of a particle with a finite *mean-life* (at rest) τ is given by $\gamma\beta c\tau$, where relativistic time dilation is accounted for by means of the factor γ. The product of the speed of light and the mean-life $c\tau$ is often tabulated. The decay length is the mean distance traveled by a particle in vacuum prior to its decay. This length must be distinguished from the *decay path*. The decay path represents a distance in space in which a given particle is permitted to decay with no or minimal competition from competing mechanisms such as scattering or absorption. Thus, the decay length is determined by the fundamental physics of the decay process, while the decay path is defined by the physical configuration of the accelerator components present.

1.7 Energy Loss by Ionization

Energy loss by ionization constitutes both a major method by which moving particles such as those produced in an accelerator lose energy and a key way in which radiation dose is delivered to living organisms, including humans. This is the stopping power due to ionization, the process in which a charged particle transfers its energy to atomic electrons in the absorbing medium. At very low energies, somewhat more complex physical mechanisms reflective of the underlying atomic and nuclear structure come into play. While the details of this topic are beyond the scope of this book, it is useful for the practitioner to be aware of their existence. The absorption of the energy of charged particles by the process of ionization is further characterized by a parameter called the *range R in material*. The range is the length of the path followed by the particle while it is coming to rest while losing all of its kinetic energy. Simplistically one might think that one could calculate the value of R by a numerical integration of the reciprocal of the stopping power; however, there are important effects at both very low energies and very high energies beyond the scope of this discussion that complicate matters.

For ions other than protons, the concept of *specific energy* is important. The specific energy is the kinetic energy of the ion divided by its mass number A, usually for this purpose taken as an integer, the sum of the number of protons and neutrons in the ion. It is approximately the kinetic energy of the "equivalent" proton. Northcliffe and Schilling (1970) have developed extensive tabulations of stopping power and range for a variety of ions and materials spanning the periodic table and specific energies up to 12 MeV/nucleon. Also, the Monte Carlo computer code SRIM (Stopping and Range of Ions in Matter) is currently easily obtained and may be used to generate similar tables as well as do simulations of protons or heavy charged ions interacting with elemental or compound materials (Biersack and Haggmark 1980; Ziegler et al. 1996). This work is extended to a much larger variety of materials and energies by the U.S. National Institute of Standards and Technology (NIST 2018a). Values of dE/dx and R are, respectively, shown for protons in Figures 1.7 and 1.8.

As a matter of technical correctness perhaps not directly relevant to this discussion in a detailed way, the plotted values of dE/dx are the *total* stopping powers including all effects,

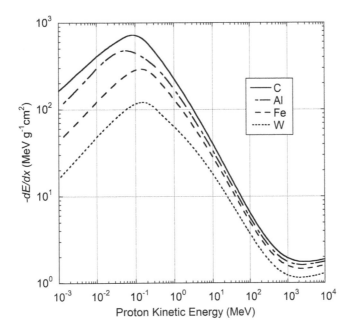

FIGURE 1.7
Values of stopping power $-dE/dx$ for energy protons in different materials as a function of proton kinetic energy. (Adapted from tabulations available from NIST. 2018a. Stopping power and range tables for protons. https://physics.nist.gov/PhysRefData/Star/Text/PSTAR.html.)

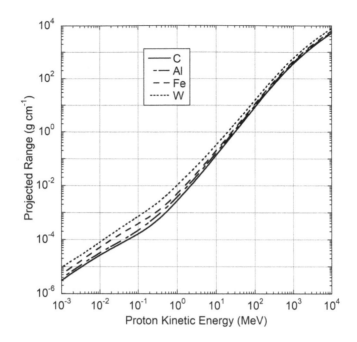

FIGURE 1.8
Values of range R for energy protons in different materials as a function of proton kinetic energy. (Adapted from tabulations available from NIST. 2018a. Stopping power and range tables for protons. https://physics.nist.gov/PhysRefData/Star/Text/PSTAR.html.)

atomic and nuclear. Likewise, the plotted values of R are the *projected* ranges representative of the deepest penetration of the material. The projected range is the average value of the depth to which a charged particle will penetrate in the course of slowing down to rest measured along the initial direction of the particle. The precise natures of the physical mechanisms invoked here are discussed in the cited references and elsewhere.

If one knows *stopping power* or rate of energy deposition dE/dx_p as a function of proton kinetic energy E_p in the material, usually expressed in units of MeV g^{-1} cm^2 of absorber material, one can obtain an *approximate* value of dE/dx_{ion} for ions of charge state z, in units of the charge of the electron, at the same numerical value of specific energy (kinetic energy per nucleon) by simply multiplying the proton value by z^2. Likewise, for an ion with a given specific energy, the *approximate* value of R will be that of the proton (g cm^{-2}) at that specific energy multiplied by the *mass* of the ion in atomic mass units (approximately equal to the number of nucleons present) divided by z^2 (Enge 1966). One can see that for highly charged ions, the value of R is much smaller than that for protons of the same specific energy.

For moderately relativistic heavy particles with rest masses much larger than that of the electron ($m_o \gg m_e$), the mean rate of energy loss, the ionization *stopping power* or rate of energy deposition, is given approximately by

$$-\frac{dE}{dx} = 4\pi N_A r_e^2 m_e c^2 z^2 \frac{Z}{A} \frac{1}{\beta^2}\left[\ln\left\{\frac{2m_e c^2 \gamma^2 \beta^2}{I}\right\} - \beta^2 - \frac{\delta}{2}\right](\text{MeV cm}^2\,\text{g}^{-1}) \qquad (1.18)$$

where N_A is Avogadro's number (atoms g-mole^{-1}), Z and A are the atomic number and weight (mass number) of the material traversed, z is the charge state of the projectile in units of electron charge, m_e and r_e are the rest mass and classical radius of the electron (see Table 1.1), and I is the ionization constant. This equation in its various forms is commonly referred to as the *Bethe Bloch formula*. For $Z > 1$, $I \approx 16Z^{0.9}$ eV. The argument of the logarithmic term in Equation 1.18 must be dimensionless, requiring that I and the rest energy of the electron $m_e c^2$ be in the same units. For diatomic hydrogen (H_2), $I = 19$ eV. β and γ are as defined in Section 1.6. δ is a small correction factor that can be approximated by $2\ln\gamma$. The initial minus sign reflects the fact that energy is being *lost* by the particle passing through the material (Hisaka et al. Particle Data Group 1992). Substituting the physical constants and taking I to be in eV,

$$-\frac{dE}{dx} = 0.3071 \, z^2 \, \frac{Z}{A} \frac{1}{\beta^2} \left[\ln \left\{ \frac{1.022 \times 10^6 \gamma^2 \beta^2}{I} \right\} - \beta^2 - \frac{\delta}{2} \right] (\text{MeV cm}^2 \, \text{g}^{-1}) \qquad (1.19)$$

In these units, the dependence on the absorbing material is rather weak given the fact that I appears only in the logarithmic term, and the ratio Z/A ranges between 0.4 and 0.5 over most of the periodic table for stable nuclides. Thus, for a given projectile charge z the value of the stopping power dE/dx is most strongly dependent on the quantity $\beta\gamma$. A broad minimum is found at an approximate value of $\gamma = 3.2$. At this value of γ, the particles are said to be *minimum ionizing*, and the corresponding minimum stopping powers are provided in Table 1.2. It can be stated that this "minimum" is in the energy domain where the phenomenon is most pure, devoid of atomic effects prominent at lower energies and below the threshold of effects prominent at higher energies. The 1992 edition of the biennial publication of the Particle Data Group (Hisaka et al. Particle Data Group 1992) was chosen as the source of these equations of stopping power and range because this version is a somewhat simpler but essentially equivalent formulation to that presented in subsequent Particle Data Group publications exemplified by more recent editions (e.g., Beringer et al. Particle Data Group 2012).

For charged particles much more massive than electrons, the trajectory through the material to first approximation is a straight line modified only by multiple Coulomb scattering (see Section 1.8), since the mass of the moving particle is so much larger than the mass of the atomic electrons.

At high energies, the situation is modified for particles such as protons that participate in the nuclear interaction. For these particles, as the kinetic energy of the particle increases, the absorption of the particles through strong (i.e., nuclear rather than atomic) interaction processes has a high probability of occurring prior to their depositing all of their energy by ionization. This is discussed further in Section 4.2.1. Figure 1.9 provides values of $\beta\gamma$ for protons, charged pions, and muons as a function of kinetic energy. Figures 1.10 and 1.11 provide stopping power and range values as a function of $\beta\gamma$ and momentum, respectively, for these same particles (Beringer et al. Particle Data Group 2012).

For a moving electron, the range is the sum of many divergent line segments through the material since its mass is identical to that of the atomic electrons encountered so that the individual angular deflections are generally much larger. As shown in Section 3.2.2, for electrons the loss of energy in matter due to the *radiation* of photons increases rapidly with electron kinetic energy and becomes much more important at relatively low energies than is the ionization stopping power or ionization range.

For muons (μ) the situation is rather unique. The muon rest energy is 105.66 MeV, its mean-life $\tau = 2.1970 \times 10^{-6}$ s, and the mean-life times the speed of light $c\tau = 658.65$ m (Beringer et al. Particle Data Group 2012). Due to their large rest mass compared to that of the electron

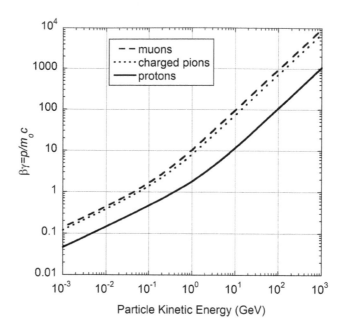

FIGURE 1.9
Value of the parameter $\beta\gamma$ as a function of particle kinetic energy for muons, pions, and protons.

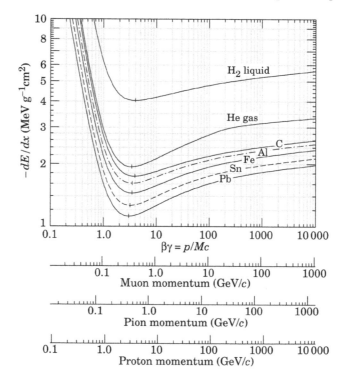

FIGURE 1.10
Values of stopping power $-dE/dx$ as a function of the parameter $\beta\gamma$ in different materials for muons, charged pions, and protons having different momenta. In this figure M denotes the rest energy of the indicated particle. (Reprinted with permission from Beringer, J. et al. [Particle Data Group] 2012. *Physical Review* D86:010001. Copyright 2012 by the American Physical Society.)

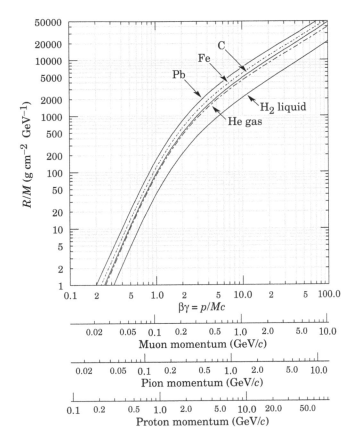

FIGURE 1.11
Values of ionization range R per particle rest energy (GeV) as a function of the parameter $\beta\gamma$ in different materials for muons, charged pions, and protons having different momenta. In this figure M denotes the rest energy of the indicated particle. (Reprinted with permission from Beringer, J. et al. [Particle Data Group] 2012. *Physical Review* D86:010001. Copyright 2012 by the American Physical Society.)

and the fact that these particles, to first order, do not participate in the strong (nuclear) interaction, muons tend to penetrate long distances in matter without being absorbed by other mechanisms. Muons, due to their heavier masses, are also far less susceptible to radiative effects than are the much lighter electrons. Thus, over a very large energy domain, the principal energy loss mechanism is that of ionization. This, as is shown later, makes the shielding of muons a matter of considerable importance at high-energy accelerators. The range-energy relation of muons is given in Figure 1.12.

At high energies ($E_\mu > 100$ GeV), the distribution of the ranges of individual muons about the mean range, called the *range straggling*, becomes important (Van Ginneken et al. 1987; Fassò et al. 1990). Also, above muon energies of several hundred GeV, radiative losses begin to dominate such that the stopping power dE/dx is given by

$$-\frac{dE}{dx} = a(E) + b(E)E \qquad (1.20)$$

where $a(E)$ is the collisional ionization energy loss (Beringer et al. Particle Data Group 2012). From Equation 1.19, in iron, $a(E)$ has an approximate value of 0.002 GeV cm^2 g^{-1}. And $b(E)$

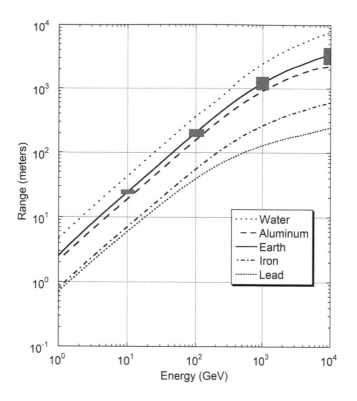

FIGURE 1.12

Range-energy curves for muons in various materials. On the curve labeled "Earth," the gray boxes are indicative of the approximate spread in the range due to range straggling at one standard deviation at the indicated muon energy. The density of "earth" was taken to be 2.0 g cm^{-2}. (Adapted from tabular data presented by Fassò, A. et al. 1990. *Landolt-Börnstein numerical data and functional relationships in science and technology new series; Group I: Nuclear and particle physics Volume II: Shielding against high energy radiation,* ed. H. Schopper. Berlin, Germany: Springer-Verlag.)

is the radiative coefficient for E in GeV. The latter parameter separated into contributions from the important physical mechanisms is plotted in Figure 1.13.

The *mean range R_μ* of a muon of kinetic energy E (GeV), is approximated by

$$R_\mu(E) = \frac{1}{b(E)} \ln\left[1 + \frac{b(E)}{a(E)} E\right] (\text{g cm}^{-2}) \qquad (1.21)$$

Muon range straggling (Van Ginneken et al. 1987; Fassò et al. 1990) chiefly results from the fact that above about 100 GeV, with increasing muon kinetic energy, electron-positron pair production, bremsstrahlung, and deep inelastic nuclear reactions become the dominant energy loss mechanisms. Although these processes have low probabilities, when they do occur they involve large energy losses. Tables 1.3 and 1.4 give fractional energy loss and comparisons of muon ranges at high energies for different physical mechanisms. Range straggling can be very important since shielding calculations based on using the mean range values can significantly underestimate the muon fluence that can penetrate a shield.

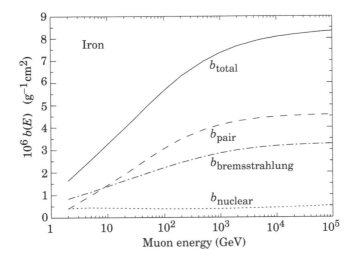

FIGURE 1.13
Contributions to the fractional energy loss by muons in iron due to e^+e^- pair production, bremsstrahlung, and photonuclear interactions. See Equation 1.20. (Reprinted with permission from Beringer, J. et al. [Particle Data Group]. 2012. *Physical Review* D86:010001. Copyright 2012 by the American Physical Society.)

TABLE 1.3

Fractional Energy Loss of Muons in Soil ($\rho = 2.0$ g cm^{-3}) due to Four Physical Mechanisms

Energy (GeV)	Ionization	Bremsstrahlung	Pair Production	Deep Inelastic Nuclear Scattering
10	0.972	0.037	8.8×10^{-4}	9.7×10^{-4}
100	0.888	0.086	0.020	0.0093
1000	0.580	0.193	0.168	0.055
10,000	0.167	0.335	0.388	0.110

Source: Adapted from Van Ginneken A. et al. 1987. Shielding calculations for multi-TeV hadron colliders. Fermi National Accelerator Laboratory: Fermilab Report FN-447. Batavia, Illinois; Fassò, A. et al. 1990. *Landolt-Börnstein numerical data and functional relationships in science and technology new series; Group I: Nuclear and particle physics Volume II: Shielding against high energy radiation*, ed. H. Schopper. Berlin, Germany: Springer-Verlag.

1.8 Multiple Coulomb Scattering

Multiple Coulomb scattering from nuclei is an important effect in the transport of charged particles through matter. A charged particle traversing a medium is deflected by many small-angle scattering events and only occasionally by ones involving large-angle scattering. These small-angle scattering events are largely due to Coulomb scattering from nuclei, hence the name of this phenomenon. This simplification is not quite completely correct for hadrons, since it ignores the contribution of strong interactions to multiple scattering. For present purposes, a Gaussian approximation adequately describes the distribution of

TABLE 1.4

Comparison of Muon Ranges (Meters) in Heavy Soil ($\rho = 2.24$ g cm^{-3}) at Selected Energies

			Mean Ranges from dE/dx in Heavy Soil (m)		
Energy (GeV)	Mean Range (m)	Standard Deviation (m)	All Processes	Coulomb Losses Only	Coulomb and Pair Production Losses
10	22.8	1.6	21.4	21.5	21.5
30	63.0	5.6	60.3	61.1	60.8
100	188	23	183	193	188
300	481	78	474	558	574
1000	1140	250	1140	1790	1390
3000	1970	550	2060	5170	2930
10,000	3080	890	3240	16,700	5340
20,000	3730	1070			

Source: Adapted from Van Ginneken A. et al. 1987. Shielding calculations for multi-TeV hadron colliders. Fermi National Accelerator Laboratory: Fermilab Report FN-447. Batavia, Illinois; Fassò, A. et al. 1990. *Landolt-Börnstein numerical data and functional relationships in science and technology new series; Group I: Nuclear and particle physics Volume II: Shielding against high energy radiation*, ed. H. Schopper. Berlin, Germany: Springer-Verlag.

deflection angles of the final trajectory compared with the incident trajectory for all charged particles. The distribution as a function of deflection angle θ is as follows:

$$f(\theta)d\theta = \left(\frac{d\theta}{\theta_0\sqrt{2\pi}}\right)\exp\left(-\frac{\theta^2}{2\theta_0^2}\right) \qquad (1.22)$$

The mean width of the projected angular distribution θ_o on a particular plane is approximated by

$$\theta_o = \frac{13.6z}{pc\beta}\sqrt{\frac{x}{X_o}}\left\{1+0.038\,\ln\left(\frac{x}{X_o}\right)\right\} \text{(radians)} \qquad (1.23)$$

where z is the charge of the projectile in units of the charge of the electron, p is the particle momentum in MeV/c, and x is the absorber thickness in the same units as the quantity X_o. X_o is a material-dependent parameter, to be discussed further in Section 3.2.2, called the *radiation length* (Hisaka et al. Particle Data Group. 1992). This description of multiple Coulomb scattering has been validated experimentally for particles having momenta up to 200 GeV/c (Shen et al. 1979). The best values of the radiation length are probably those of Tsai (1974), the values tabulated in Table 1.2. A compact, approximate formula for calculating the value of X_o as a function of atomic number Z and atomic weight A of the material medium is (Hisaka et al. Particle Data Group 1992)

$$X_o = \frac{716.4A}{Z(Z+1)\ln(287/\sqrt{Z})}(\text{g cm}^{-2}) \qquad (1.24)$$

PROBLEMS

1. Provide answers to the following:

   a. Express 1.0 kilowatt (1.0 kW) of beam power in GeV s^{-1}.

   b. To how many singly charged particles per second does 1.0 ampere of beam current correspond?

   c. Express an absorbed dose of 1.0 Gy in GeV kg^{-1} of energy deposition.

2. In the 1973 System, which has the higher quality factor, a 10 MeV (kinetic energy) α-particle or a 1.0 MeV neutron? Write down the quality factors for each particle. What is the quality factor of a 1.0 MeV neutron in the 1990 System?

3. Calculate the number of ^{12}C and ^{238}U atoms in a cubic centimeter of solid material.

4. Calculate the velocity and momenta of a 200 MeV electron, proton, iron ion, π^+, and μ^+. The 200 MeV is kinetic energy, and the answers should be expressed in units of the speed of light (velocity) and MeV/c (momenta). Iron ions have an isotope-averaged mass (rest energy) of 52,019 MeV ($A = 55.845 \times 931.494$ MeV/amu). The π^+ mass is 140 MeV, and the μ^+ mass = 106 MeV. Do the same calculation for 20 GeV protons, iron ions, and muons. It is suggested that these results be presented in tabular form. Make general comments on the velocity and momenta of the particles at the two energies. (The table may help you notice any algebraic errors that you may have made.)

5. Calculate the mass stopping power of a 20 MeV electron (ionization only) and a 200 MeV proton in ^{28}Si.

6. Calculate the fluence of minimum ionizing muons necessary to produce an effective dose of 1.0 mrem assuming a radiation weighting factor of unity and that *tissue* is equivalent to *water* for minimum ionizing muons. (Hint: Use Table 1.2.) Compare with the results given in Figure 1.4 for high energies.

2

General Considerations for
Accelerator Radiation Fields

2.1 Introduction

In this chapter general properties of the radiation fields at accelerators are discussed. For this discussion, the concept of particle yield into a given solid angle is introduced. Following that, a generalized theoretical approach to particle transport is introduced. The Monte Carlo technique is described and illustrated by simple examples. The manipulation of charged particles using electromagnetic fields is reviewed due to its importance in understanding the handling of the charged particle beams.

2.2 Primary Radiation Fields at Accelerators: General Considerations

Accelerated charged particles, except in the singular phenomenon of synchrotron radiation discussed in Chapter 3, do not produce radiation unless there is some interaction with matter. The charged particles directly accelerated, and otherwise manipulated by the electromagnetic fields within the accelerator, are referred to as the *primary particles* or *primary beam*. All other particles that are produced from this beam either result from interactions of these primary particles in matter or are due to synchrotron radiation and are referred to as *secondary particles* or *secondary beam*. Sometimes one finds references to *tertiary particles* or *tertiary beam* that result from interactions in matter of the secondary particles or their radioactive decay. Confusion can also arise when secondary or tertiary particles are collected into beams of their own and sometimes even accelerated. When this is done and the secondary or tertiary particles are employed at some location separated from the place where they were initially produced, they can obviously play the role of primary particles.

If one considers primary particles incident upon material such as a target, the *yield* Y of secondary particles is a crucial parameter. In this book, photons are considered to be "particles" for purposes of general discussions such as this one. For a given type of secondary particle, the yield is typically a function of primary particle type and energy, the angle of emission, and the secondary particle energy. It is defined according to Figure 2.1.

The coordinate system shown in Figure 2.1 is a two-dimensional projection of a more general spherical coordinate system shown in Figure 2.2.

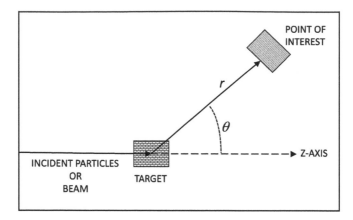

FIGURE 2.1
Conceptual interaction of incident beam with material (the "target") that produces radiation at the point of interest located at polar coordinates (r, θ).

Scattered reaction products are found at a "point of interest" located at radius r and polar angle θ relative to the direction of the incident particle conventionally taken to be along the positive z-axis. In general, particle *differential yields* are expressed in terms of particles per unit solid angle at the point of interest and are commonly normalized to the number of incident particles or to the beam current or total delivered charge. Such particle yields, dependent on both target material and thickness, are reported in terms of particle type, energy, and angular distribution. The rate of production of the desired reaction products and their energy spectra are generally a strong function of both θ and the incident particle energy E_o. Almost always, discussions of particle energy mean, more precisely, the kinetic energy (see Section 1.6). Usually there is no dependence on the azimuthal angle ψ measured in the xy-plane. However, there are exceptions. One is the situation in which the spins of the target nuclei and/or the incident particles are oriented along some chosen direction in a polarized beam or polarization experiment. Interactions of colliding beams involving spin-polarized particles have azimuthal dependencies. Secondary particles resulting from multipole emission/deexcitation processes from excited atomic or nuclear states will also have a dependence on azimuthal angle. Physical processes involving parity violation constitute other examples.

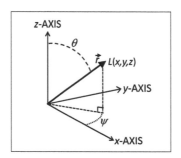

FIGURE 2.2
Coordinate system used to describe radiation transport. The three Cartesian axes x, y, and z are shown along a radius vector $\vec{r}$ connected to a particular point of location $L(x,y,z)$ in space making polar angle θ with the z-axis. The projection of $\vec{r}$ on the xy-plane makes an azimuthal angle ψ with the x-axis.

In principle, the particle yield could be obtained directly from differential cross sections for given incident particle kinetic energy E; $d\sigma(E,\theta)/d\Omega$, where $\sigma(E,\theta)$ is the cross section as a function of energy and angle, and Ω is the solid angle into which the secondary particles are directed. For example, Y could in principle be obtained from an integration of this cross section as it varies with energy while the incident particle loses energy in passing through the target material.

Calculations of the radiation field that directly use the cross sections may not always be practical because targets hit by beam are not really "thin." One cannot always ignore energy loss or secondary interactions in the target. Furthermore, the knowledge of cross sections at all energies is often incomplete with the unfortunate result that one cannot always perform an integration over θ and E to get the total yield.

For many applications, the details of the *angular distributions* of *total secondary particle yield* $dY(\theta)/d\Omega$ and the *angular dependence of the emitted particle energy spectrum* $d^2Y(E,\theta)/dEd\Omega$ are very important.

Often, the particle fluence is needed at a particular location at coordinates (r,θ) from a known *point source* of beam loss while the angular distributions of $dY/d\Omega$ are generally expressed in units of particles steradian^{-1} incident particle^{-1}. A point source is one in which the dimensions of the source are small compared with the distance to some other location of interest. While care should be taken, it is common for "point source" conditions to be a satisfactory approximation.

To obtain the total fluence $\Phi(\theta)$, for example, particles cm^{-2} (incident particle)$^{-1}$, or differential fluence $d\Phi(E,\theta)/dE$, for example, particles cm^{-2} MeV^{-1} (incident particle)$^{-1}$, at a given distance r (cm) at a specified angle θ from such a point source, one must simply multiply the yield values by $1/r^2$ since an area dA subtends a solid angle $d\Omega$ of dA/r^2 at distance r from an arbitrary point in space:

$$\Phi(\theta) = \frac{1}{r^2}\frac{dY(\theta)}{d\Omega} \quad \text{and} \quad \frac{d\Phi(E,\theta)}{dE} = \frac{1}{r^2}\frac{d^2Y(E,\theta)}{dEd\Omega} \tag{2.1}$$

Given the fact that secondary, as well as primary, particles can create radiation fields, it is clear that the transport of particles through space and matter can be a complex matter. In the following section, the advanced techniques for handling these issues are described.

2.3 Theory of Radiation Transport

The theoretical material in this section largely follows O'Brien (1980). It is included here to show clearly the mathematical basis of the contents of shielding codes, especially those that use the Monte Carlo method. Vector notation is used in this section.

2.3.1 General Considerations of Radiation Transport

Stray and direct radiations at any location are distributed in particle type, direction, and energy. Figure 2.2 shows the coordinate system used in this discussion.

To determine the amount of radiation present for radiation protection purposes, one must assign a magnitude to this multidimensional quantity. This is done by forming a double integral over energy and direction of the product of the flux density and an approximate dose per unit fluence conversion factor summed over particle type:

$$\frac{dH(\vec{r},t)}{dt} = \sum_i \int_{4\pi} d\vec{\Omega} \int_0^\infty dE f_i(\vec{r},E,\vec{\Omega},t) P_i(E) \tag{2.2}$$

where the summation index i is over the various particle types, $\vec{\Omega}$ is the direction *vector* of a given particle passing through point location L at coordinates $\vec{r}$, the coordinate *vector* of the point in space where the quantities exemplified by dose rate dH/dt are to be calculated, E is the particle energy, t is time, and i is the particle type. $P_i(E)$ is the dose per fluence conversion factor for the radiation dosimetry quantity desired (e.g., dose equivalent, effective dose, etc., here *generically* denoted by H) expressed as a function of energy and particle type for the ith type of particle. The inner integral is over all energies present while the outer integral is over all spatial directions from which contributions to the radiation field at the location specified by $\vec{r}$ originate. The summation is over the particle types present. The result of the integration is $dH(\vec{r},t)/dt$, the dose rate at location $\vec{r}$ and time t. Values of $P_i(E)$ are given in Figures 1.4 through 1.6.

The *angular flux density* $f_i(\vec{r},E,\vec{\Omega},t)$ is the number of particles of type i per unit area, per unit energy, per unit solid angle, per unit time at location $\vec{r}$, with energy E, at a time t, traveling in a direction $\vec{\Omega}$. It is related to the total flux density $\phi(\vec{r},t)$ by integrating over direction and particle energy:

$$\phi(\vec{r},t) = \sum_i \int_{4\pi} d\vec{\Omega} \int_0^\infty dE f_i(\vec{r},E,\vec{\Omega},t) \tag{2.3}$$

$f_i(\vec{r},E,\vec{\Omega},t)$ is connected to the total fluence $\Phi(\vec{r})$ by integrating over a relevant interval of time (from t_i to t_f), as well as direction and energy by

$$\Phi(\vec{r}) = \sum_i \int_{4\pi} d\vec{\Omega} \int_0^\infty dE \int_{t_i}^{t_f} dt f_i(\vec{r},E,\vec{\Omega},t) \tag{2.4}$$

$f_i(\vec{r},E,\vec{\Omega},t)$ is connected to the energy spectrum expressed as a flux density for particle type i at point $\vec{r}$ at time t, $\phi_i(\vec{r},t,E)$ by

$$\phi_i(\vec{r},t,E) = \int_{4\pi} d\vec{\Omega} f_i(\vec{r},E,\vec{\Omega},t) \tag{2.5}$$

To determine the proper dimensions and composition of a shield, the amount of radiation, expressed in terms of the dose (i.e., dose equivalent, effective dose, etc.) that penetrates the shield and reaches locations of interest must be calculated. This quantity must be compared with the maximum allowed by the design objectives or by applicable regulation requirements. If the calculated result is too large, either the conditions associated with the source of the radiation or the physical properties of the shield must be changed. The latter could be a change in shield materials, dimensions, or both. If the shield cannot be adjusted, then the amount of beam loss allowed by the beam control instrumentation, the amount of residual gas in the vacuum system, or the amount of beam accelerated must be reduced. It is difficult and expensive, especially in the case of the larger accelerators, to alter permanent shielding or operating conditions if the determination of shielding dimensions and composition has not been done correctly. The methods for determining these quantities mathematically have been described by several workers with only a summary of this important work given in the next section.

2.3.2 The Boltzmann Equation

The primary tool for determining the amount of radiation reaching a given location is the *stationary form of the Boltzmann equation*, or more simply the Boltzmann equation. The solution of this equation yields the angular flux density f_i. The angular flux density is then converted to dose rate by means of Equation 2.2. This section reviews the theory that determines the distribution of radiation in matter and discusses some of the methods for extracting detailed numerical values for elements of this distribution, such as the particle flux, or related quantities, such as dose, radioactivation, or instrument response. The Boltzmann equation includes all the processes that the particles of various types that comprise the radiation field can undergo. More details are given by O'Brien (1980), and a summary has been provided by the NCRP (2003).

The Boltzmann equation was originally derived by Ludwig Boltzmann in 1872 to describe the properties of gases but is well-adapted to address radiation fields. The equation is an integral-differential equation describing the behavior of a dilute assemblage of moving particles. It is a continuity equation of the angular flux density f_i in a phase space made up of the three space coordinates of Euclidean geometry, the three corresponding direction cosines (i.e., the cosines of the angles between the trajectory vector and each of the three standard axes of Cartesian coordinates), the kinetic energy, and the time. The density of radiation in a volume of phase space may change in the following five ways:

- *Uniform translation*: Where the spatial coordinates change, but the energy-angle coordinates remain unchanged
- *Collisions*: As a result of which the energy-angle coordinates change, but the spatial coordinates remain unchanged, or the particle may be absorbed and disappear altogether
- *Continuous slowing down*: In which uniform translation is combined with continuous energy loss
- *Decay*: Where particles are changed through radioactive transmutation into particles of another kind
- *Introduction*: Involving the direct emission of a particle from a source into the volume of phase space of interest: electrons or photons from radioactive materials, neutrons from an α-neutron source, the "appearance" of beam particles, or particles emitted from a collision at another (usually higher) energy

Combining these five elements yields

$$\tilde{B}_i f_i(\vec{x}, E, \vec{\Omega}, t) = Q_{ij} + Y_i \qquad (2.6)$$

where the mixed differential-integral *Boltzmann operator* for particles of type i, $\tilde{B}_i$, is

$$\tilde{B}_i = \nabla \cdot \vec{\Omega} + \sigma_i + d_i + \frac{\partial S_i}{\partial E} \tag{2.7}$$

$$Q_{ij} = \sum_j \int_{4\pi} d\vec{\Omega}' \int_0^{E_{max}} dE_B \sigma_{ij} (E_B \rightarrow E, \vec{\Omega}' \rightarrow \vec{\Omega}) f_j(\vec{x}, E, \vec{\Omega}', t) \tag{2.8}$$

and

$$d_i = \frac{\sqrt{1 - \beta_i^2}}{\tau_i \beta_i c} \tag{2.9}$$

The *gradient operator* ∇ inherently results in a vector, not scalar, quantity. Not surprisingly, except for a few special cases the Boltzmann equation is difficult to solve.

In Equations 2.6 and 2.7,

- Y_i is the number of particles of type i introduced by a source per unit area, time, energy, and per unit solid angle.

- σ_i is the absorption cross section for particles of type i. To be dimensionally correct, this is really the *macroscopic cross section* or linear absorption coefficient $\mu = N\sigma$ as defined in Equation 1.6.

- d_i is the decay probability per unit flight path of radioactive particles (such as muons, pions, or radionuclides) of type i.

- S_i is the stopping power for charged particles of type i (assumed to be zero for uncharged particles including photons) with its partial derivative combining that continuous energy loss with uniform translation.

- Q_{ij} is the "scattering-down" integral; the production rate of particles of type i with a direction $\vec{\Omega}$, an energy E at a location $\vec{x}$, by collisions with nuclei or decay of j-type particles having a direction $\vec{\Omega}'$ at a higher energy E_B.

- Rarely, there are analogous "scattering-up" integrals of similar form for exothermic processes such as thermal neutron capture and some others.

- σ_{ij} is the doubly differential inclusive cross section for the production of i-type particles with energy E and a direction $\vec{\Omega}$ from nuclear collisions or decay of j-type particles with an energy E_B and a direction $\vec{\Omega}'$.

- β_i is the velocity of a particle of type i divided by the speed of light c; τ_i is the mean-life of a radioactive particle of type i in its rest frame of reference.

2.4 The Monte Carlo Method

2.4.1 General Principles of the Monte Carlo Technique

The Monte Carlo method is the most common approach in radiation physics to solving the Boltzmann equation for realistic geometries that are difficult to characterize using analytic techniques (i.e., with equations in closed form). The method proceeds by constructing a series of *trajectories* or *histories*, choosing each step at random from a distribution of applicable processes described as realistically as possible. In perhaps its most widely used form in physics applications, the *inverse transform method*, one starts with calculating a set of travel distances between collisions using known cross sections. Then, the cross sections for changing energy, particle type, or direction are used to create a set of possible outcomes of each collision. The result of the interaction may be a number of particles of varying types, energies, and directions, each of which will be followed in turn. Along the way, decay processes are included. The results of many histories are tabulated, leading typically to mean values and standard deviations of particle types, locations, directions, energies, and other quantities of interest.

If $p(x)dx$ is the *differential probability* of an occurrence at $x \pm 1/2dx$ within the mathematical interval $[a,b]$, then the integration

$$P(x) = \int_a^x dx' p(x') \qquad (2.10)$$

gives $P(x)$, the *cumulative probability* of the event occurring in the interval $[a, x]$. The cumulative probability increases monotonically with x and must satisfy the conditions $P(a) = 0$, $P(b) = 1.0$. If a *random number* R uniform on the interval $[0, 1]$ is chosen,

$$R = P(x) \qquad (2.11)$$

corresponds to a random choice of the value of x, since the distribution function for the cumulative $P(x)$ can, in principle, be inverted as a unique one-to-one mapping:

$$x = P^{-1}(R) \qquad (2.12)$$

To illustrate, in determining when an uncharged particle undergoes a reaction in a one-dimensional system with no decays $(d = 0)$, no competing processes $(S = 0)$, and no "in-scattering" $(Q = 0)$, one recognizes from Equations 1.3, 2.6, and 2.7 that a simple application of the Boltzmann equation is evident:

$$\tilde{B}\Phi = \{\nabla \cdot \vec{\Omega} + \sigma_i\}\Phi \qquad (2.13)$$

This simple situation reduces to the following, when one understands σ_i to be the *macroscopic* cross section otherwise denoted by $N\sigma$:

$$\tilde{B}\Phi = \frac{d\Phi}{dx} + N\sigma\Phi = 0 \qquad (2.14)$$

The solution to this equation is the familiar one:

$$\Phi = \Phi_0 \exp(-x/\lambda) \qquad (2.15)$$

where $\lambda = 1/N\sigma$ as in Equation 1.5. Replacing x/λ with r, the number of mean-free-paths the particle travels in the medium, the differential probability per unit mean-free-path for an interaction is given by

$$p(r) = \exp(-r) \qquad (2.16)$$

then,

$$P(r) = \int_0^r dr' \exp(-r') = -\exp(-r')\big|_0^r = 1 - \exp(-r) = R \Rightarrow r = -\ln(1-R) \qquad (2.17)$$

Selecting a random number R then determines a depth r that has the proper distribution. Obviously, identical results apply to other processes described by an exponential function such as radioactive decay. In this simple situation, it is clear that one can solve the above for r as a function of R and thus obtain individual values of r from a corresponding set of random numbers. For many processes, an inversion this simple is not possible analytically. In those situations, other techniques exemplified by successive approximations and table look-up procedures can be employed.

In a real Monte Carlo calculation, further sampling processes might select the scattering of the particles being followed, particle production, decays, etc. Deflections by magnetic fields can be readily included for each segment of the trajectory.

The Monte Carlo result is the number of times the event of interest occurred in a given number of trajectories or histories. As a counting process it has an uncertainty. The reciprocal of the relative error will tend to be roughly proportional to the square root of the number of trajectories calculated. Thus, high probability processes are more accurately simulated than are low probability ones. This becomes a problem for calculations of radiation field properties external to a thick shield in which the particle fluences are attenuated over many orders of magnitude. Sophisticated techniques are often used that temporarily give enhanced probabilities, called *weights,* to the low-probability events during the calculation in order to study them. The correct probabilities are then restored at the end of the calculation by removing the weights. It is commonly a matter of best practice to evaluate the accuracy of a Monte Carlo result by repeating calculations using different initial values of the random number generator, the "seed."

2.4.2 Monte Carlo Example: A Sinusoidal Angular Distribution of Beam Particles

Suppose one has a distribution of beam particles such as exhibited in Figure 2.3.

For this distribution, $p(\theta) = A \cos \theta$ for $0 < \theta < \pi/2$, where A is a constant. The fact that the integral of $p(\theta)$ over the relevant interval $0 \leq \theta \leq \pi/2$ to get the cumulative probability $P(\theta = \pi/2)$ must be unity implies $A = 1$, since

$$P(\pi/2) = \int_0^{\pi/2} d\theta A \cos\theta = A \sin\theta \ \Big|_0^{\pi/2} \overset{\text{def}}{=} 1 \qquad (2.18)$$

Thus, $p(\theta) = \cos \theta$. The cumulative probability $P(\theta)$ is then given by

$$P(\theta) = \int_0^\theta d\theta' p(\theta') = \int_0^\theta d\theta' \cos\theta' = \sin\theta' \ \Big|_0^\theta = \sin\theta \qquad (2.19)$$

If R is a random number, then $R = P(\theta)$ determines a unique value of θ; hence,

$$\theta = \sin^{-1}(R) \qquad (2.20)$$

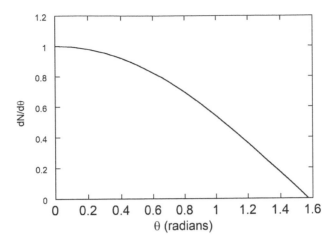

FIGURE 2.3
Hypothetical angular distribution of particles obeying a distribution proportional to cos θ.

It is instructive to do a simple Monte Carlo calculation with, say, 50 random numbers to illustrate the basic features of the method. One can proceed by setting up a table such as Table 2.1 populated by a particular set of 50 random numbers where the results have been "tallied" (2nd column) into "bins" of ranges of θ-values (first column).

The tally marks in the second column collect those "events" for which an individual random number R results in a value of θ within the associated first column bin. θ_{mid} is the midpoint of the bin (0.1, 0.3,...). The fourth column is the normalized result determined from the following:

$$N = \frac{\text{Number found in Monte Carlo bin}}{(\text{Total number of events}) \times (\text{bin width})}$$
$$= \frac{\text{Number found in bin in Monte Carlo}}{(50) \times (0.2\,\text{radians})} \tag{2.21}$$

TABLE 2.1

Tally Sheet for Monte Carlo Example

θ (Radians)	Total Events in Bin	N (Normalized Number)	θ_{mid}
0.0–0.199	11	1.1	0.1
0.2–0.399	13	1.3	0.3
0.4–0.599	11	1.1	0.5
0.6–0.799	4	0.4	0.7
0.8–0.999	7	0.7	0.9
1.0–1.199	4	0.4	1.1
1.2–1.399			1.3
1.4–1.57			1.49

One can calculate the exact value of the mean value of θ for the specified distribution $\langle\theta\rangle$:

$$\langle\theta\rangle = \frac{\int_0^{\pi/2}\theta p(\theta)d\theta}{\int_0^{\pi/2}p(\theta)d\theta} = \frac{\int_0^{\pi/2}\theta\cos(\theta)d\theta}{1} = [\cos\theta + \theta\sin\theta]_0^{\pi/2} \qquad (2.22)$$

$$\langle\theta\rangle = \left[(0-1) + \left(\frac{\pi}{2} - 0\right)\right] = 0.57$$

To calculate the same quantity from the Monte Carlo result, one proceeds first by multiplying the frequency of Monte Carlo events for each of the eight angular bins from the table by the midpoint value of the bins. Then one sums over the eight bins and divides by the number of incident particles (50 in this example). Thus one can determine the average value of θ, $\langle\theta\rangle_{MC}$, calculated by the Monte Carlo technique in this extremely simple example:

$$\langle\theta\rangle_{MC} = [(11)(0.1) + (13)(0.3) + (11)(0.5) + (4)(0.7) + (7)(0.9) + (4)(1.1)]/50 = 0.48 \qquad (2.23)$$

Despite the very coarse bins and an *extremely* small number of histories used in this example, the agreement is perhaps surprisingly good. This example also illustrates that the statistical errors are generally larger for the more rare events here represented by large values of θ (i.e., perhaps for $\theta > 1$ radian).

Practical Monte Carlo calculations generally involve the need to follow a huge number of histories. While historically some Monte Carlo calculations predating the advent of computers were performed successfully using wheels of chance and hand-tallying techniques (e.g., Wilson 1952), modern results rely on the use of computers to perform useful calculations in nearly all circumstances. As computer technology has advanced, the ability to model the physical effects in more detail and with ever-improving statistical accuracy has resulted. In subsequent chapters, results obtained using some specific codes will be presented. Brief, qualitative descriptions of these codes, accurate as of this writing, are given in Appendix. It should be noted that most of these codes are being constantly improved with the result that the wisest practice in using them is to consult with their sponsoring entities or authors directly.

2.5 Review of Magnetic Deflection and Focusing of Charged Particles

2.5.1 Magnetic Deflection of Charged Particles

Particle accelerators of all types utilize *electromagnetic forces* to accelerate, deflect, and focus charged particles. The physics has been well-described in textbooks such as those by Carey (1987), Chao et al. (2013), Edwards and Syphers (1993), and Lee (2012). A review of this topic along with the physics of particle acceleration tailored to the needs of radiation physics at accelerators has been given by Cossairt (2008). In accelerator radiation protection, an understanding of these forces is motivated by the need to be able to determine the deflection of particles by electric or magnetic fields to be able to assure that particles in a deflected particle beam either interact with material where such interactions are desired or avoid such beam loss. Doing this is interconnected with the design of the accelerator and is discussed in the advanced texts referenced above. This is especially important when radiofrequency (RF) electromagnetic fields are applied to the particle beams where a full treatment using electrodynamics is needed. As the reader should recall, *Maxwell's equations* interconnect the electric and magnetic fields when they vary with time. However, some of the simpler issues are discussed in this section for static, or slowly varying electric and magnetic fields.

The *Lorentz force* $\vec{F}$ (Newtons) on a given charge q (Coulombs) in SI units at any point in space is

$$\vec{F} = q(\vec{v} \times \vec{B} + \vec{E}) = \frac{d\vec{p}}{dt} \tag{2.24}$$

where the electric field $\vec{E}$ is in Volts m^{-1}, the magnetic field $\vec{B}$ is in Tesla (1.0 Tesla = 10^4 Gauss), $\vec{v}$ is the velocity of the charged particle in m s^{-1}, $\vec{p}$ is the momentum of the particle in kg m s^{-1}, and t is the time in seconds. The direction of the force due to the cross product in Equation 2.24 is, of course, determined by the usual *right-hand* rule. Static electric fields (i.e., $d\vec{E}/dt = 0$), if present, serve to accelerate or decelerate the charged particles. Electrostatic deflection according to Equation 2.24 is used and is of considerable importance even at large accelerators. This has been discussed in the references cited.

Due to the cross product in this equation, in a uniform magnetic field without the presence of an electric field, any component of $\vec{p}$ that is parallel to $\vec{B}$ will not be altered by the magnetic field. Typically, charged particles are deflected by *dipole magnets* in which the magnetic field is, to high order, spatially uniform and constant with time, or slowly varying compared with the time during which the particle is present. For this situation, if there is no component of $\vec{p}$ that is parallel to $\vec{B}$, the motion is a circular orbit, or segment thereof, and the magnetic force serves to supply the requisite centripetal acceleration. The presence of a component of $\vec{p}$ that is parallel to $\vec{B}$ results in a trajectory that is a spiral rather than a circle. Figure 2.4 illustrates the condition of circular motion.

Equating the centripetal force to the magnetic force and recognizing that $\vec{p}$ is perpendicular to $\vec{B}$ leads to

$$\frac{mv^2}{R} = qvB \tag{2.25}$$

where m is the *relativistic* mass (see Equation 1.14). Solving for the radius of the circle R (meters), recognizing that the magnitude of the momentum $p = mv$, and changing the units of measure for momentum,

$$R(\text{meters}) = \frac{p}{qB} \text{ (SI units)} = \frac{p(\text{GeV}/c)}{0.29979qB} \tag{2.26}$$

where q in the denominator of the far-right-hand side is now the *number* of electronic charges carried by the particle, and B remains expressed in Tesla. The numerical factor in the denominator is just the exact value of the speed of light in SI units divided by 10^9.

At large accelerators, one is often interested in the angular deflection of a magnet of length L that provides such a uniform field orthogonal to the particle trajectory. Such a situation

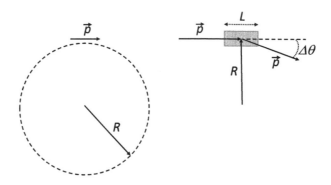

Magnetic field $\vec{B}$ is perpendicular to the Fig. and directed toward the reader.

FIGURE 2.4

A particle of positive charge q having momentum $\vec{p}$ follows a circular path when directed perpendicular to a static, uniform magnetic field $\vec{B}$. (*Left*) This condition for a complete circle. (*Right*) A particle of momentum $\vec{p}$ enters a magnet of length L that has field integral value of BL. For this example, $L \ll R$ and the particle experiences a small angular deflection $\Delta\theta$. The angular deflection is exaggerated in this figure for clarity. (Reprinted with permission from Cossairt, J. D. 2008. In *Topics in accelerator health physics*, eds. J. D. Cossairt, V. Vylet, and J. W. Edwards, 1–45. Madison, WI: Medical Physics Publishing. https://www.medicalphysics.org/SimpleCMS.php?content=default.html.)

is also shown in Figure 2.4. If L is only a small piece of the complete circle (i.e., $L = R$), one can consider the circular path over such a length to be two straight line segments "bent" by the deflection. Doing this, the change in direction in the small-angle approximation $\Delta\theta$ is given by

$$\Delta\theta = \frac{L}{R} = \frac{0.29979qBL}{p} \text{ (radians)} \tag{2.27}$$

where the product BL (Tesla-meters) is commonly referred to the *field integral* of the magnet system, and p remains in GeV/c. BL could just as well be obtained by integrating a nonuniform field over the length of a particular magnet system. This angle of deflection can be used to determine if a particle beam will interact with or "cleanly" avoid some solid object near its path, a matter commonly of practical importance for radiation protection.

2.5.2 Magnetic Focusing of Charged Particles

Charged particle beams can be *focused* by a variety of devices including the edge fields of dipole magnets and electrostatic quadrupoles. The references cited in Section 2.5.1 describe in much more detail these systems and those of higher order that focus particle beams. Mathematical methods analogous to those found in the study of geometrical optics are used to describe the optics of charged particles because of what will shortly be seen to be useful analogies. Where time-varying electric and magnetic fields are involved, the full complement of Maxwell's equations must, of course, be used to describe the motion of charged particles. The application of higher-order multipole fields and the employment of RF electromagnetic fields to accelerate, decelerate, and otherwise manipulate charged particle beams is left to the specialized texts.

While electrostatic quadrupoles and other types of magnetic focusing devices are used at accelerators (e.g., Cossairt 2008), magnetic quadrupoles are the most common method

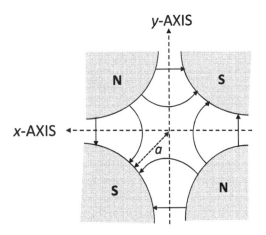

FIGURE 2.5
Cross section of a typical quadrupole magnet. The pole pieces are of opposite magnetic polarities, denoted **N** and **S**, and are, ideally, of hyperbolic shapes. A Cartesian coordinate system is used in which x and y denote transverse coordinates, while z is along the desired beam trajectory, the optic axis of the beam optical system. In this figure, a beam carrying positive electric charge enters the quadrupole *into* the paper along the positive z-axis. The curves with arrows denote magnetic field "lines." Magnetostatics determines that the magnetic fields as illustrated by the magnetic field "lines" are perpendicular to the surface of the pole pieces. The parameter a is the *radius* or gap radius of the quadrupole. (Reprinted with permission from Cossairt, J. D. 2008. In *Topics in accelerator health physics*, eds. J. D. Cossairt, V. Vylet, and J. W. Edwards, 1–45. Madison, WI: Medical Physics Publishing. https://www.medicalphysics.org/SimpleCMS.php?content=default.html.)

of focusing charged particle beams and are the only type discussed here. An idealized quadrupole magnet has the transverse cross section shown in Figure 2.5, which also defines the Cartesian coordinate system to be used in the remainder of this discussion. As one can see, the polarities of the pole pieces alternate. Following the usual convention, the incoming beam direction is taken to be along the longitudinal coordinate z, in this case, "into the paper" along the *optic axis* of the quadrupole. Positive values of the y-coordinate measure upward deviations from the optic axis, while positive values of the x-coordinate measure deviations from the optic axis to "beam left," for consistency with the right-hand rule as applied to the Cartesian coordinates (x,y,z).

Often in the accelerator magnets themselves and nearly always in beam lines transmitting extracted particles, the electromagnetic fields vary only slowly with time or are static compared with the particle transit times. Under these conditions, it is shown in the references that if the shapes of the pole pieces are hyperbolae described by equations of form $xy = \pm k$, where k is a constant, and if the pole pieces are uniformly magnetized, then the components of the magnetic field within the gap containing the beam are

$$B_x = -\frac{B_o}{a}y = -gy \tag{2.28}$$

and

$$B_y = -\frac{B_o}{a}x = -gx \tag{2.29}$$

Here a is the gap *radius* defined in Figure 2.5, and B_o is the magnitude of the magnetic field strength at the pole pieces. The parameter g is, quite naturally, called the *gradient* of the quadrupole and has units of Tesla meter^{-1}. This ideal quadrupole has length L along the z-coordinate, the *optic axis* of the system.

Now examine qualitatively what happens to a particle having positive charge that enters this magnet parallel to the z-axis. If the particle trajectory is on the optic axis, then it will not be deflected at all since $B_x = B_y = 0$. If, however, a particle enters the magnet parallel to the optic axis but with some finite *positive* value of y, it will receive a deflection toward *smaller* values of y in accordance with the right-hand rule and Equation 2.24. It should also be clear that the amount of the deflection is proportional to the distance "off axis," a direct analogy with what happens optically with glass lenses and visible light. Likewise, if it enters with a finite *negative* value of y, it will receive a deflection toward *less negative* values of y. Thus, a beam of such particles is said to be *focused* in the yz plane since it is moved closer to the center of the magnet. However, if the particle enters with a finite positive value of x, it will be deflected toward a *larger* value of x, away from the optic axis. Finally, a particle incident with a finite negative value of x will similarly be deflected away from the optic axis. Thus, a beam of such particles is, logically, said to be *defocused* in the xz plane. Even qualitatively, it is clearly evident that more than one quadrupole is needed to achieve a net focusing effect on all particles in the beam. Figure 2.6 shows several features of focusing/defocusing by magnetic quadrupoles.

Considering just the situation in the yz plane, it is easy to see that the analogy with geometrical optics is instructive even in mathematical detail. For a particle entering with coordinate y, one can substitute into Equation 2.27 and find the angular deflection to be

$$\Delta\theta = \frac{0.29979qLgy}{p} \text{ (radians)} \tag{2.30}$$

where the same units as Equation 2.27 are used, with g (Tesla meter^{-1}) and y (meters) inserted and the negative sign of Equation 2.28 now implicit. If the incident particle trajectory is parallel with the z-axis, the situation is schematically shown in Figure 2.6a. In schematic drawings of beam optics, it is customary to show *convex lenses* to denote focusing elements and *concave lenses* to represent defocusing elements pertinent to a given plane. Bending magnets are similarly represented by *prisms* in such drawings.

Applying trigonometry, one finds that after deflection in this situation, the particle trajectory will intercept the z-axis at a distance f, which is

$$f = \frac{y}{\tan \Delta\theta} \approx \frac{y}{\Delta\theta} = \frac{p}{0.29979qLg} \tag{2.31}$$

since the deflection $\Delta\theta$ is small and the employment of the small-angle approximation thus valid. This approximation is called the *thin lens* approximation. Recognizing that f is independent of the y coordinate, f is called the *focal length* of the quadrupole. By analogy with optical thin lenses, the *thin lens equation* connecting the *image distance* z_i, with the object distance z_o for other rays is as follows:

$$\frac{1}{z_o} + \frac{1}{z_i} = \frac{1}{f} \tag{2.32}$$

In this equation, z_o and z_i are both greater than zero if the object is to the left of the lens and the image is to the right of the lens, forming a *real* image, for a focusing lens

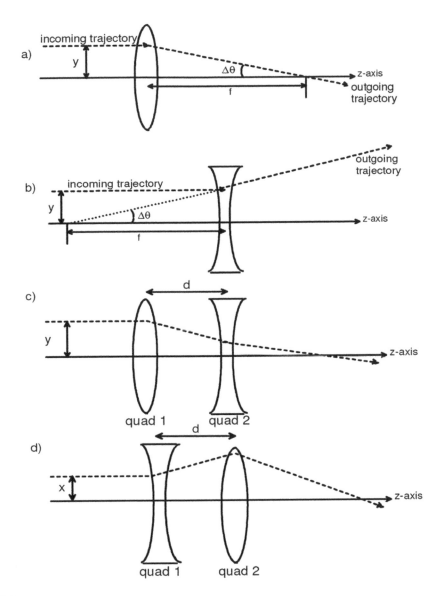

FIGURE 2.6
Configurations of quadrupole lenses are shown with the symbolism explained in the text: (a) Representation of focusing in the yz plane of a beam trajectory incident from the left parallel to the z-axis. A *real* image is formed at the focal length f from the lens. (b) Representation of defocusing in the yz plane. The parallel beam is deflected so that it appears to emerge from a point a distance f before the lens, thus forming a *virtual* image. (c) Representation of a particle trajectory in the yz plane of a quadrupole doublet. The particle enters a quadrupole doublet parallel to the z-axis from the left. First a focusing quadrupole (quad 1) is encountered and then a defocusing quadrupole (quad 2) follows. (d) Representation of a particle trajectory in the xz plane of the same doublet. The particle enters the doublet parallel to the z-axis. In this plane, the defocusing quadrupole is encountered first. (Reprinted with permission from Cossairt, J. D. 2008. In *Topics in accelerator health physics*, eds. J. D. Cossairt, V. Vylet, and J. W. Edwards, 1–45. Madison, WI: Medical Physics Publishing. https://www.medicalphysics.org/SimpleCMS. php?content=default.html.)

with $f > 0$. The situation for the defocusing plane, here the xz plane, is shown in Figure 2.6b as a concave lens. For that plane, the equations are still workable if one applies a negative sign to the value of f and understands that a value of $z_i < 0$ describes a *virtual* image, another analogy with visible light optics.

The simplest configuration of quadrupole magnets is a pair of two. In a given plane, say the yz, the first might be focusing while the second is defocusing. In the orthogonal plane, here the xz, the defocusing quadrupole would thus be encountered first. Commonly, but not exclusively, these magnets are of identical dimensions and have the same or similar gradients. Such a *quadrupole doublet* is shown in Figure 2.6c and d for the yz and xz planes, respectively.

Equation 2.32 can now be used to see how a quadrupole doublet can focus a parallel beam in both the xz and yz planes in a simple example. For this general discussion, the two quadrupoles, quad 1 and quad 2, have different focal lengths f_1 and f_2, respectively, and are separated by distance d. Quad 1 is focusing in the yz plane. As in geometrical optics, for an incoming parallel beam, the object distance relative to quad 1 is $z_{yo1} \to \infty$. Thus, the image distance from quad 1 is at $z_{yi1} = f_1$. The object distance of this image from quad 2 is then $z_{yo2} = d - f_1$. From Equation 2.32, relative to quad 2, the final image will be at z_{yi2}:

$$\frac{1}{z_{yi2}} = \frac{1}{-f_2} - \frac{1}{d - f_1} \tag{2.33}$$

where the negative coefficient of f_2 explicitly incorporates the fact that lens 2 is *defocusing* in the yz plane. Solving,

$$z_{yi2} = \frac{f_2(f_1 - d)}{f_2 - f_1 + d} \tag{2.34}$$

If the quadrupoles are identical ($f = f_1 = f_2$), then

$$z_{yi2} = \frac{f(f - d)}{d} \tag{2.35}$$

Following the same procedure for the xz plane to obtain the corresponding image distance z_{xi2},

$$z_{xi2} = \frac{f_2(f_1 + d)}{f_1 - f_2 + d} \tag{2.36}$$

With identical quadrupoles, this becomes

$$z_{xi2} = \frac{f(f + d)}{d} \tag{2.37}$$

One notices that with identical quadrupoles the focal positions on the z-axis are not identical:

$$z_{xi2} - z_{yi2} = 2f \tag{2.38}$$

For many common applications this is an undesirable result and an example of the limitations of the analogy with thin lens optics of visible light. A nicer result would be for $z_{xi2} - z_{yi2} = 0$, with planes in focus at the same point. However, this nonzero result is expected since the particles in the xz plane are first subject to *defocusing*, and thus become *more* divergent, prior to their being focused. The average focal length of the system for both the xz and yz planes is thus f^2/d. More sophisticated schemes like quadrupole triplets and nonidentical magnets can be used to obtain a specialized beam envelope. Quadrupole triplets can, in fact, achieve the desired simultaneous image points in both transverse planes. These advanced methods are discussed in great detail in the specialized references.

In quadrupoles, the beam axis should coincide with the optic axis. If a beam enters a quadrupole significantly off-axis, the entire beam will be deflected nearly as if a quadrupole were a dipole magnet of equivalent field strength and length L. Beams that are deflected in this manner by a quadrupole are said to have suffered *steering*. The steering of beams can constitute significant loss points in the beam transport system and is reflective of beam lines not perfectly aligned or operational errors in beam tuning.

In this simple exposition, some significant effects have been ignored. First, a typical particle beam will contain some spread in particle momenta. The derivation given ignores the fact that *dispersion* will occur in the magnetic fields in the same way that a prism disperses a visible beam of "white" light into the various colors. There also may be *aberrations* or distortions of an image. An example is *chromatic aberration*, analogous to its namesake encountered in geometrical optics. For particle beams chromatic aberration is due to the dependence of focal length on particle momentum in Equation 2.31. Also, no particle beam is ever completely parallel or emergent from a geometrical point.

All particle beams possess a property called *transverse emittance* ε. This quantity is expressed as the product of angular divergence and transverse dimensional size, typically in units of π mm-milliradian. The explicit display of the factor π is a matter of custom and arises from the mathematical derivations intrinsic in the precise definition of this quantity. Emittance is used to describe both longitudinal and transverse phenomena, but the discussion here is limited to transverse emittance. During the acceleration process, the beam emittance becomes smaller because the normalized transverse emittance, the product $\gamma\beta\varepsilon$, is an *invariant*. Thus, as velocity and $\gamma\beta$ increase with energy, ε must decrease. There are exceptions to this generalization not discussed here. Once a beam is no longer undergoing acceleration (e.g., in an extracted beamline), the value of ε remains constant or generally increases due to processes such as multiple Coulomb scattering, ionization losses, space charge effects, etc. Under some conditions not discussed further in this text, synchrotron radiation can, in fact, reduce the transverse emittance.

When ε remains constant, the product of the angular divergence of the beam envelope and the transverse size of the beam envelope are conserved. This is a consequence of *Liouville's theorem* of classical dynamics as applied to the coordinate system defined here, which requires that the volume of the phase space defined by the transverse spatial coordinates, x and y, and their corresponding "conjugate" momentum components, p_x and p_y is conserved. Since in this Cartesian coordinate system, the momentum components p_x and p_y are, in small-angle approximation, directly proportional to the angles (in radians) between the momentum vector and the x and y coordinate axes, respectively, the assertion made directly follows. The specialized references describe this in more detail.

Thus, efforts made to focus the beam into a smaller cross-sectional size will unavoidably result in a beam with a correspondingly larger angular spread. Likewise, attempts to create a parallel beam (one with essentially no angular spread) will result in a correspondingly larger beam size.

PROBLEMS

1. This problem gives two elementary examples of Monte Carlo techniques that are almost "trivial." In this problem, obtaining random numbers from a standard table or from a hand calculator should be helpful.

   a. First, use a random number table or random number function on a calculator along with the facts given about the cumulative probability distribution for exponential attenuation to demonstrate that, even for a sample size as small as, say, 15, the mean value of paths traveled is "within expectations" if random numbers are used to select those path lengths from the cumulative distribution. Do this, for example, by calculating the mean and standard deviation of your distribution.

   b. An incident beam is subjected to a position measurement in the coordinate x. It is desirable to "re-create" incident beam particles for a shielding study using Monte Carlo. The x distribution as measured is given in the following table:

x	#
0	0
1	1
2	2
3	3
4	4
5	5
6	4
7	3
8	2
9	1
10	0

   Determine, crudely, $p(x)$ and $P(x)$, and then use 50 random numbers to "create" particles intended to represent this distribution. Then compare with the original one which was measured in terms of the average value of x and its standard deviation. Do *not* take the time to use interpolated values of x, simply round off to integer values of x for this demonstration.

2. A beam of protons having a kinetic energy of 100 GeV is traveling down a beam line. The beam is entirely contained within a circle of diameter 1 cm. All of the beam particles have the same kinetic energy. An enclosure farther downstream must be protected from the beam or secondary particles produced by the beam by shielding it with a large-diameter iron block that is 20 cm in radius centered on the beam line. The beam passes by this block by being deflected by a uniform field magnet that is 3 m long, the longitudinal center of which is located 30 m upstream of the iron block. Calculate the magnetic field B that is needed to accomplish this objective.

3

Prompt Radiation Fields due to Electrons

3.1 Introduction

In this chapter the major features of the prompt radiation fields produced by electrons are described. The development of the electromagnetic cascade and the shielding of photoneutrons and high-energy particles that result are discussed. The utilization of Monte Carlo calculations in electron shielding problems is also addressed. The material presented in this chapter is useful for understanding electron, photon, and photoneutron radiation from electron accelerators used in medicine and in research. A section on synchrotron radiation completes this chapter.

3.2 Unshielded Radiation Produced by Electron Beams

At all energies photons produced by *bremsstrahlung* dominate the unshielded radiation field aside from the hazard of the direct beam. As the energy increases, neutrons become a significant problem. For electrons having kinetic energy E_o approaching 100 MeV or higher, the *electromagnetic cascade* is of great importance. A useful rule of thumb is that electrons have a finite ionization range R in any material that monotonically increases with the initial kinetic energy E_o (MeV). For $2 < E_o < 10$ MeV,

$$R \approx 0.6E_o \text{ (g cm}^{-2}) \tag{3.1}$$

In air at standard temperature and pressure (STP) over this energy domain, R(meters) $\approx 5E_o$ (MeV). At energies above 10 MeV or so, an approximate threshold that is dependent on the absorbing medium, the loss of energy begins to be dominated by *radiative processes*, whereby the emission of photons begins to dominate over those losses of energy due to collisions, a phenomenon discussed further in Section 3.2.2.

3.2.1 Dose Rate in a Direct Beam of Electrons

At all accelerators, the dose rate in the direct particle beam is generally larger than in any purely secondary radiation field. This is certainly true at electron accelerators. Swanson (1979a) has given a rule of thumb, said to be "conservative," meaning one that should lead to an overestimate, for electrons in the energy domain of $1.0 < E_o < 100$ MeV;

$$\frac{dH_{equiv}}{dt} = 1.6 \times 10^{-6} \phi \tag{3.2}$$

where dH_{equiv}/dt is the dose equivalent rate (Sv h^{-1}), and ϕ is the flux density (cm^{-2} s^{-1}) in the electron beam. One of the problems at the end of this chapter examines the domain of validity of this approximation. The coefficient is 1.6×10^{-4} if dH_{equiv}/dt is to be in rem h^{-1} with ϕ still expressed as cm^{-2} s^{-1}. For electrons in an external radiation field, there is no significant difference between dose equivalent and effective dose.

3.2.2 Bremsstrahlung

Bremsstrahlung is the radiative energy loss of charged particles, especially electrons, as they interact with materials. It appears in the form of photons. Some of the information in this section is also useful in understanding the radiation produced by certain accelerator components, such as radiofrequency (RF) cavities operated apart from the main accelerator, as pointed out by Silari et al. (1999).

An important parameter when considering the radiative energy loss of *electrons* in matter is the *critical energy* E_c above which the radiative loss of energy exceeds that due to ionization. There are several alternative formulae used to calculate E_c with representative ones given here. For electrons, the value of E_c is a smooth function of atomic number, approximated by

$$E_c = \frac{800\,(\text{MeV})}{Z+1.2} \tag{3.3}$$

where Z is the atomic number of the material. *Muons* behave much like "heavy" electrons in matter and are subject to the same radiative phenomena that become much more important at a much higher energy due to the much larger rest energy of the muon. For muons (see Sections 1.7 and 3.2.4) the corresponding critical energy $E_{c,muon}$ is much larger and differs for solid and gaseous media (Beringer et al. Particle Data Group 2012);

$$E_{c,muon} = \frac{5700\,\text{GeV}}{(Z+1.47)^{0.838}}\,(\text{solids}), \text{ and } E_{c,muon} = \frac{7980\,\text{GeV}}{(Z+2.03)^{0.879}}\,(\text{gases}) \tag{3.4}$$

The transition from dominance by ionization to dominance by radiation is a smooth one. The total stopping power for electrons or muons may be written as the sum of collisional and radiative components, respectively;

$$\left(\frac{dE}{dx}\right)_{tot} = \left(\frac{dE}{dx}\right)_{coll} + \left(\frac{dE}{dx}\right)_{rad} \tag{3.5}$$

Another parameter of significant importance is the radiation length X_0, the mean thickness of material in which a high-energy electron loses all but $1/e$ of its energy by bremsstrahlung. The radiation length also plays a role in the description of multiple Coulomb scattering for all charged particles and was discussed in that context in Section 1.8. X_0 is the approximate scale length for describing high-energy electromagnetic cascades, supplanting the ionization range for even moderate electron energies. The radiation length is approximated by Equation 1.24. For electrons with energies well above E_c, the *radiative fractional energy* loss of the electron dE_{rad}/E is equal to the fraction of a radiation length the electron penetrates;

$$\frac{dE_{rad}}{E} = -\frac{dx}{X_0}, \text{ thus } \left(\frac{dE}{dx}\right)_{rad} = -\frac{E}{X_0} \tag{3.6}$$

so that under these conditions (i.e., where loss by ionization can be neglected), by integration, the energy of the electron E as a function of thickness of shield penetrated x is given by

$$E(x) = E_o \exp\left(-\frac{x}{X_o}\right) \tag{3.7}$$

Figure 3.1 gives the percentage of energy E_o that appears as radiation as a function of energy for electrons stopped in various materials as plotted by Swanson (1979a) based on the calculations of Berger and Seltzer (1964, 1966).

Bremsstrahlung develops as a function of target thickness. As the thickness increases, the intensity of the radiation increases until reabsorption begins to take effect. Then, self-shielding begins to take over. One talks about the maximal conditions as being a "thick-target" bremsstrahlung spectrum, with $x \approx X_o$. This phenomenon becomes dominant above energies of about 100 MeV for low atomic number ("low-Z") materials and above 10 MeV for high atomic number ("high-Z") materials.

The energy spectrum of the radiated photons ranges from zero to the energy of the incident electron, and the number of photons in a given energy interval is approximately inversely proportional to the photon energy. The amount of energy radiated per energy interval is practically constant according to Fassò et al. (1990). Detailed spectral information for bremsstrahlung photons has been provided by various workers. Figures 3.2 and 3.3 are provided as examples of such spectra at moderate electron beam energies.

Bremsstrahlung spectra are noticeably more energetic, in the vernacular such spectra are said to be "harder," at forward angles.

For thin targets of thickness x ($x \ll X_o$), the spectrum of photons of energy k per energy interval dk, dN/dk, can be approximated by

$$\frac{dN}{dk} \propto \frac{x}{X_o k} \tag{3.8}$$

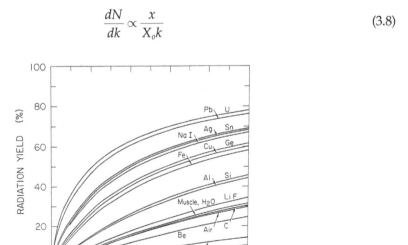

FIGURE 3.1
Bremsstrahlung efficiency for electrons stopped in various materials. This is the percentage of the kinetic energy of incident electrons converted to radiation as a function of incident energy E_o based on calculations of Berger and Seltzer (1964, 1966). The remainder of the kinetic energy is transferred to the medium by ionization. (Reprinted from Swanson, W. P. 1979a. *Radiological safety aspects of the operation of electron linear accelerators.* International Atomic Energy Agency. IAEA Technical Reports Series No. 188. Vienna, Austria. Used with permission.)

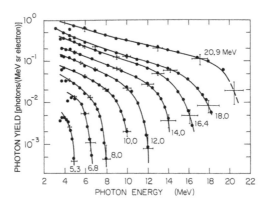

FIGURE 3.2

Bremsstrahlung spectra measured at $\theta = 0°$ from intermediate thickness ($0.2X_o$) targets of high atomic number (Z) material. (Reprinted from *Nuclear Instruments and Methods*, 6, O'Dell, A. A. et al, Measurements of absolute thick-target bremsstrahlung spectra, 340–346, Copyright 1968, with permission from Elsevier.)

For thick targets, those with $x \gg X_0$, the intensity falls with energy approximately as $1/k^2$ at $\theta = 0$ and even more dramatically at larger angles (Swanson 1979a). Thick targets may require consideration of the electromagnetic cascade, discussed later in this chapter.

A more detailed parameterization of the normalized total photon differential yield per incident electron $dN/d\Omega$ for photons of all energies is (National Council on Radiation Protection and Measurements [NCRP] 2003) as follows:

$$\frac{1}{E_o}\frac{dN}{d\Omega} = 4.76E_o \exp(-\theta^{0.6}) + 1.08\exp\left(-\frac{\theta}{72}\right) \text{(photons sr}^{-1}\text{GeV}^{-1}\text{electron}^{-1}) \quad (3.9)$$

where E_o is in gigaelectron volt (GeV), and θ is in degrees. As is shown in the following text, this semiempirical parameterization is especially useful as a source term in thick shields and is particularly valid for production angles around $\theta = 90°$.

The three *Swanson's Rules of Thumb* parameterize this behavior for the absorbed dose rates dD/dt normalized to 1.0 kW of incident beam power for E_o in megaelectron volt (MeV), expected at 1.0 m from a point "target" of high atomic number Z (Swanson 1979a):

Swanson's Rule of Thumb 1:

$$\frac{dD}{dt} \approx 20E_o^2 \text{ (Gy m}^2) \text{ (kW}^{-1}\text{h}^{-1}) \text{ at } \theta = 0°, E_o < 15\,\text{MeV} \quad (3.10)$$

Swanson's Rule of Thumb 2:

$$\frac{dD}{dt} \approx 300E_o \text{ (Gy m}^2) \text{ (kW}^{-1}\text{h}^{-1}) \text{ at } \theta = 0°, E_o > 15\,\text{MeV} \quad (3.11)$$

Swanson's Rule of Thumb 3:

$$\frac{dD}{dt} \approx 50 \text{ (Gy m}^2) \text{ (kW}^{-1}\text{h}^{-1}) \text{ at } \theta = 90°, E_o > 100\,\text{MeV} \quad (3.12)$$

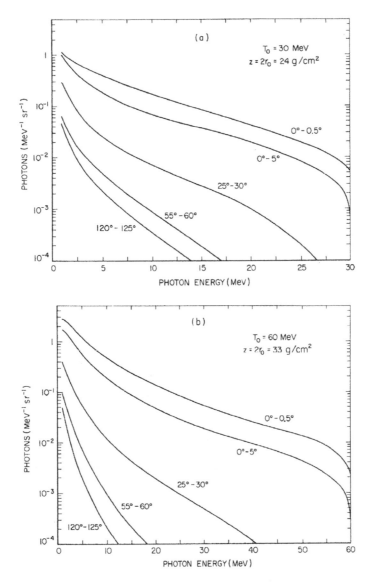

FIGURE 3.3
Spectra of bremsstrahlung photons, per incident electron, emerging in various directions θ from thick targets irradiated by normally electron beams that are monoenergetic and incident perpendicularly. T_0 is the incident kinetic energy of the electrons. Frame (a) shows results for $T_0 = 30$ MeV, while frame (b) shows results for $T_0 = 60$ MeV. The target thickness z at both energies, indicated in each frame of the figure, is twice the mean electron ionization range r_0. (Reprinted with permission from Berger, M. J., and S. M. Seltzer. 1970. *Physical Review* C2:621–631. Copyright 1970 by the American Physical Society.)

Higher absorbed dose rates at $\theta = 90°$ can arise in certain circumstances due to the presence of softer (i.e., less energetic) radiation components. In Equation 3.12, the coefficient of 50 is sometimes increased to 100 to better describe measurements (Fassò et al. 1984). For point-like sources, one can scale these results to other distances (in meters) using the inverse square law. Figure 3.4 shows the behavior for a high-Z target. The forward intensity is a slowly varying function of target material except at very low values of Z.

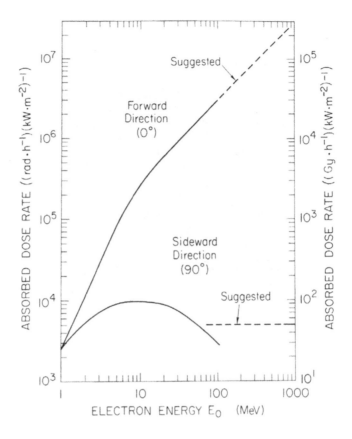

FIGURE 3.4

Thick target bremsstrahlung from a high atomic number target. Absorbed dose rates at 1.0 m per unit incident electron beam power (kW) are given as a function of incident electron kinetic energy E_o. The dashed lines represent a reasonable extrapolation. The dose rates measured in the sideward direction (smoothed for this figure) depend strongly on target and detector geometry and can vary by more than a factor of two. (Reprinted from Swanson, W. P. 1979a. *Radiological safety aspects of the operation of electron linear accelerators.* International Atomic Energy Agency. IAEA Technical Reports Series No. 188. Vienna, Austria. Used with permission.)

$\theta_{1/2}$ is the value of θ where the intensity in the forward lobe has half of its maximum intensity and is approximately given by a relationship with E_o (MeV) due to Swanson (1979a):

$$\theta_{1/2} = \frac{100}{E_0} \text{ (degrees)} \tag{3.13}$$

Alternatively, according to Fassò et al. (1990) the *average angle of emission* is of the order of m_e/E_o (radians), where m_e is the rest mass (in energy units, e.g., MeV) of the electron. At higher energies (E_o greater than approximately 100 MeV), the electromagnetic cascade development in accelerator components is very important and can result in a forward "spike" of photons of *characteristic angle* of $\theta_c = 29.28/E_o$ (degrees, for E_o in MeV). At $\theta = \theta_c$ the intensity of the spike has fallen to $1/e$ of its value at $\theta = 0$.

A formula for calculating the unshielded bremsstrahlung dose equivalent at 1.0 m H_{brem} from a point source is needed. An approximation found to be in agreement with Equations 3.11 through 3.13 for all angles for $E_0 \geq 0.1$ GeV is

$$H_{brem} = E_0 \begin{cases} (1.33\times10^6)E_0\exp(-1000E_0\theta/2.51) \\ +(1.33\times10^5)\exp(-\theta/0.159) \\ +(3.0\times10^3)\exp(-\theta/0.834) \end{cases} (Sv\,m^2\,electron^{-1})\times10^{-17} \qquad (3.14)$$

where E_o is in GeV and θ is in radians (NCRP 2003). Equation 3.14 includes the "doubling" suggested for Equation 3.12 (Fassò et al. 1984).

3.2.3 Neutrons

3.2.3.1 Giant Photonuclear Resonance Neutrons

Neutron production can be expected to occur in any material irradiated by electrons in which bremsstrahlung photons above the material-dependent threshold are produced. This *neutron production threshold* varies from 10 to 19 MeV for light nuclei and 4–6 MeV for heavy nuclei. Thresholds of 2.23 MeV for deuterium and 1.67 MeV for beryllium are exceptions. Other exceptions exist for isolated target nuclei and should be verified individually, perhaps using the online resources of the U.S. National Nuclear Data Center (NNDC 2018). Between this threshold and approximately 30 MeV, a production mechanism known as the giant photonuclear resonance, a (γ,n) reaction, is the most important source of neutron emission from material. The notation (γ,n) is that common in nuclear physics where the symbol for the incoming particle, here γ for a photon, is on the left while the symbol representative of the outgoing particle, here n for the neutron, is on the right within the parentheses.

Swanson (1979a) has given a detailed description of this process that is summarized here. A simple picture of this phenomenon is that the electric field of the photon produced by bremsstrahlung transfers its energy to the nucleus by inducing an oscillation in which the protons as a group move opposite to the neutrons as a group. This process has a broad maximum cross section at photon energies k_o between about 20 and 23 MeV for materials having mass numbers A less than about 40. For heavier targets, the peak is at an energy of approximately $k_o = 80A^{-1/3}$ MeV. Fassò et al. (1990) have provided a great deal of data on the relevant cross sections. It turns out that the yield Y of giant resonance neutrons at energies above approximately $2k_o$ is nearly independent of energy and nearly proportional to the beam power.

This process may be thought of as one in which the entire target nucleus is excited by the incident electron and then decays somewhat later by means of neutron emission. The directionality of the incident electron or photon becomes "lost" so that these emissions are *isotropic*, with no dependence on θ. Because of this isotropicity, and assuming that "point source" conditions are present, the inverse square law may be used to estimate the flux density at any given distance r. The spectrum of neutrons of energy E_n is similar to that seen in a fission neutron spectrum and can be described as a *Maxwellian distribution*:

$$\frac{dN}{dE_n} = \frac{E_n}{T^2}\exp\left(-\frac{E_n}{T}\right) \qquad (3.15)$$

where T is a nuclear "temperature" characteristic of the "excited" (i.e., "heated") target nucleus. Its excitation energy T in energy units is generally in the range $0.5 < T < 1.5$ MeV. For this distribution, the most probable value of $E_n = T$ and the average value of $E_n = 2\,T$. This process generally is the dominant one for incident photon kinetic energies $E_o < 150$ MeV.

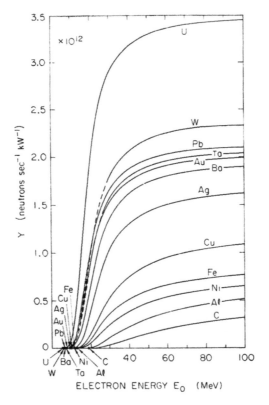

FIGURE 3.5

Neutron yield rate dY_n/dt from infinitely thick targets per kilowatt of electron beam power as a function of electron beam energy E_o, ignoring target self-shielding. (Reprinted with permission from Swanson, W. P. 1979b. *Health Physics* 37:347–358. https://journals.lww.com/health-physics/pages/default.aspx.)

The energy dependences, sometimes called *excitation functions*, of total *neutron yields* in various materials are plotted in Figure 3.5.

Table 3.1 gives the high energy limits for total yield rate dY_n/dt of giant resonance neutrons per watt of beam power (s^{-1} W^{-1}), the isotropic differential neutron yield $dY_n/d\Omega$ (GeV^{-1} sr^{-1}) per unit of beam energy per electron, and a recommended dose equivalent source term S_n (Sv cm^2 GeV^{-1}) per unit beam energy per electron to be used as follows:

$$H = \frac{S_n}{r^2} E_o I \tag{3.16}$$

where H is the dose equivalent in sieverts, r is the radial distance from the target in centimeters, E_o is in GeV, and I is the total number of beam particles incident (e.g., during some time interval).

In Table 3.1 no energy dependence "near threshold" is assumed. The neutrons are distributed isotropically over all directions. These results are best used for $E_o \geq 0.5$ GeV. For electron energies below 0.5 GeV, appropriate values can be obtained by scaling the Table 3.1 entries according to the Figure 3.5 curves. The agreement with various experiments is quite good according to Fassò et al. (1990). The use of these "saturation" values can support reasonable, but conservative, estimates.

TABLE 3.1

Total Neutron Yield Rate per Unit Beam Power dY_n/dt (s^{-1} W^{-1}), Differential Yield per Unit Electron Energy $dY_n/d\Omega$ (GeV^{-1} sr^{-1}) per Electron, and Source Term S_n per Unit Electron Energy (Sv cm^2 GeV^{-1}) per Electron for Giant Resonance Neutrons in an Optimum Target

Material	Total Neutron Production dY_n/dt (s^{-1} W^{-1})	Differential Neutron Yield $dY_n/d\Omega$ (GeV^{-1} sr^{-1}) per Electron	Recommended Source Terms,[a] S_n (Sv cm^2 GeV^{-1}) per Electron
C	4.4×10^8	5.61×10^{-3}	4.3×10^{-12}
Al[b]	6.2×10^8	7.90×10^{-3}	6.0×10^{-12}
Fe	8.18×10^8	1.04×10^{-2}	7.7×10^{-12}
Ni	7.36×10^8	9.38×10^{-3}	6.9×10^{-12}
Cu	1.18×10^9	1.50×10^{-2}	1.1×10^{-11}
Ag	1.68×10^9	2.14×10^{-2}	1.5×10^{-11}
Ba	1.94×10^9	2.47×10^{-2}	1.8×10^{-11}
Ta	2.08×10^9	2.65×10^{-2}	1.8×10^{-11}
W	2.36×10^9	3.01×10^{-2}	2.0×10^{-11}
Au	2.02×10^9	2.58×10^{-2}	1.8×10^{-11}
Pb	2.14×10^9	2.73×10^{-2}	1.9×10^{-11}
U	3.48×10^9	4.44×10^{-2}	3.0×10^{-11}

Source: Adapted from Swanson, W. P. 1979a. Radiological safety aspects of the operation of electron linear accelerators. International Atomic Energy Agency: IAEA Technical Report No. 188. Vienna, Austria; Fassò, A. et al. 1990. *Landolt-Börnstein numerical data and functional relationships in science and technology new series; Group I: Nuclear and particle physics Volume II: Shielding against high energy radiation*, ed. H. Schopper. Berlin, Germany: Springer-Verlag.

[a] To get Sv cm^2 h^{-1} kW^{-1}, multiply the entries in this column by 2.25×10^{16}.
[b] The value for aluminum is also recommended for concrete.

3.2.3.2 Quasi-Deuteron Neutrons

At energies above the giant resonance, the dominant neutron production mechanism is one in which the photon interacts with a neutron-proton pair within the nucleus rather than with the whole nucleus. The *quasi-deuteron effect* is so-named because for $E_o \approx 30$ MeV the photon wavelength nearly matches the average internucleon distance so that the photon interactions tend to occur with "pairs" of nucleons. Only neutron-proton pairs have a nonzero electric dipole moment, a fact that favors interactions of photons with such pairs (i.e., quasi-deuterons). This mechanism is important for $30 < E_o < 300$ MeV as described by Swanson (1979a). An important effect due to this mechanism is to add a tail of higher-energy neutrons to the giant resonance spectrum. For $5 < E_n < E_o/2$ (MeV), the nearly isotropic spectrum of quasi-deuteron neutrons is given by

$$\frac{dN}{dE_n} \approx E_n^{-\alpha} \text{ where, approximately, } 1.7 < \alpha < 3.6 \tag{3.17}$$

The slope becomes steeper as E_o, the kinetic energy of the incident electron, is approached. Equation 3.17 is for *thin* targets, those thinner than approximately one radiation length. For *thick* target situations, the fall-off with E_n is generally steeper. Since the reaction is (γ,np) and the neutron and the proton are nearly identical in mass, they share the available energy equally so that the yield of neutrons due to this mechanism is essentially zero for neutrons having kinetic energy $E_n > E_o/2$. In general the quasi-deuteron neutrons are fewer in number and generally less important than are the giant resonance neutrons. Shielding against the latter will usually provide adequate protection against the former for shielding purposes, but should not be neglected when the fluence of particularly energetic neutrons may be important.

3.2.3.3 High-Energy Particles

There are interactions in which the production of other elementary particles, perhaps best typified by charged pions, $\pi^{\pm}$, becomes energetically possible at still higher energies—those well above the $\pi^{\pm}$ rest energy of 139.57 MeV (Beringer et al. Particle Data Group 2012). These particles can then produce neutrons, and other particles, through secondary interactions as discussed in Chapter 4. The neutrons from this source tend to dominate the lateral shielding requirements in the GeV region. Fassò et al. (1990), based on the earlier work of DeStaebler (1965), give a parameterization of the measured yield per incident electron on iron for high-energy particles of

$$\frac{d^2Y_n}{dEd\Omega} = \frac{7.5\times10^{-4}}{\left(1-0.75\cos\theta\right)^2 A^{0.4}} (\text{GeV}^{-1}\text{sr}^{-1}), \tag{3.18}$$

where A is the atomic mass (g mole^{-1}) of the target material, and θ is the production angle. It is conservative to use a dose per fluence factor of approximately 10^{-13} Sv m^2 (10^{-3} μSv cm^2) for these particles that are mostly high-energy neutrons (see Figure 1.5). Obviously, these neutrons are forward-peaked, not isotropic.

In general photons are produced more copiously, while the neutrons are more difficult to shield.

3.2.3.4 Production of Thermal Neutrons

This discussion would not be complete without mentioning *thermal neutrons*. At all accelerators capable of producing neutrons, the neutrons produced can scatter from walls, components, etc., and *thermalize*; that is, come into equilibrium with an energy spectrum characteristic of a thermal gas at the ambient absolute temperature T present. As discussed in more detail in Section 9.5.1, at $T = 293°$, the most probable kinetic energy of such *thermal neutrons* is 0.025 eV. Such neutrons can participate in exothermic *thermal neutron capture reactions* and thus emit energetic photons, protons, or ions such as α-particles. The particles and ions produced can represent an important source of ionizing radiation that merits further consideration. As subsequently explained in later chapters, some thermal neutron capture cross sections are quite large, enhancing the importance of these neutrons.

3.2.4 Muons

With electron beams, muons become of significance above an electron energy of approximately 211 MeV, the threshold of the process in which a μ^+ and a μ^- are produced in a *pair production* process quite analogous to the more familiar one in which an electron-positron pair results from photon interactions. Muons can also be produced with much smaller yields at electron accelerators by means of the decay of $\pi^{\pm}$ and K$^{\pm}$ particles. The latter are due to secondary particle production processes exemplified by photo-pion creation. Such *decay muons*, more prominent at proton and ion accelerators, are discussed in Section 4.2.4. A detailed theoretical treatment of muon production by incident electrons is given by Nelson (1968), Nelson and Kase (1974), and Nelson et al. (1974). Figure 3.6 shows the

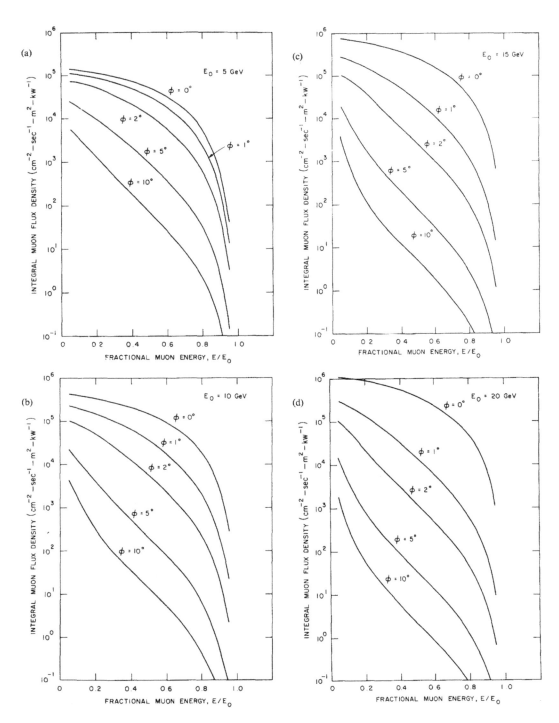

FIGURE 3.6
Integrated muon flux density at 1.0 m per kilowatt of electron beam power as a function of muon energy four different values of incident electron energy incident on a thick iron target at several values of ϕ. Here ϕ denotes the polar angle usually denoted θ in this book. The integral of the flux density over energy includes all muons that have energies that exceed the value of the abscissa at the specified value of ϕ. (Reprinted from *Nuclear Instruments and Methods*, 66, Nelson, W. R., The shielding of muons around high energy electron accelerators: Theory and measurement, 293–303, Copyright 1968, with permission from Elsevier.)

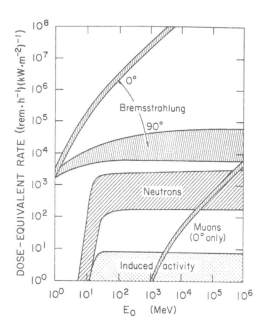

FIGURE 3.7
Qualitative sketch of dose equivalent rates per unit primary electron beam power at 1.0 m produced by various types of "secondary" radiation from a high-Z target as a function of primary beam energy, if no shielding were present. The width of the bands suggests the degree of variation found, depending on such factors as target material and thickness. The angles at which the various processes are most important are indicated. Dose due to neutrons and induced activity have essentially no angular dependence. Induced activity is discussed in Chapters 7 and 8. (Reprinted from Swanson, W. P. 1979a. *Radiological safety aspects of the operation of electron linear accelerators*. International Atomic Energy Agency. IAEA Technical Reports Series No. 188. Vienna, Austria. Used with permission.)

angular dependence of the integrated muon flux density at four incident electron energies E_o and a variety of angles for electrons incident on an iron target (Nelson 1968).

3.2.5 Summary of Unshielded Radiation Produced by Electron Beams

The broad features of the radiation field due to the unshielded initial interactions of electrons are illustrated in Figure 3.7, a diagram intended to only be *qualitative*. As one can see, at large angles, from the standpoint of the dose equivalent, the unshielded field is always dominated by photons. At small angles, the field is dominated by photons at the lower energies with muons increasing in importance as the energy increases to large values. The production of induced radioactivity is discussed in Chapters 7 and 8.

3.3 Electromagnetic Cascade: Introduction

As a prelude to discussing the electromagnetic cascade process, one must look a bit more at the dose equivalent due to thick target bremsstrahlung at large values of θ for targets surrounded by cylindrical shields. The situation is given in Figure 3.8.

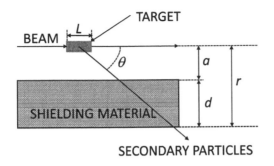

FIGURE 3.8
Target and shielding geometry for the estimation of dose equivalent due to electron beam interactions with a target surrounded by a cylindrical shielding. L is the length of the target, and the other parameters specify the geometry.

Following Equation 3.9, the NCRP (2003) gives the photon absorbed dose per incident electron D external to such a shield:

$$D(\theta) = (1 \times 10^{-11})\left\{10.2E_o \exp(-\theta^{0.6}) + 2.3\exp\left(-\frac{\theta}{72}\right)\right\}E_o\left(\frac{\sin\theta}{a+d}\right)^2 \exp\left(-\frac{\mu}{\rho}\frac{\rho d}{\sin\theta}\right) (\text{Gy electron}^{-1})$$

(3.19)

As was the case for Equation 3.9, this expression is normalized to results involving thick iron and copper targets at $E_0 = 15$ GeV. Here, E_o is the electron energy in GeV, θ is in degrees, a is the target-to-shield distance (cm), d is the shield thickness (cm), ρ is the shielding material density (g cm^{-3}), and μ/ρ is the value of the attenuation coefficient at the *Compton minimum*, the energy where the total photon cross section is at a minimum and the *photon mean free path* λ_γ is thus a maximum. The use of the term *Compton minimum* is somewhat inaccurate since the cross section for the Compton scattering process *monotonically* decreases with energy. However, the contribution of the pair production mechanism to the total photon cross section *increases* with energy, thus creating a "minimum" in the total attenuation coefficient. For concrete $\mu/\rho = 2.4 \times 10^{-2}$ cm^2 g^{-1} at the Compton minimum. Figure 3.9 gives values of the photon mean free path for a variety of materials as a function of energy.

Values for more materials, including mixtures and compounds, and energies are available from the U.S. National Institute of Standards and Technology (NIST 2018b). The major feature that needs to be considered in the shielding design at electron accelerators is the electromagnetic cascade.

One should recall the definitions of critical energy E_c and radiation length X_0 that were given in Equations 3.3, 3.4, and 1.24 (Fassò et al. 1990).

A related parameter of importance for describing the electromagnetic cascade is the *Molière radius* X_m:

$$X_m = \frac{X_o E_s}{E_c}$$

(3.20)

where

$$E_s = \left(\sqrt{\frac{4\pi}{\alpha}}\right)m_e c^2 = 21.2\,\text{MeV}$$

(3.21)

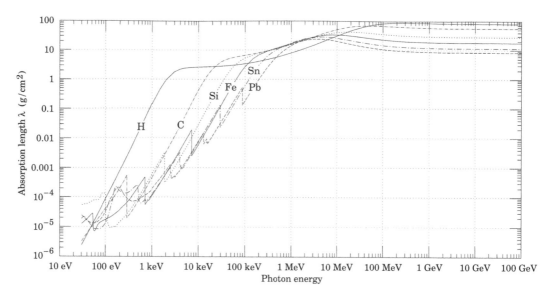

FIGURE 3.9
Photon absorption length, or mean free path, plotted as a function of photon energy in various materials. (Reprinted with permission from Beringer, J. et al. [Particle Data Group] 2012. *Physical Review* D86:010001. Copyright 2012 by the American Physical Society.)

Here α is the fine structure constant of atomic physics (see Table 1.1), and m_e is the mass of the electron. X_m is a good characteristic length for describing *radial* distributions in electromagnetic showers. Two additional dimensionless scaling variables are commonly introduced to describe electromagnetic shower behavior:

$$t = \frac{x}{X_o} \text{ (for longitudinal distance scaling)} \tag{3.22}$$

and

$$y = \frac{E}{E_c} \text{ (for energy scaling)} \tag{3.23}$$

For mixtures of n elements, these quantities and also the stopping power dE/dx scale according to the elemental fractions by mass f_i as follows, exemplified for the stopping power:

$$\frac{dE}{dx} = \sum_{i=1}^{n} f_i \left(\frac{dE}{dx}\right)_i \tag{3.24}$$

For photons of energies E_o greater than about 1.0 GeV, the total $e^+ e^-$ *pair production* cross section σ_{pair} is approximately given, for a single constituent element, by

$$\sigma_{pair} = \frac{7}{9} \left(\frac{A}{X_0 N_A}\right) (\text{cm}^2) \tag{3.25}$$

where A is the atomic weight, N_A is Avogadro's number, and X_o is the radiation length expressed in units of g cm^{-2}. For energies larger than a few MeV, the pair production process dominates the total photon attenuation. The mean free path length for pair production λ_{pair} is thus given by

$$\lambda_{pair} = \frac{\rho}{N\sigma_{pair}} = \frac{\rho}{(\rho N_A/A)(7/9)(A/X_0 N_A)} = \frac{9}{7}X_0 \quad (3.26)$$

The energy-independence and near-equality of λ_{pair} and X_o lead to the most important fact about the electromagnetic cascade:

> The electrons radiatively produce photons with almost a characteristic length only a bit shorter than that with which the photons produce more e$^\pm$ pairs.

This first-order approximation is important because it means that the "size" in physical space of individual electromagnetic interactions is approximately independent of energy. For *hadronic cascades* (see Chapter 4) the results are considerably different and can likely be viewed as more complicated.

3.4 Electromagnetic Cascade Process

Figure 3.10 conceptually illustrates the electromagnetic cascade process.

As illustrated in Figure 3.10, the electromagnetic cascade at an electron accelerator proceeds qualitatively according to the following steps:

Step 1: At point **B** a high-energy electron ($E_o \gg m_e c^2$) produces a high-energy photon by means of bremsstrahlung after traveling an average distance of X_o.

Step 2: At point **P** this photon produces an e$^+$ e$^-$ pair after traveling an average distance of 9/7 X_o. Each member of the pair will have, on average, half the energy of the original photon.

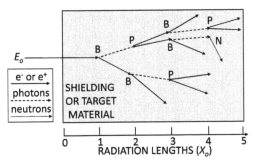

FIGURE 3.10

Conceptual view of the development of an electromagnetic cascade in a semi-infinite medium with no magnetic or electric fields present. The solid lines represent electrons or positrons, the dashed lines represent photons, and the dotted lines represent neutrons. The shower is initiated by an electron or positron of energy E_o incident on the medium from the left. In this figure the spreading in the transverse direction is greatly exaggerated for clarity. Bremsstrahlung and pair production events are denoted by **B** and **P**, respectively. Compton scattering, ionization, and the production of other hadrons in addition to neutrons are not shown but also play a role in the dispersal of energy. Photonuclear reactions, as illustrated by the (γ,n) reaction at point **N** also play a role, albeit much more infrequently than inferred from this illustration. The process could just as well be initiated by a photon.

Step 3: After traveling an average distance of X_0, the electrons and positrons will produce bremsstrahlung photons at other points **B**.

Step 4: Each electron or positron may continue on to interact again at points **B** and **P** and release yet more photons and the photons will produce more $e^\pm$ pairs until the incoming energy is totally absorbed.

Step 5: Occasionally, a photon produces a neutron by means of photonuclear processes at point **N**.

This chain of events can equally well be initiated by a high-energy photon, even one produced in secondary interactions at a proton or ion accelerator. Eventually, after a number of generations, the individual energies of the electrons and positrons will be degraded to values below E_c so that ionization processes then begin to dominate and terminate the shower. Likewise, the photon energies eventually are degraded so that Compton scattering and the photoelectric effect compete with further $e^+ e^-$ pair production. Both phenomena work to end the cascade.

Of course, there are subtleties representing many different physical processes, such as the production of other particles, which must be taken into account and are best handled by Monte Carlo calculations. A general discussion of the use of Monte Carlo techniques for such problems has been given by Rogers and Bielajew (1990). A widely used code incorporating the Monte Carlo method applied to electromagnetic cascades is the successive versions of Electron Gamma Shower (EGS) code (Nelson et al. 1985) (see Appendix, Section A.2.1). Van Ginneken (1978) developed the Monte Carlo program called AEGIS, which was very effective in calculating the propagation of such cascades through thick shields. Analytical approximations have been developed and are summarized elsewhere (e.g., Swanson 1979a, Fassò et al. 1990). The results of published calculations are used in the following discussion to aid in improving the reader's understanding of electromagnetic cascades. For detailed calculations, readers should obtain the latest version of the modern Monte Carlo code of their choice.

3.4.1 Longitudinal Shower Development

The dosimetric properties of the calculations of an electromagnetic cascade may be summarized in curves that give fluence, dose, or other quantities of interest as functions of shower depth or distance from the axis. Figure 3.11 shows the fraction of total energy deposited (integrated over all radii about the shower axis) versus longitudinal depth as calculated by Van Ginneken and Awschalom (1975). They introduced a longitudinal scaling parameter ζ given by

$$\zeta = 325(\ln Z)^{-1.73} \ln E_0 \ (\text{g cm}^{-2}) \tag{3.27}$$

where E_0 is in MeV and Z is the atomic number of the absorber. When the longitudinal coordinate is expressed in units of ζ, all curves approximately merge into this universal one and are rather independent of target material.

In their well-referenced classic development of analytical shower theory, Rossi and Griesen (1941) and Rossi (1952), using their *Approximation B*, a more advanced formalism than their parallel *Approximation A* not further discussed here, predicted for an *electron-initiated* shower that the total number of electrons and positrons at the *shower maximum* N_{show} is proportional to the primary energy as follows:

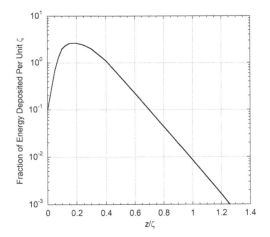

FIGURE 3.11
Fraction of total energy deposited by an electromagnetic cascade versus longitudinal depth x integrated over all radii about the shower axis. See Equation 3.27. (Adapted from Van Ginneken, A., and M. Awschalom. 1975. *High energy particle interactions in large targets: Volume I, hadronic cascades, shielding, and energy deposition.* Fermi National Accelerator Laboratory. Batavia, IL.)

$$N_{show} = \frac{0.31 E_o/E_c}{\sqrt{\ln(E_o/E_c) - 0.37}} \qquad (3.28)$$

For a *photon-initiated* shower, a value of 0.18 should replace that of 0.37 in the denominator of Equation 3.28. This distinction related to the initiator of the shower (electron/positron or photon), and others reflect the deeper penetration of an initiating photon implied by the 9/7 factor in Equation 3.26. The result embodied in the mathematical language of this equation is intuitively sensible, since the final outcome of the shower is to divide the energy at the shower maximum among a number of particles with energies near E_c. One can obtain the maximum energy deposited per radiation length from Equation 3.28 as the product $E_c N_{show}$ (Fassò et al. 1990).

Also from the Rossi-Griesen work, the location of the *shower maximum* t_{max} along the longitudinal coordinate in units of radiation length, as determined by Equation 3.22, is approximated by

$$t_{max} = 1.01 \ln\left(\frac{E_o}{E_c}\right) - C_{show}, \text{ with } C_{show} = 1 \qquad (3.29)$$

Experimentally Bathow et al. (1967) found that values of $C_{show} = 0.77$ for copper and $C_{show} = 0.47$ for lead fit experimental data better. Not surprisingly, photon-initiated showers penetrate about 0.8 radiation lengths deeper than do the electron-initiated showers. Fassò et al. (1990) simply give values of $C_{show} = 1.0$ and $C_{show} = 0.5$ for electron- and photon-initiated showers, respectively.

The longitudinal "center of gravity" (i.e., the mean depth in the shield) $\bar{t}$ of all the shower electrons in the scaled unit t is given by

$$\bar{t} = 1.009 \ \ln\left(\frac{E_o}{E_c}\right) + 0.4 \text{ (electron-induced shower)} \qquad (3.30)$$

and

$$\bar{t} = 1.012 \ \ln\left(\frac{E_o}{E_c}\right) + 1.2 \ \text{(photon-induced shower)} \tag{3.31}$$

Fassò et al. (1990) give the mean-squared *longitudinal* spread τ^2 (squared standard deviation in units of X_o^2) about $\bar{t}$ in this same scaled unit system to be

$$\tau^2 = 1.61 \ \ln\left(\frac{E_o}{E_c}\right) - 0.2 \ \text{(electron-induced shower)} \tag{3.32}$$

and

$$\tau^2 = 1.61 \ \ln\left(\frac{E_o}{E_c}\right) + 0.9 \ \text{(photon-induced shower)} \tag{3.33}$$

There are other, perhaps less important, differences between photon- and electron-induced showers. EGS4 results tabulated by Fassò et al. (1990) have been parameterized to determine *source terms* S_i for longitudinal distributions of absorbed dose in various materials and for the associated dose equivalent within shields composed of these materials over the energy region of 1.0 GeV $< E_o <$ 1.0 TeV. This has been done for the dose on the z-axis (subscripts "*a*") and for the dose averaged over a 15 cm radius about the z-axis (subscripts "15"). Table 3.2 gives parameters for calculating dose equivalent H_{long} (Sv per electron), at the end of a beam absorber of length L (cm) of density ρ (g cm^{-3}), and gives fitted values of the various "attenuation lengths" λ_i (g cm^{-2}) to be used with the corresponding tabulated values of S_i.

For absorbed dose calculations, the factor C, which is the ratio of dose equivalent in tissue (Sv) to absorbed dose in the material (not tissue) (Gy), should be set to unity. The formula in which these parameters from Table 3.2 are to be used is as follows:

$$H_{long} = C S_i \exp\left(-\frac{\rho L}{\lambda_i}\right) \tag{3.34}$$

TABLE 3.2

Source Terms S_a and S_{15} and Corresponding Recommended Longitudinal Attenuation Lengths, λ_a and λ_{15} for Doses on the Axis and Averaged over a Radius of 15 cm in the Forward Direction for Beam Absorbers and End-Stops, Respectively, for Use in Equation 3.34

Material	C (Sv Gy^{-1})	S_a (Gy electron^{-1})	λ_a (g cm^{-2})	S_{15} (Gy electron^{-1})	λ_{15} (g cm^{-2})
Water	0.95	$1.9 \times 10^{-10}E_o^{2.0}$	58	$1.5 \times 10^{-11}E_o^{2.0}$	59.9
Concrete	1.2	$1.9 \times 10^{-9}E_o^{1.8}$	44	$2.2 \times 10^{-11}E_o^{1.8}$	45.6
Aluminum	1.2	$2.3 \times 10^{-9}E_o^{1.7}$	46	$3.4 \times 10^{-11}E_o^{1.7}$	46.3
Iron	1.3	$2.9 \times 10^{-8}E_o^{1.7}$	30	$1.8 \times 10^{-10}E_o^{1.7}$	33.6
Lead	1.8	$1.9 \times 10^{-7}E_o^{1.4}$	18	$4.6 \times 10^{-10}E_o^{1.4}$	24.2

Source: Adapted from Fassò, A. et al. 1990. *Landolt-Börnstein numerical data and functional relationships in science and technology new series; Group I: Nuclear and particle physics Volume II: Shielding against high energy radiation*, ed. H. Schopper. Berlin, Germany: Springer-Verlag.

These results are most valid in the region of incident electron energy E_o from 1.0 GeV to 1.0 TeV. Conversion factors C from absorbed dose in the shielding material to dose equivalent within the shield are given. E_o is the beam kinetic energy in GeV. Equation 3.34 is valid in the longitudinal region *beyond the shower maximum*.

3.4.2 Lateral Shower Development

Figure 3.12 shows the fraction U/E_o of the incident electron energy that escapes laterally from infinitely long cylinders made of various materials as a function of cylinder radius R for showers caused by electrons of various energies that bombard the front face of the cylinder. On this graph R is in units of the Molière radius X_m.

Beyond about four Moliere radii the slope of this curve can be taken to be the photon mean free path at the Compton minimum for the absorbing material of interest. An analytic function that fits data between 100 MeV and 20 GeV, a rather wide energy domain, for electrons incident on targets ranging from aluminum to lead is (Swanson and Thomas 1990)

$$\frac{U(R/X_m)}{E_o} = 0.8\exp\left[-3.45\left(\frac{R}{X_m}\right)\right] + 0.2\exp\left[-0.889\left(\frac{R}{X_m}\right)\right] \qquad (3.35)$$

As was done for the longitudinal situation, EGS4 (Fassò et al. 1990) has been similarly used to give the maximum energy deposition (and by extension, the maximum absorbed dose

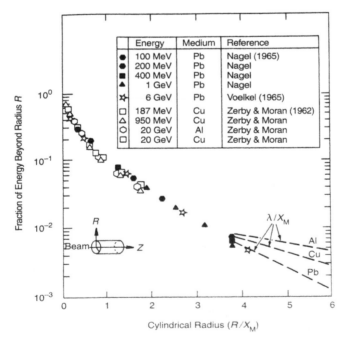

FIGURE 3.12

Fraction of total energy deposited beyond a cylindrical radius R/X_m, as a function of radius R for showers caused by 0.1–20 GeV electrons incident on various materials. The original results as designated by the author names originate from Nagel (1965), Völkel (1965), and Zerby and Morgan (1963 *sic*). (Reprinted from DeStaebler, H. et al. 1968. In *The Stanford two mile accelerator*, ed. R. B. Neal, 1029–1067. New York, NY: Benjamin Publishing Co. The SLAC National Accelerator Laboratory has declared this reference including this figure to be public domain and not subject to copyright: http://www.slac.stanford.edu/library/2MileAccelerator/copyrightinfo.htm)

TABLE 3.3

Conversion Factors C from Absorbed Dose in Shielding Material to Dose Equivalent, Source Terms S_{lat} for the Maximum of the Electromagnetic Component, and Recommended Lateral Attenuation Lengths λ_{lat} for the Electron Energy Range E_o from 1.0 GeV to 1.0 TeV Laterally for Beam Absorbers or End-Stops for Use in Equation 3.36

Material	C (Sv Gy^{-1})	S_{lat} (Gy cm^2 GeV^{-1} electron^{-1})	λ_{lat} (g cm^{-2})
Water	0.95	2.5×10^{-12}	26
Concrete	1.2	3.6×10^{-12}	27
Aluminum	1.2	3.4×10^{-12}	29
Iron	1.3	4.7×10^{-11}	33
Lead	1.8	1.3×10^{-10}	26

Source: Adapted from Fassò, A. et al. 1990. *Landolt-Börnstein numerical data and functional relationships in science and technology new series; Group I: Nuclear and particle physics Volume II: Shielding against high energy radiation*, ed. H. Schopper. Berlin, Germany: Springer-Verlag.

and dose equivalent) as a function of radius r. Over the energy range 1.0 GeV $< E_o <$ 1.0 TeV, there is direct scaling with energy in the formula for maximum dose equivalent at $\theta \approx 90°$:

$$H_{lat} = CE_0 S_{lat} \frac{\exp(-\rho d / \lambda_{lat})}{r^2} \tag{3.36}$$

where H_{lat} is the maximum dose equivalent laterally (Sv per electron), C is the same as in Equation 3.34, E_o is the electron kinetic energy in GeV, S_{lat} is the source term from the EGS4 calculations, d is the lateral dimension of the shield (shield thickness) in centimeters, ρ is the material density (g cm^{-3}), λ_{lat} is the attenuation length (g cm^{-2}), and r is the distance from the axis (cm), where the dose equivalent is desired (see Figure 3.8). Table 3.3 gives the parameters needed for Equation 3.36.

3.5 Shielding of Hadrons Produced by Electromagnetic Cascade

3.5.1 Neutrons

As discussed before, neutrons are produced by high-energy electrons and photons. These neutrons must be accounted for in order to properly shield electron accelerators. Tesch (1988) summarized shielding against these neutrons by developing simple analytical relations for cases where thick targets are struck by the electron beam. Figure 3.8 defines the shielding geometry. For lateral concrete shielding of density ρ (g cm^{-3}), the maximum dose equivalent outside of shield thickness d (cm) which begins at radius a (cm) from a thick iron or copper target struck by electrons having primary energy E_o (GeV) per incident electron is

$$H(d,a) = \frac{4 \times 10^{-13}}{[(a+d)]^2} E_o \exp\left(-\frac{\rho d}{100}\right) \text{(Sv electron}^{-1}) \tag{3.37}$$

This equation is valid for E_o greater than about 0.4 GeV and values of the product ρd greater than about 200 g cm^{-2}. The maximum dose equivalent will be found at $\theta \approx 90°$. For other target materials one can scale this equation in the following way: According to Tesch the neutron production is proportional to the photoproduction cross section, the track length in centimeters, and the atom number density (cm^{-3}). The interaction cross section is generally proportional to the atomic weight A. Since the track length is proportional to the radiation length X_0, the production becomes proportional to the radiation length in units of $g\,cm^{-2}$. Thus, for rough estimates of the dose equivalent in the environs of targets of materials other than iron, one can obtain results by multiplying these values by the scaling factor f determined from the ratio of the values of the radiation lengths expressed in units of $g\,cm^{-2}$:

$$f = \frac{X_o(\text{material } g\,cm^{-2})}{X_o(\text{iron, } g\,cm^{-2})} \tag{3.38}$$

Furthermore, for shields composed of other materials, one can simply adjust the implicit attenuation length (i.e., the value of 100 g cm^{-2} in the exponential function in Equation 3.37) to that appropriate to the material.

Fassò et al. (1990) give a more detailed treatment separately handling the giant resonance neutrons and high-energy particle components of dose while deriving "source terms" and appropriate formulae. The formula for the dose equivalent H_n due to the giant resonance neutrons given as follows is held to be valid for $1.0\text{ GeV} < E_o < 1.0\text{ TeV}$ and for $30 < \theta < 120°$. For the giant resonance neutrons,

$$H_n = \eta_n S_n E_o \left(\frac{\sin\theta}{a+d}\right)^2 \exp\left(-\frac{\rho d}{\lambda_n \sin\theta}\right), \text{ (Sv electron}^{-1}) \tag{3.39}$$

where E_o is the beam energy (GeV), ρ (g cm^{-3}) is the density, and the quantities a (cm) and d (cm) are defined in Figure 3.8. S_n (Sv cm^2 GeV^{-1}) is the source term from Table 3.1, while λ_n (g cm^{-2}) is the attenuation length λ_n recommended for giant resonance neutrons for representative materials listed in Table 3.4. This formula is regarded as being valid for $30 < \theta < 120°$.

In Equation 3.39 dimensionless factor η_n, where $\eta_n \leq 1$, gives an estimate of the fractional "efficiency" for the production of neutrons by the target. It is generally connected with details of a given configuration. For "conservative" calculations, it can be taken to have a

TABLE 3.4

Recommended Attenuation Lengths for Use in Equation 3.39 for Various Materials

Material	λ_n (g cm^{-2})
Water	9
Concrete	42
Iron	130
Lead	235

Source: Adapted from Fassò, A. et al. 1990. *Landolt-Börnstein numerical data and functional relationships in science and technology new series; Group I: Nuclear and particle physics Volume II: Shielding against high energy radiation*, ed. H. Schopper. Berlin, Germany: Springer-Verlag.

TABLE 3.5

Attenuation Lengths λ_h in g cm^{-2} for the High-Energy Particle Component for Use in Equation 3.40

Material	Energy Limit >15 MeV or >25 MeV (g cm^{-2})	Energy Limit >100 MeV	Nuclear Interaction Length (g cm^{-2})	Recommended λ_h (g cm^{-2})
Water			84.9	86
Aluminum			106.4	128
Soil (sand)	101,104[a]	117	99.2	117
	102,105[b]	96		
Concrete	101,105[a]	120	99.9	117
	91	105		
	82,100[b]	100		
Iron	139[b]		131.9	164
Lead	244[b]		194	253

Source: Adapted from Fassò, A. et al. 1990. *Landolt-Börnstein numerical data and functional relationships in science and technology new series; Group I: Nuclear and particle physics Volume II: Shielding against high energy radiation,* ed. H. Schopper. Berlin, Germany: Springer-Verlag.

[a] Attenuation lengths for the indicated values are slightly dependent on angle with the higher value at $\theta = 0°$ and the smaller value in the backward direction for $E > 15$ MeV.

[b] Attenuation lengths for the indicated values are slightly dependent on angle with the higher value at $\theta = 0°$ and the smaller value in the backward direction for $E > 25$ MeV.

value of unity. It smoothly increases from very small values to unity as the target thickness approaches X_o.

3.5.2 High-Energy Particles

When considering the production of high-energy particles by electrons, no correction for target thickness is generally employed. These particles tend to drive the shielding requirements of large electron accelerators. The following formula for the dose equivalent per incident electron external to such a shield is based on Equation 3.18 augmented with exponential attenuation in the shielding and an inverse square law dependence;

$$H_h = \frac{7.5 \times 10^{-13} E_o}{(1 - 0.75 \cos\theta)^2 A^{0.4}} \left[\frac{\sin\theta}{a+d} \right]^2 \exp\left[-\frac{\rho d}{\lambda_h \sin\theta} \right] (\text{Sv electron}^{-1}) \qquad (3.40)$$

The cylindrical geometry is as defined in Figure 3.8. H_h is the dose equivalent due to these particles (Sv), E_o is the beam energy (GeV), A is the atomic weight of the target, and λ_h (g cm^{-2}) is the attenuation length typical of these particles. Table 3.5 gives values of λ_h for representative materials. Fassò et al. (1990) go further and describe a variety of special cases.

3.6 Synchrotron Radiation

While the physics of synchrotron radiation has long been understood, its importance in both basic and applied modern research has grown dramatically with the advent of electron accelerators designed for use as photon sources. Basic electromagnetic theory provides a sound theoretical basis for the phenomena. Konopinski (1981); Jackson (1998); Wiedemann (2003. 2013) and Margaritondo (1988) provide much detailed information on both the synchrotron radiation and the modern facilities that have been built to utilize it. Important reviews of radiation protection considerations at synchrotron radiation facilities have been provided by Swanson and Thomas (1990) and Liu and Vylet (2001).

3.6.1 General Discussion of the Phenomenon

The movement of electrons in a curved orbit results in their centripetal acceleration. This gives rise to emission of photons. At nonrelativistic energies this radiation is largely isotropic. However, for relativistic energies, a condition readily achieved by accelerated electrons, the photons emerge in a tight bundle along a tangent to any point on a circular orbit. For a single electron, or a small bunch of electrons orbiting together, the photon beam will sweep around like a searchlight. As a function of the relativistic parameter γ defined in Equation 1.13, there is a *characteristic angular frequency* ω_c (radians s^{-1}) given by

$$\omega_c = \frac{3}{2}\frac{c}{R}\gamma^3 \text{ (radians s}^{-1}) \tag{3.41}$$

where c is the speed of light (m s^{-1}), and R is the radius of the orbit (m). The angular frequency is equal to 2π times the ordinary frequency (i.e., cycles s^{-1} or Hz). This corresponds to a *characteristic energy* ε_c of the synchrotron radiation power spectrum, the median energy of this distribution, that is given by

$$\varepsilon_c = \frac{3}{2}\frac{\hbar c}{R}\gamma^3 = \frac{2.218}{R}W^3 \text{ (keV)} \tag{3.42}$$

in which $\hbar$ is *Planck's constant reduced*, that is, Planck's constant divided by 2π (see Table 1.1); and W is the total energy of the electrons. Some references use the term *critical energy* for this quantity, a choice of nomenclature avoided here to reduce confusion with the completely different quantity of the same name discussed in connection with electromagnetic cascades. The characteristic angular half-width of this "searchlight" beam θ_c as a function of emitted photon energy k is

$$\theta_c = \frac{1}{\gamma}\left(\frac{\varepsilon_c}{k}\right)^{1/3} \tag{3.43}$$

Figure 3.13 shows this bundle at the characteristic energy, that is, at $k = \varepsilon_c$, where $\theta_c = 1/\gamma$. According to Equation 3.43, for $k < \varepsilon_c$ the bundle is broader, while for $k > \varepsilon_c$ the bundle is narrower.

The energy spectrum of the photons emitted by electrons captured in such a circular orbit turns out to be a standard mathematical function, the shape of which is independent of the electron beam energy. It is an integral of a modified Bessel function of the third kind with numerical tabulations of its values available in the published literature. Figure 3.14 shows this function.

This photon spectrum is called the *bending magnet spectrum*. In this spectrum, half of the radiated energy is in photons with $k \leq \varepsilon_c$, while the remainder is in photons with $k \leq \varepsilon_c$. The peak value is at 0.3 ε_c.

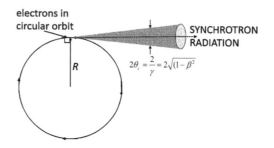

FIGURE 3.13
Synchrotron radiation pattern for relativistic particles at the instantaneous orbit location denoted by "electrons." Twice the characteristic opening angle θ_c for photons with energy $k = \varepsilon_c$ is shown as the conical shaded region.

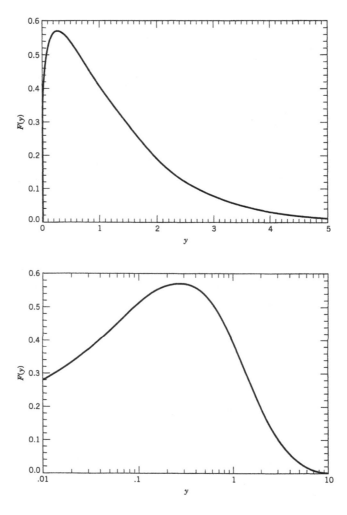

FIGURE 3.14
Universal synchrotron radiation spectrum plotted on two different sets of scales. The graph gives the intensity of radiation, here $F(y)$, as a function of the ratio $y = k/\varepsilon_c$. (Jackson, J. D: *Classical electrodynamics*, third edition. New York, NY: John Wiley and Sons. 1998. Copyright Wiley-VCH Verlag GmbH & Co. KGaA. Reproduced with permission.)

For singly charged particles other than electrons of rest mass m_x, the characteristic energy is obtained by multiplying the result of Equation 3.42 by a factor of $(m_e/m_x)^3$, where m_e is the rest mass of the electron. The characteristic energy for synchrotron radiation for *protons* having the same total energy as electrons is thus far lower. As will become obvious, it is often convenient to specify these and other quantities as functions of the magnetic field B (Tesla) that creates the circular orbit of radius R for particles of momentum p (GeV/c) by recalling Equation 2.26:

$$R = \frac{p}{0.29979qB} \text{ (meters)} \tag{3.44}$$

If one substitutes for R, recall that $q = 1$ for electrons (ignoring the negative sign of the charge); disregard the small distinction between kinetic energy, momentum, and total energy for *relativistic* electrons; and combine constants:

$$\varepsilon_c = 0.6649W^2B \text{ (keV)} \tag{3.45}$$

For relativistic conditions (i.e., $\gamma \gg 1$), the mean number of photons emitted per complete revolution is

$$N_\gamma = \frac{5\pi}{\sqrt{3}}\alpha\gamma \tag{3.46}$$

where α is the *fine structure constant* of atomic physics ($\approx 1/137$, see Table 1.1). Since the bending magnet spectrum has considerable "skewness," again for $\gamma \gg 1$, the *mean energy per photon* $\langle \varepsilon \rangle$ is

$$\langle \varepsilon \rangle = \frac{8}{15\sqrt{3}}\varepsilon_c \tag{3.47}$$

As an electron circulates in this circular orbit, the *energy loss per revolution* is given by

$$\delta E = \frac{0.08846}{R}W^4 \text{ (MeV)} \tag{3.48}$$

with W in GeV and R in meters. An alternative form commonly used arises from substituting for R employing Equation 3.44

$$\delta E = 0.02652W^3B \text{ (MeV)} \tag{3.49}$$

with B in Tesla.

If the orbit is a circle with continuous, uniform bending around the circumference and with straight sections of "negligible" length, it should be clear that a circulating current I (milliamperes) can be connected with the *radiated power* P (watts). First, one needs to determine the number of electrons s^{-1} per milliampere current by making the simple unit conversion:

$$I(\text{mA}) = I\left(\frac{10^{-3}\text{Coulombs}}{\text{sec}}\right) \times \left(\frac{1 \text{ electron}}{1.602 \times 10^{-19}\text{Coulombs}}\right) = I\left(\frac{\text{electrons}}{\text{sec} \times \{1.602 \times 10^{-16}\}}\right)$$

Then one can derive the radiated power from Equation 3.48;

$$P = \frac{0.08846W^4}{R} \frac{\text{MeV}}{\text{electron}} I \left(\frac{\text{electrons}}{\text{sec} \times \{1.602 \times 10^{-16}\}} \right) \times \frac{1.602 \times 10^{-13} \text{ Joule}}{\text{MeV}}$$

$$= \frac{88.46W^4 I}{R} \frac{\text{Joules}}{\text{sec}} = \frac{88.46W^4 I}{R} \text{ watts} \qquad (3.50)$$

Again using Equation 3.44, this can be expressed in terms of the magnetic field:

$$P = 26.52W^3 BI \text{ watts} \qquad (3.51)$$

For singly charged particles other than electrons of rest mass m_x, the corresponding radiated power P is obtained by multiplying this result by a factor of $(m_e/m_x)^4$, where m_e is the rest mass of the electron. Again, one can see why all synchrotron radiation facilities (i.e., "light sources") are based on circulating electrons, not protons or ions. However, for ultra-high-energy proton accelerators (e.g., the Large Hadron Collider at CERN, France and Switzerland), the need to replenish the energy lost through synchrotron radiation to maintain circular orbits can be a significant electrical power demand. The energy lost by synchrotron radiation is replaced by the RF system of the accelerator or storage ring, in order to maintain a specified circular beam orbit.

Synchrotron radiation possesses an additional property not further discussed in detail here: the fact that the photons are polarized to a rather high degree, greater than 80% is typical, in the plane of the ring in which they orbit. These large polarizations can be further manipulated and are extremely beneficial to the users of light sources. They also can result in asymmetries in the radiation production by these accelerators by favoring photons or neutrons emitted in one range of azimuthal angle ψ (see Figure 2.2) over others.

3.6.2 Insertion Devices

The researchers who use modern light sources are not limited to the broad band of photons obtained from the general bending of the electron beam around its circular orbit. It was realized at an early stage in the development of this technology that if one were to insert a set of bending magnets of alternating polarities into a straight section of a ring, *smaller* bending radii over a localized region of the orbit could be produced that would result in radiation of *higher* energy photons according to Equation 3.42 or Equation 3.45. Figure 3.15 shows such a *wavelength shifter* schematically.

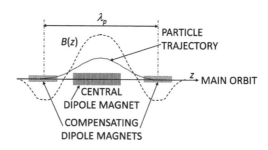

FIGURE 3.15
A wavelength shifter showing the three magnets involved, the magnetic field strength $B(z)$ as a function of longitudinal coordinate z, and the electron trajectory.

In Figure 3.15 λ_p is the length of a *period* of a group of such magnets. As discussed further in the following text, the pole pieces are typically short; the magnetic field strength may have an approximately sinusoidal dependence on z as depicted here. The compensating magnets are required to restore the electrons to the main circulation orbit. It is clear that while more energetic photons would be emitted, their intensities would be limited by the short fraction of time of each orbit that the individual beam electrons are deflected by this higher magnetic field.

This alignment of a series of such magnets of alternating polarities in a row can be used to increase the intensity of the desired photons. If there are N_m such pairs of magnet poles (i.e., "periods") in the system, then one will get $2N_m$ times the photons provided by one of them, neglecting end effects. These magnets could be dipoles of alternating polarities lined up in any plane. In practice, they are generally set to bend charged particles in the bending plane of the storage ring. This avoids some complications for accelerator functionality, since it limits the coupling between horizontal and vertical phase space (i.e., "betatron") oscillations in the storage ring. Such betatron oscillations are described in numerous references on accelerator physics (e.g., Edwards and Syphers 1993; Cossairt 2008; Chao et al. 2013).

The magnetic field strength in these magnets can be of any value if there is no net deviation of the overall orbit of the electrons, aside from corrections that might be needed to compensate for additional dispersion and aberrations introduced by this "device" (see Section 2.5.2).

Components of this type placed in storage rings to create specialized photon energy spectra are called *insertion devices*. Typical modern light sources contain multiple sets of such devices designed to create particular photon beam properties desired by specific scientific users of the facilities. Some employ permanent magnets, while others may utilize electromagnets including superconducting magnets, to achieve the desired magnetic fields. The design of such devices is an ongoing development at such facilities. Figure 3.16 is a schematic picture of an insertion device.

The definition of a special parameter is useful in this discussion. Consider a device consisting of a large number of alternating magnet poles. The spacing of each *pair* of poles λ_p as defined in Figure 3.15 constitutes the length of the *period*. Because the lengths of the individual pole pieces are often short compared with the dimensions of the field gaps, truly *uniform* dipole field conditions are commonly not fully achieved if this is not necessary for the desired performance. Instead, the magnetic field component perpendicular to the bending plane, here denoted B_y, is often approximated by a *sinusoidal* dependence on the longitudinal coordinate z as shown in Figure 3.15:

$$B_y = B_o \sin \frac{2\pi z}{\lambda_p} \tag{3.52}$$

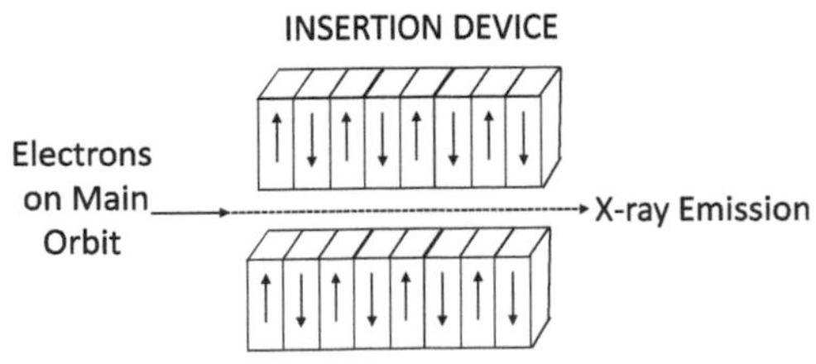

FIGURE 3.16
A typical insertion device. The vertical arrows show the orientation of the magnetic field in the individual gaps of the components of the device.

Now one can calculate the angle α_m, the maximum deflection of the electrons away from the central axis, as they proceed along the insertion device using Equation 3.44 by performing an integration over the longitudinal coordinate z. Given the size of practical insertion device pole pieces, it is useful to work with z and λ_p in centimeters. Performing the integration after making the unit conversion,

$$\alpha_m = \frac{2.9979 \times 10^{-3} B_o}{p(\text{GeV}/c)} \int_0^{\lambda_p/4} dz \sin \frac{2\pi z}{\lambda_p} = \frac{\lambda_p}{2\pi} \frac{2.9979 \times 10^{-3} B_o}{p(\text{GeV}/c)} \left[-\cos \frac{2\pi z}{\lambda_p} \right]_0^{\lambda_p/4}$$

$$= (4.771 \times 10^{-4}) \frac{B_o \lambda_p}{p} \text{ (radians)} \tag{3.53}$$

One multiplies this by the relativistic parameter of the circulating electrons γ to define a new dimensionless parameter K, the *wiggler strength parameter* or *deflection parameter*. Since the electrons that produce useful synchrotron radiation are highly relativistic, working in energy units as was done in Equation 3.53,

$$\gamma = \frac{W}{m_o} \approx \frac{p}{m_o}, \text{ and } K = \gamma \alpha_m = (4.771 \times 10^{-4}) \frac{B_o \lambda_p}{m_o} = 0.934 B_o \lambda_p \tag{3.54}$$

with B_o in Tesla, λ_p in centimeters, and the rest energy of the electron ($m_o c^2$) in GeV. K is called the *wiggler strength parameter* or *deflection parameter*. Its role can be better understood with the help of Figure 3.17.

Reminded by Figure 3.17 that the cone of emission of the synchrotron radiation has an approximate half-width of $1/\gamma$, for K > 1 the maximum deflection α_m is thus *larger* than the cone of emission. Under such conditions, the insertion device is called a *wiggler*, and the synchrotron radiation produced has the bending magnet energy spectrum shape appropriate for the magnetic fields of the insertion device.

For K < 1, the divergence due to the magnetic deflections is *smaller* than the intrinsic cone of emission, and the device is then called an *undulator*. In an undulator, since the deflections occur *within* the cone of emission, interference effects as can occur with visible light result. In fact, these are exploited to provide approximately monochromatic photons or spectra with other desired properties.

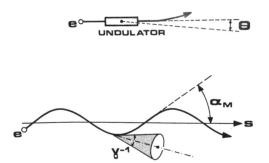

FIGURE 3.17
Comparison of the half-angle of the cone of emission of radiation $\approx \gamma^{-1}$ and maximum deflection angle of the electron trajectory caused by an insertion device α_m. (Margaritondo, G, *Introduction to synchrotron radiation*, 1988, New York, NY, by permission of Oxford University Press. www.oup.com.)

While wigglers are useful for making the energy spectrum of the photons more energetic (i.e., "hardening" the spectrum), undulators can be used to create high-intensity (i.e., "bright") beams of nearly monoenergetic photons or a spectrum of photons delivered in a few narrow spectral bands. The increased brightness is due to the smaller dispersion due to the bending magnet deflections. Foregoing the details of a somewhat complicated derivation, the *undulator frequency* ν_u of the photons produced is given by Wiedemann (2003, 2013):

$$\nu_u \approx \frac{2c\gamma^2}{\lambda_p} \frac{1}{(1+K^2/2+\gamma^2\theta^2)} \text{ (Hz)} \tag{3.55}$$

for small, but not negligible, values of K, angles of emission θ in radians and λ_p in centimeters. Since K is a function of magnetic field strength and magnet pole spacing, this frequency can be adjusted to some degree by altering those parameters. At "intermediate" values of K, other spectral peaks at harmonics of the previously mentioned frequencies become possible. It is, of course, easy to obtain the corresponding photon energy E_u by applying Planck's constant h:

$$E_u = h\nu_u \approx \frac{(2.480\times10^{-7})\gamma^2}{\lambda_p} \frac{1}{(1+K^2/2+\gamma^2\theta^2)} \text{ (keV)} \tag{3.56}$$

One should consider the power that can be emitted in the tightly focused undulator beam. For an undulator of N_m periods, the power emitted (Margaritondo 1988) is given by

$$P_{tot} = \frac{(1.9\times10^{-12})N_m\gamma^2K^2I \text{ (mA)}}{\lambda_p \text{ (cm)}} \text{ (kW)} \tag{3.57}$$

Unlike for the bending magnet situation, this power would be emitted into a very small $1/\gamma$ cone, not in the "pancake-shaped" distribution around the entire circumference representative of the bending magnet situation.

Figure 3.18 shows some qualitative examples of spectra emitted by different types of insertion devices. Obviously, these devices continue to evolve, and more complicated ones are being developed to address specific applications. Collimation is often used to select desired portions of these spectra, optimized for their intended use.

3.6.3 Radiation Protection Issues Specific to Synchrotron Radiation Facilities

Obviously, all the radiation protection concerns discussed elsewhere in this text pertinent to electron accelerators of the same energies and intensities apply to synchrotron radiation facilities. These include the production of bremsstrahlung photons, the production of neutrons and high-energy particles, the development of electromagnetic cascades, and the production of induced radioactivity. However, there are unique phenomena prominent at these facilities. Though summarized here, these and related topics are discussed in more detail elsewhere (e.g., Rindi 1982; Ban et al. 1989; Tromba and Rindi 1990; Ipe and Fassò 1994; Liu et al. 1995; Liu et al. 2005).

3.6.3.1 Operating Modes

Synchrotron radiation sources largely operate as storage rings. To accommodate insertion devices and experimental apparatus, these storage rings often have multiple straight

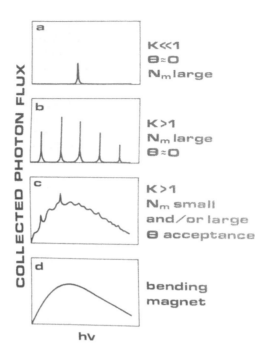

FIGURE 3.18

Different examples of insertion device emission, compared to bending magnet radiation. *Case a* is the line emission from an undulator, seen through a pinhole that limits the angular acceptance. *Case b* is a strong field device (not further described here), again seen through a pinhole so as to limit the angular acceptance. In *case c*, a limited number of periods are used to create a spectrum of small peaks atop the bending magnet spectrum. *Case d* is the pure bending magnet spectrum. All four of these are plotted on a *logarithmic* abscissa akin to that of the lower frame of Figure 3.14. (Margaritondo, G, *Introduction to synchrotron radiation*, 1988, New York, NY, by permission of Oxford University Press. www.oup.com.)

sections. Operating modes need to be considered. Typically, the electron beam is produced by an injector accelerator of some type and injected into the main storage ring in an *injection event*. Following injection the beam is typically smoothed out for several thousand turns before being added to the stored beam. During the injection process, electrons having trajectories that are undesired in the storage ring are typically absorbed on limiting apertures designed to "clean-up" the beam for storage or are otherwise lost around the ring. Then, the beam is used for the intended research purposes for long periods of time. At the end of a defined period of storage ring operations usually called a *store*, the beam is generally disposed of in a beam absorber, and the cycle is repeated. An advancement made to improve the efficiency of operations is the capability to *replenish* the beam without ending a store by delivering additional electrons from the injector to the storage ring in a *top-off mode*.

Often the personnel protection requirements and beamline access restrictions imposed on the researchers are considerably different during injection events and storage ring operations due to the differences in the levels and types of radiological hazards presented.

Typically, radiological problems are most prominent during the injection events because at that time, errors of beam tuning may result in point losses, and the rate of beam delivery can be large, perhaps up to the output of the injector. These considerations cannot be ignored during the top-off mode of operation. During pure storage ring operations, since accelerator orbits have been established to achieve a useful beam lifetime, inductive time constants render sudden, large losses due to mistuning or collapse of the magnetic field (e.g., during a power failure) much less probable. Even a slow deterioration of the vacuum would cause beam losses to occur over numerous orbits; thus, such losses would not be highly localized. Point losses can, however, occur due to other types of events such as the sudden closure of vacuum valves or some other unintended movement of material into the beam. However, while the stored beam *current* may be significant, during the storage ring mode (i.e., not during injection or top-off), the total *number* of stored electrons is limited and serves as an upper limit to beam loss events. Devices called beam stops, shutters, or injection stoppers are inserted into the front end of beamlines to protect personnel and equipment from the consequences of beam losses during injection events. Obviously, it is imperative to fully understand the beam loss characteristics at every stage of operation.

3.6.3.2 Gas Bremsstrahlung: Straight Ahead

At these facilities, the decay of the stored beam will be dominated by scattering from the residual gas particles. Though the vacuum can be made to be very good, the path lengths of the electrons in a storage ring mode are extremely long when huge numbers of orbits are taken into account. Also, the synchrotron radiation photons can induce outgassing in certain materials that may *increase* residual pressures within the vacuum chamber. The process of beam interaction in the residual gas is obviously a "thin target" phenomenon, otherwise the electron beam could not be stored. Under some circumstances, equipment damage concerns are important. Following scattering events the electrons will spiral radially inward and be lost. Workers at various laboratories have developed computational methods using both analytical and Monte Carlo techniques to address these matters. In the discussion here, results using both methods are used as illustrations.

From Equation 3.8 *gas bremsstrahlung* has a nearly $1/k$ energy spectrum (with k denoted as the photon energy to distinguish it from the electron beam energy). The spectrum extends essentially from zero up to almost the kinetic energy of the stored electrons, an energy far higher than that seen for the synchrotron radiation photons as discussed. The angular distribution is highly forward-peaked, having a characteristic angle (i.e., a "1/e" angle) of $0.511/E$ in radians for electron beam energy E (MeV). The dose is approximately proportional to $E^{2.5}$ and of course the mass thickness of the air column in the ring section through which the electrons pass. It is obvious that the photons from the gas bremsstrahlung are far more energetic and hence more difficult to shield than are the synchrotron radiation photons of much lower energies (i.e., x-rays). To better understand this, Tromba and Rindi (1990) performed Monte Carlo calculations with the code EGS4 for the geometry shown in Figure 3.19.

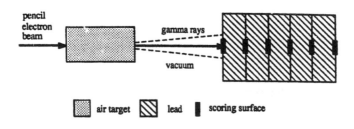

FIGURE 3.19
Geometry considered by Tromba and Rindi in their Monte Carlo calculations of gas bremsstrahlung. (Reprinted from *Nuclear Instruments and Methods*, A292, Tromba G., and A. Rindi, Gas bremsstrahlung from electron storage rings: A Monte Carlo evaluation and some useful formulae, 700–705, Copyright 1990, with permission from Elsevier.)

In this geometry, an electron pencil beam crosses an air target. The bremsstrahlung photons are attenuated in lead. The number of photons and the relative dose are scored on a small surface (about 1.0 cm²), smaller than the photon-beam angular opening, at different depths in the lead. The Monte Carlo program tabulated "histories" on the locations labeled "scoring surfaces." As a result of their calculations, these authors proposed the following expression for the dose rate $dD/dt_{10\,meters}$ at 10 m "on axis" from the end of the straight section (i.e., the "air target" in Figure 3.19):

$$\frac{dD}{dt}_{10\,meters} = 3.32 \times 10^{-9} E^{2.43} \frac{p}{p_{atm}} \frac{dN}{dt} L \, (\text{Gy h}^{-1}) \tag{3.58}$$

where E is the electron energy (GeV), dN/dt is the number of electrons s⁻¹ passing through the straight section, p/p_{atm} is the ratio of the residual pressure to atmosphere pressure, and L is the length of the straight section in meters. It is often convenient to work with the beam current I (mA) and residual pressure P (torr). Units of measure of pressure are 1.0 torr = 1.0 mm of Hg pressure = 1/760 of a standard atmosphere = 133.3 Pa. Equation 3.58 then becomes

$$\frac{dD}{dt}_{10\,meters} = 2.72 \times 10^{4} E^{2.43} PIL \, (\text{Gy h}^{-1}) \tag{3.59}$$

For other distances r(m) measured from the center of the straight section, inside of this narrow radiation cone, one should scale this result by the inverse square factor:

$$\frac{dD(r)}{dt} = \frac{dD}{dt}_{10\,meters} \left\{ \frac{10 + L/2}{r} \right\}^{2} \tag{3.60}$$

Equation 3.60 is valid as long as one is within the radiation emission cone and $r > L/2$.

Of course one will need to calculate the thickness of shielding needed to attenuate this source of radiation to some desired level. Fortunately, as exhibited in Figure 3.9, the photon mean free path is very weakly dependent on energy over several orders of magnitude in the energy domain of interest. Tromba and Rindi (1990) found that for lead shielding, the attenuation, after some initial buildup region of a few centimeters, can be characterized as an exponential one having an attenuation coefficient of ≈ 0.6 cm⁻¹, a parameter only very weakly dependent on beam energy.

3.6.3.3 Gas Bremsstrahlung: Secondary Photons

Another consequence of gas bremsstrahlung is the generation of secondary photons by interactions of the gas bremsstrahlung photons with materials. This applies when the bremsstrahlung photons are incident on some sort of absorber or beam "shutter" in the absence of the beam electrons that have been deflected somewhere else by bending magnets. A useful prescription has been presented by Liu and Vylet (2001) to estimate this effect. For a thin piece of material, the lateral photon dose is largest at somewhat forward angles. For more massive objects, the maximum in the lateral photon dose is at "backward" (i.e., $\theta > 90°$) angles. From Equation 3.6, the fractional energy transferred from an electron to the photons dE/E is equal to the ratio of the mass thickness t of the column of residual gas to its radiation length X_o. Thus, the fractional energy or power (with units of time included) transferred to gas bremsstrahlung photons from the circulating electrons is t/X_o, where the radiation length of air from Table 1.2 is 36.62 g cm^{-2}. One can multiply this ratio by the stored power of the electron beam to determine the *bremsstrahlung power*; that is, the power that is delivered to the bremsstrahlung photons. To illustrate how this can be used, consider an example for a 3.0 GeV storage ring that stores 500 mA of electron beam current. Further assume that $L = 5.0$ meters and the residual gas pressure is 1.0×10^{-9} torr. For this, applying the atmospheric density at normal temperature and pressure (NTP), one can calculate mass thickness t by simple unit conversion:

$$t = 500\,\text{cm} \times \frac{10^{-9}\,\text{torr} \times \text{atmosphere}}{760\,\text{torr}}\, \frac{1.205\,\text{g}}{1000\,\text{cm}^3\,\text{atmosphere}} = 7.928 \times 10^{-13}\,(\text{g cm}^{-2})$$

Thus the fraction of the total beam power diverted into gas bremsstrahlung by interactions in the residual gas of this particular straight section is

$$F_{brem} = \frac{t}{X_o} = \frac{7.928 \times 10^{-13}}{36.62} = 2.165 \times 10^{-14} \tag{3.61}$$

Under these conditions the stored beam power is 1.5×10^9 watts. Applying the result of Equation 3.61, 3.24×10^{-5} watts is transformed into bremsstrahlung radiation at this particular location. When this bremsstrahlung bombards a solid object in a beamline, an electromagnetic cascade is initiated. For simplicity, this "object" will be taken to be a cylinder several radiation lengths long characterized by a Molière radius X_m as defined in Equation 3.20. Equation 3.35 gives the fraction of the incident bremsstrahlung beam power that escapes a thickness R of this shield *laterally* F_{esc} $(0 \leq F_{esc} \leq 1)$:

$$F_{esc} = \frac{U(R/X_m)}{E_o} = 0.8\exp\left[-3.45\left(\frac{R}{X_m}\right)\right] + 0.2\exp\left[-0.889\left(\frac{R}{X_m}\right)\right] \tag{3.62}$$

Thus, in our example $3.24 \times 10^{-5}F_{esc}$ watts will escape laterally. Making a unit conversion, this is equivalent to $2.022 \times 10^8 F_{esc}$ MeV s^{-1}. For these laterally produced photons, it is reasonable to take their average energy to be about 1.0 MeV. Thus, this configuration represents a finite, uniform *line source* of strength $S_L = 4.044 \times 10^5 F_{esc}$ photons cm^{-1} s^{-1}. It has been shown by others (e.g., Jaeger et al. 1968), that the flux density at a distance a away from a line source length L on the perpendicular bisector is

$$\phi(a) = \frac{S_L}{2\pi a}\text{Tan}^{-1}\left(\frac{L}{2a}\right) \tag{3.63}$$

with the value of Tan^{-1}(L/2a) obtained in radians. Evaluating this at $a = 1.0$ meter,

$$\phi(100\,\text{cm}) = \frac{4.045 \times 10^5}{2\pi(100\,\text{cm})}\left(\frac{\text{photons}}{\text{cm s}}\right)F_{esc}\,\text{Tan}^{-1}\left(\frac{500}{200}\right) = 766.3 F_{esc}(\text{photons cm}^{-2}\,\text{s}^{-1}) \tag{3.64}$$

Multiplying by an appropriate dose per fluence value of about 4×10^{-6} μSv cm^2 (see Figure 1.5) gives a dose rate $dH/dt = 3.07 \times 10^{-3}F_{esc}$ μSv s^{-1} = 11.0 F_{esc} μSv h^{-1}, a measurable value even with the extremely good vacuum postulated if F_{esc} is at all significant.

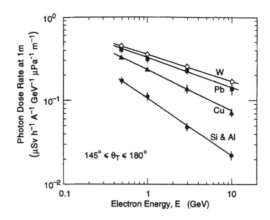

FIGURE 3.20

Secondary photon dose rate at 1.0 m lateral to large targets of various materials as a function of electron beam energy. The values are normalized to a beam current of 1.0 A, per GeV beam energy, and an air path of 1.0 m with a pressure of 1 µPa (=7.50 × 10⁻⁹ torr). (Reprinted with permission from Liu, J. C. et al. 1995. *Health Physics* 68:205–213. https://journals.lww.com/health-physics/pages/default.aspx.)

Liu and Vylet (2001) report that estimates of this type agree with measurement to within a factor of two or three. This methodology is somewhat overly simplistic; it ignores some forward peaking. For thick targets, the dose equivalent rate will be larger at *backward* angles. For small objects struck by the beam, a point source approximation may be a better choice.

Liu et al. (1995) have performed a more sophisticated calculation of such photon dose rates using the EGS4 code for backward angles (i.e., "upstream" of the beam shutter). Their results for various materials are given in Figure 3.20.

These results exhibit an energy dependence, because at the higher energies, the location of the shower maximum is located *deeper* in the absorber. Thus, at higher energies the photons will be *more* attenuated as they move backward out of the target. Liu et al. (1995) report parametrizations of the energy dependencies of H_s, the photon surface dose rate (µSv h⁻¹ A⁻¹ GeV⁻¹ µPa⁻¹ m⁻¹) at 1.0 m lateral distance from the beam line:

$$H_s = 0.35E^{-0.33} \text{ (tungsten)},$$
$$H_s = 0.32E^{-0.36} \text{ (lead)},$$
$$H_s = 0.23E^{-0.49} \text{ (copper), and} \tag{3.65}$$
$$H_s = 0.11E^{-0.69} \text{ (silicon and aluminum)}$$

3.6.3.4 Gas Bremsstrahlung: Neutron Production Rates

The prodigious production of photons can lead to correspondingly significant photoneutron production. Using a methodology similar to that employed previously, Equation 3.61 can be used to calculate the bremsstrahlung power at the end of a given straight section. These photons may then be incident on some device such as a beam shutter and produce neutrons. For example, when the straight section as previously discussed is operational, then 3.24 × 10⁻⁵ watts goes into the bremsstrahlung at that location. Typical beam shutters might be made of tungsten. For tungsten Figure 3.5 and Table 3.1 give a total photoneutron yield of 2.36 × 10⁹ neutrons s⁻¹ W⁻¹, so 7.65 × 10⁴ neutrons s⁻¹ will be emitted if this

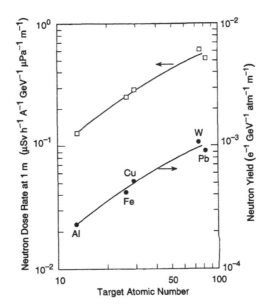

FIGURE 3.21
Neutron dose rate at 1.0 m away from a beamline device struck by gas bremsstrahlung and the neutron yield within the device as a function of the atomic number of the device. The values are normalized to a beam current of 1.0 A, per GeV beam energy, and an air path of 1.0 m with a pressure of 1.0 μPa. (Reprinted with permission from Liu, J. C. et al. 1995. *Health Physics* 68:205–213. https://journals.lww.com/health-physics/pages/default.aspx.)

bremsstrahlung power is incident. Since the photoneutrons are isotropic and "point source" conditions are a good approximation, a flux density of 0.61 neutrons cm^{-2} s^{-1} is found at a distance of 1.0 m without including the effects of any intervening shielding. For these giant resonance neutrons, a conservative dose equivalent per fluence value is 3.2×10^{-4} μSv cm^2. When applied here, a dose equivalent rate of $dH/dt = 1.95 \times 10^{-4}$ μSv $s^{-1} = 0.70$ μSv h^{-1} is calculated. Neutrons from quasi-deuteron and photopion reactions are ignored in this estimate. According to Liu and Vylet (2001), these values are in reasonable agreement with measurement and more sophisticated calculations.

Liu et al. (1995) have given more detailed results for neutron production by gas bremsstrahlung incident on various materials provided in Figures 3.21 and 3.22.

Figure 3.21 presents both the normalized neutron dose rates at 1.0 m from the target and the neutron yield with the target as a function of target atomic number. These results are for targets with sufficient size ($\approx 30X_o$ long and $30X_m$ in diameter) to generate near maximal neutron yields.

Figure 3.22 can be used to estimate results for shorter targets made of lead.

3.6.3.5 Importance of Ray Tracing

Figure 3.23 shows a schematic plan view layout of a typical synchrotron radiation light source facility showing the main synchrotron ring, insertion devices, and the experimental apparatus housings commonly called *shacks* or *hutches*. At such facilities a great many components, beamlines, shielding, and experimental apparatus are placed in a small space and are generally tangential to the main synchrotron ring. Given this plethora of equipment elements mounted both within a light source storage ring and associated with its beam

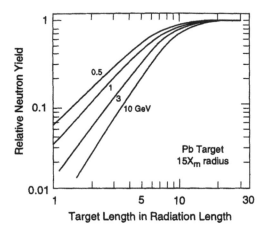

FIGURE 3.22
Relative neutron yield as a function of the target length in units of radiation length for a cylindrical lead target 15 Molière units in radius struck by gas bremsstrahlung for four electron beam energies. (Reprinted with permission from Liu, J. C. et al. 1995. *Health Physics* 68:205–213. https://journals.lww.com/health-physics/pages/default.aspx.)

lines and experiments, it is generally important to do careful ray tracing studies for both the ring shielding and beamline shielding designs to be sure that secondary radiation from electron losses in the ring (normal and abnormal) as well as synchrotron radiation, gas bremsstrahlung "beams," and photoneutrons in beamlines are effectively prevented from reaching undesired locations. All operational modes including injection events, top-offs, and storage ring runs need to be considered. One must be sure that no shielding voids, cracks, or holes are present that could result in significant radiation exposures, especially since many penetrations are commonly needed to support operation of the apparatus located in the hutches. It is evident that all three spatial dimensions must be considered, especially given the fact that such facilities will commonly have numerous such experimental stations so that personnel must be watchful for radiation sources impinging on their neighbors.

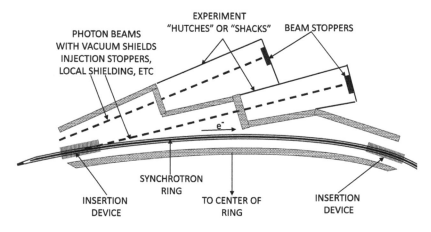

FIGURE 3.23
Plan view schematic layout of a portion of a typical synchrotron radiation source showing the beamline elements and the location experimental stations, called "shacks" or "hutches."

PROBLEMS

1. An electron accelerator has a beam profile in the form of a 2.0 mm diameter circle uniformly illuminated by the beam. Make a crude plot of the value of the dose equivalent rate *in the beam* as the energy increases from 1.0 MeV to 10 GeV. The average beam current is 1.0 microamp (1.0 μA). Assume the beam profile is unchanged during acceleration. Compare the results with Swanson's simple formula, said to be a "conservative" value. Is his formula "conservative" above 100 MeV? (Hint: Use Figure 1.4.)

2. Calculate for electrons the critical energy and length of material that corresponds to the radiation length for carbon and for lead. What does this say about the effectiveness of low-Z versus high-Z shielding materials for electrons?

3. A 100 MeV electron accelerator produces a 1.0 μA beam incident on a high-Z (thick) target. Estimate the bremsstrahlung absorbed dose rates at $\theta = 0°$ and 90° at $r = 2.0$ m from the target using Swanson's rules of thumb. Compare the 0° result with the "in-the-beam dose equivalent rate" found in Problem 1. How do the bremsstrahlung and in-beam dose rates compare?

4. Suppose the enclosure of the former Tevatron at Fermilab is used to house an electron synchrotron. The radius of the synchrotron is 1000 m. If the circulating beam is 10^{12} electrons, calculate the characteristic energy ε_c of the synchrotron radiation photons for $E_0 = 100$ GeV. Also find θ_c of the "lobe."

5. For the accelerator of Problem 3, calculate the neutron flux density at $r = 2.0$ m from giant resonance neutrons at large angles using the values in Table 3.1 for a high-Z (tungsten) target. Also use Table 3.1 to estimate the dose equivalent rate at $r = 2.0$ m. Check this result by making an "educated guess" that the average neutron energy is between 1.0 and 10 MeV, and use the curve in Figure 1.5. Compare this neutron dose with the bremsstrahlung dose at large angles obtained in Problem 3.

6. For a 20 GeV electron accelerator operating at 1.0 kW, the electron beam strikes a beam stop made of aluminum or iron. How *long* (in z) does the beam stop have to be to range out all of the muons for either aluminum or iron based on the mean range? Compare the dose equivalent rates at the immediate downstream ends of each material if 10% of the muons leak through due to straggling and multiple Coulomb scattering can be neglected. Assume the production of muons from iron is approximately equal to that from Al. Recall the inverse square law.

7. In the discussion of the longitudinal development of electromagnetic showers, there are three different formulations (Rossi-Griesen, Bathow, and Van Ginneken). Using Van Ginneken's scaling method, calculate the value of the parameter ζ (g cm^{-2}) for $E_0 = 1000$ MeV, 10 GeV, and 100 GeV for copper and lead. Determine the number of radiation lengths to which ζ corresponds for each material at each energy.

8. Compare the results of Van Ginneken for the location of the longitudinal shower maximum with Bathow's result for copper and lead at the three energies given in Problem 7 for incident electrons. Is the agreement better or worse as the energy increases?

9. A hypothetical electron accelerator operates at either 100 MeV or 10 GeV and delivers a beam current of 1.0 μA. Using Table 3.2, calculate the dose equivalent rates in both Sv s^{-1} and rem h^{-1} at the end of a 300 cm long aluminum beam stop, averaged over a 15 cm radius that are due to bremsstrahlung. (The beam stop is

a cylinder much larger than 15 cm in radius.) Then assume that, in order to save space, a high-Z beam stop is substituted. How long of a high-Z beam stop is needed to achieve the same dose rates? (Assume lead is a suitable high-Z material.) Why is the length of high-Z shield different for the two energies? In this problem, assume the values in Table 3.2 are valid for energies as low as 0.1 GeV.

10. In the accelerator and beam stop of Problem 9, if the radius of the beam stop is 30 cm, what is the maximum dose equivalent rate (Sv s^{-1} and rem h^{-1}) on the lateral surface (at contact at $r = 30$ cm) of the beam stop for both energies, 100 MeV and 10 GeV, and both materials? Again assume approximate validity at 100 MeV of the results.

11. Calculate the dose equivalent rate due to neutrons outside a 1.0 m thick concrete shield surrounding a cylindrical tunnel (inner radius 1 m) in which is located a copper target stuck by 1.0 μA beam of 100 GeV electrons. The geometry should be assumed to be optimized for producing giant resonance photoneutrons, and the calculations should be performed at $\theta = 30°$, 60°, and 90°. ($\rho = 2.5$ g cm^{-3} for concrete.) Express the result as Sv s^{-1} and rem h^{-1}. For $\theta = 90°$, use Equation 3.37 as a check.

4

Prompt Radiation Fields due to Protons and Ions

4.1 Introduction

In this chapter the major features of development of prompt radiation fields and the shielding of these fields as they are produced at proton and ion accelerators are addressed. Emphasis is placed on the shielding of neutrons in view of their general dominance of the radiation fields. The shielding against muons at such accelerators is also covered. Methods for utilizing the results of both semi-empirical and Monte Carlo methods in the solution of practical shielding problems are presented.

4.2 Radiation Production by Proton Beams

4.2.1 The Direct Beam: Radiation Hazards and Nuclear Interactions

Measured by the connection of dose with fluence, the direct beams at proton accelerators nearly always present a greater hazard than do any secondary beams since the primary beam intensities are generally larger and their cross-sectional areas smaller. Figure 1.4 gives the dose equivalent per fluence P as a function of proton energy. This quantity has a prominent transition at about 200 MeV. Below that approximate energy threshold the proton ionization range in tissue is less than the typical thickness of the human body. Those protons thus range out in the body and deposit nearly all of their energy in its tissues. Above 200 MeV, the fraction of the proton's energy that escapes from the body gets larger, with less dose thus delivered.

The ionization range of a proton increases monotonically with energy. Since the mass of the proton is so much larger than that of the electron, the radiative processes of bremsstrahlung and synchrotron radiation are usually negligible at proton accelerators. As discussed in Section 4.5, at high energies the cross sections for inelastic interactions become nearly independent of energy with approximately the values tabulated in Table 1.2. Thus, as an individual proton passes through a material medium, the probability of it participating in an inelastic nuclear reaction before it loses its remaining energy to ionization becomes significant and, as the energy increases, the dominant means by which protons are absorbed. Tesch (1985) has summarized this effect, and the results are shown in Figure 4.1 for various materials and energies.

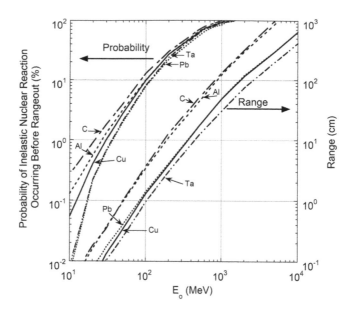

FIGURE 4.1
Ionization range of protons (curves on right and right-hand scale) and probability of inelastic nuclear interaction within the range (curves on left and left-hand scale) for various materials. (Reproduced with permission from Tesch, K. 1985. *Radiation Protection and Dosimetry* 11:165–172.)

4.2.2 Neutrons and Other Hadrons at High Energies

The production and behavior of neutrons at proton and ion accelerators have different characteristics as the energy of the beam particles E_o is increased. Phenomena in a sequence of domains of proton energy of approximate boundaries are discussed.

4.2.2.1 $E_o < 10$ MeV

For a nuclear reaction, the *Q-value* Q_v is the energy released by the reactions and is defined in terms of the rest masses m_i of the participating particles or ions as

$$Q_v = [(m_1 + m_2) - (m_3 + m_4)]c^2 \qquad (4.1)$$

for the nuclear reaction: $m_1 + m_2 \rightarrow m_3 + m_4$ exemplified by, say, $p + {}^{12}C \rightarrow n + {}^{12}N$. A more compact notation commonly used for the same nuclear reaction is $m_2(m_1,m_3)m_4$; for example, ${}^{12}C(p,n){}^{12}N$ for the example given. In using this notation, it is conventional but not required for the "lighter" projectile to be represented by m_1, while generally the less massive emitted particle is represented by m_3. A value of $Q_v > 0$ implies an *exothermic* nuclear reaction. Endothermic ($Q_v \leq 0$) reactions are characterized by a *nuclear reaction threshold energy E_{th}* given by

$$E_{th} = \frac{m_1 + m_2}{m_2}|Q_v| \qquad (4.2)$$

where the absolute value of the negative value of Q_v is taken. Below a kinetic energy of about 10 MeV, (p,n) reactions are important for some materials because these reactions commonly have very low threshold energies (e.g., $E_{th} < 5$ MeV). For example, the reaction ^{7}Li(p,n)^{7}Be has a threshold energy of 1.9 MeV and a reaction cross section σ that quickly rises as a function of energy to a value of about 300 mb. When considering such thresholds, one should keep in mind that nuclear reactions are intrinsically quantum mechanical processes rendering these thresholds to be "blurred" rather than sharp points of transition. For example, quantum-tunneling effects can allow for reactions to proceed that are slightly below the calculated nuclear reaction threshold.

The cross sections of low energy nuclear reactions are highly reflective of the details of the structure of the target nuclei including its excited states. This results in dependencies on target material, scattering angle, and energy. A large repository of related data is maintained by the National Nuclear Data Center (NNDC 2018).

As discussed in Section 3.2.3.4, all accelerators capable of producing neutrons, thermal neutrons can be present and represent an important source of ionizing radiation that requires further consideration.

4.2.2.2 $10 < E_o < 200$ MeV

For protons having kinetic energies of this magnitude and higher, neutrons are usually the dominant feature of the radiation fields resulting from the interactions of the protons. In this energy domain, the neutron yields are smoother functions of energy than was the case at lower energies but are also more forward peaked. For this domain Tesch (1985) has summarized the total neutron yields Y per incident proton for different materials as a function of energy in Figure 4.2.

In Figure 4.2 the smooth curves agree with the original primary data obtained from a myriad of experiments to within about a factor of two. An important feature is that for low energies $Y \propto E_o^2$, while for $E_o > 1.0$ GeV, $Y \propto E_o$. Especially at lower energies, many *evaporation neutrons* are produced. These neutrons can be viewed as "boiling" off of a nucleus that has been "heated" by absorption of energy from the incident particle. They consequently have an isotropic spatial distribution. Other neutrons that are produced are *cascade neutrons* that result directly from individual nuclear reactions that occur within the target nucleus. In contrast to the evaporation neutrons, cascade neutrons are likely to have a directionality that is at least mildly "forward peaked". In this region, angular distribution data are available from nuclear physics research. Representative examples are given in Figures 4.3 and 4.4 for 52 and 200 MeV protons, respectively. Additional examples are provided by the National Council on Radiation Protection and Measurements (2003) and by the NNDC (2018).

In Figure 4.4, the target thicknesses were chosen to exceed the ionization ranges of the protons by approximately 15%, just sufficient to stop all of the protons with this approximate allowance for range straggling.

4.2.2.3 200 MeV $< E_o < 1.0$ GeV: "Intermediate" Energy

In this region, many more reactions become energetically available. Also, the number of protons emitted gradually becomes approximately equal to the number of neutrons produced. In fact, at the highest energies the radiation effects of protons and neutrons are essentially identical. Both sources are important, and reliance on the values shown in Figure 4.2 could underestimate radiation effects by as much as a factor of two. Also, at these energies, cascade neutrons become much more important than evaporation

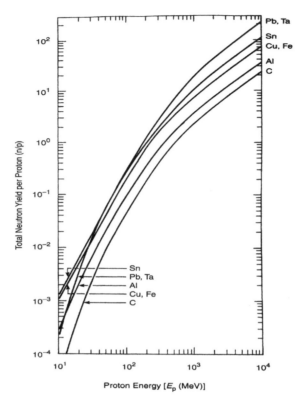

FIGURE 4.2

Total neutron yield per proton for different target materials as a function of incident proton energy E_o. These values apply to relatively thick targets and include some degree of shower development. (Reproduced with permission from Tesch, K. 1985. *Radiation Protection and Dosimetry* 11:165–172.)

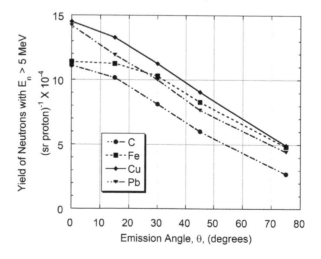

FIGURE 4.3

Measured angular distributions of total neutron yield above 5.0 MeV for carbon, iron, copper, and lead bombarded by 52 MeV protons. The measurements were normalized at $\theta = 15°$. The curves are drawn to guide the eye. (Reprinted from *Nuclear Instruments and Methods*, 151, Nakamura, T., M. Yoshida, and T. Shin. Spectra measurements of neutrons and photons from thick targets of C, Fe, Cu, and Pb by 52 MeV protons, 493–503. Copyright 1978, with permission from Elsevier.)

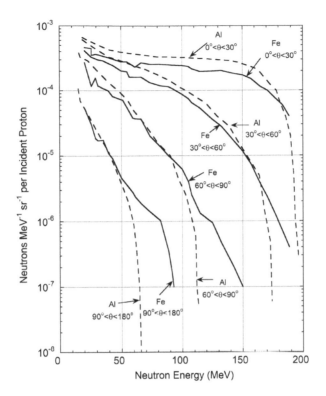

FIGURE 4.4
Calculated energy spectra of neutrons emitted by iron and aluminum targets bombarded by 200 MeV protons for four ranges of the value of θ. (Adapted from the results of Hagan, W. K. et al. 1988. *Nuclear Science and Engineering* 98:272–278 for an iron target and Alsmiller, Jr., R. G. et al. 1975. *Particle Accelerators* 7:1–7 for an aluminum target.)

neutrons. This makes the radiation field more sharply forward peaked with increasing primary particle energy.

4.2.2.4 $E_o > 1.0$ GeV: "High"-Energy Region

In this domain the situation is more complex. Figures 4.5 through 4.8 show representative results for incident protons of energies 14, 26, 22, and 225 GeV, respectively. These results should be regarded as *thin target* values. "Thin" target in this context means a target shorter than the *mean free path* for *removal* of the high-energy protons. Table 4.1 summarizes common removal mean free paths based on the nuclear collision lengths of Table 1.2.

The yield Y or fluence Φ is often normalized either to the number of protons *incident* on a given target N_o or to the number of *interacting protons* N_{int}. The distinction between the two different normalizations must be understood. In view of Equation 1.5, in a target of thickness x and mean free path λ, $N_{int} = N_o[1 - \exp(-x/\lambda)]$ so that $Y/N_o = \{Y/N_{int}\}[1 - \exp(-x/\lambda)]$.

Considerable efforts have been made to semi-empirically fit the distributions of the yields of secondary particles produced by proton interactions. These were done both to supply the needs of the particle physics community and address radiation safety issues. They began in the early days of radiation protection, and the results were embodied in the continual development of Monte Carlo programs designed to calculate the properties of hadronic cascades as discussed in Section 4.6. An example of a particularly successful early model

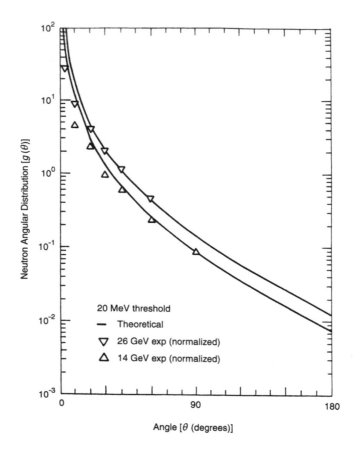

FIGURE 4.5

Measurements of the angular distribution of the yield of neutrons $dY/d\Omega$, here labeled $g(\theta)$, having kinetic energies above 20 MeV produced by 14 and 26 GeV protons incident on a thin beryllium target. The yield is plotted per interacting proton. The symbols represent measurements, while the lines are theoretical calculations using the Ranft formula, Equation 4.3. (Reprinted from *Accelerator health physics*, Patterson, H. W., and R. H. Thomas. New York, NY: Academic Press. Copyright 1973, with permission from Elsevier.)

is the *Ranft formula* expressed as the following formula for the yield $Y(p)$ of protons or neutrons of momentum p (GeV/c) (Ranft 1967):

$$\frac{d^2Y(p)}{d\Omega dp} = \left\{ \frac{A}{p_o} + \frac{Bp}{p_o^2} \left[1 + f(p_o) \left\{ 1 - \frac{p_o}{p} \right\} \right] \right\} \times \left\{ 1 + f(p_o) \left[1 - \frac{pp_o}{m_p^2} \right] \right\} p^2 \exp(-Cp^2\theta) \tag{4.3}$$

$$[\text{protons or neutrons sr}^{-1}(\text{GeV/c})^{-1}\text{per interacting proton}]$$

where p_o is the primary proton momentum (GeV/c), m_p is the proton rest energy (GeV/c²), $f(p_o) = \{1 + (p_o/m_p)^2\}^{1/2}$, and θ is the production angle (radians). The parameters A, B, and C of the Ranft model are material-dependent. Selected values are given in Table 4.2.

4.2.3 Sullivan's Formula

For simple radiation protection estimates, Sullivan (1989, 1992) developed a formula for the fluence $\Phi(\theta)$ of hadrons with $E_o > 40$ MeV that will be produced at 1.0 m from a

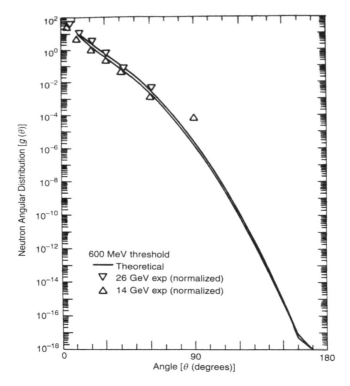

FIGURE 4.6
Measurements of the angular distribution of the yield of neutrons $dY/d\Omega$, here labeled $g(\theta)$, having kinetic energies above 600 MeV produced by 14 and 26 GeV protons incident on a thin beryllium target. The yield is plotted per interacting proton. The symbols represent measurements, while the lines are theoretical calculations using the Ranft formula, Equation 4.3. (Reprinted from *Accelerator health physics*, Patterson, H. W., and R. H. Thomas. New York, NY: Academic Press. Copyright 1973, with permission from Elsevier.)

copper target struck by protons in the energy region $5 < E_o < 500$ GeV per interacting proton:

$$\Phi(\theta) = \frac{1}{2\left[\theta + \left(35/\sqrt{E_o}\right)\right]^2} \text{ (cm}^{-2}\text{per interacting proton)} \tag{4.4}$$

where E_o is in gigaelectron volt (GeV), and θ is in degrees. At proton energies below 2.0 GeV, this formula also approximately accounts for the angular dependence of fluence of all *neutrons* per interacting proton. This equation is plotted in Figure 4.9, for "lateral" ($\theta \approx 90°$) and "forward" ($\theta \approx 0°$) directions.

Of course, the effective dose or dose equivalent is more directly germane to radiation protection than is the fluence. In principle, the dose equivalent can be obtained by integrating the product of the fluence and the dose equivalent or effective dose per fluence $P(E)$ over the spectrum discussed in Section 1.4.3:

$$H = \int_{E_{min}}^{E_{max}} dE P(E) \Phi(E) \tag{4.5}$$

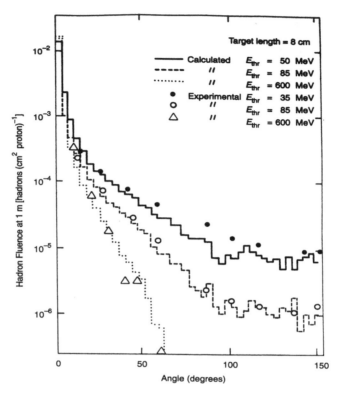

FIGURE 4.7

Measured angular distributions of hadron fluence (particles cm^{-2}) at 1.0 m from a copper target bombarded by 22 GeV protons. Several choices of hadron energy thresholds are shown. The measured fluence is normalized to the number of interacting protons. (Ranft, J., and J. T. Routti. 1972. *Particle Accelerators* 4:101–110. Reprinted with permission from Taylor and Francis Ltd. http://www.tandfonline.com.)

or by discrete summation, taking into account the "coarseness" of available data, and/or calculations:

$$H = \sum_{j=1}^{m} \Delta_j(E) P_j(E) \Phi_j(E) \tag{4.6}$$

Tesch (1985) has done this obtaining the dose equivalent at $\theta = 90°$ and 1.0 m from a copper target bombarded by protons of various energies. The result is plotted in Figure 4.10.

Above about 1.0 GeV the dose equivalent is approximately proportional to E_o. Levine et al. (1972) have measured the angular distribution of absorbed dose for 8.0 and 24 GeV/c protons incident on a Cu target. Results are in approximate agreement with those of Tesch.

4.2.4 Muons

At proton accelerators muons arise from two principal mechanisms: from pion and kaon decay and from "direct" production. Direct muon production is important at only very high–energy proton accelerators. Production by means of pion and kaon decay proceeds as follows where rest energies of the parent particles, the branching fractions (the percentage of time

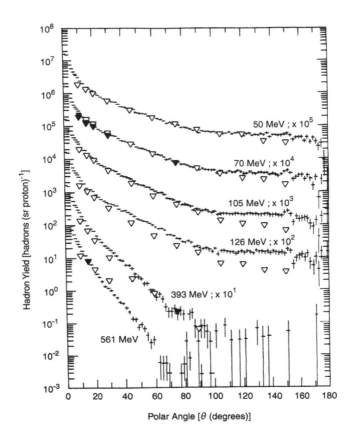

FIGURE 4.8
Measurements of hadron yields $dY/d\Omega$ above different energy thresholds as a function of production angle θ around a 15 cm long copper target bombarded by 225 GeV protons. The data have been multiplied by the indicated powers of 10 prior to plotting. The ∇ symbols denote the results of measurements while the $+$ symbols denote calculations. The measured yield is normalized to the number of protons incident on this particular target. (Reprinted from *Nuclear Instruments and Methods in Physics Research*, A245, Stevenson, G. R. et al., Comparison of measurements of angular hadron energy spectra, induced activity, and dose with FLUKA82 calculations, 323–327, Copyright 1986, with permission from Elsevier.)

the parent particle decays by the stated reaction), the mean-lives τ, and the product of the speed of life c and the mean-life $c\tau$ are also given (Beringer et al. Particle Data Group 2012):

$$\pi^\pm \to \mu^\pm + \nu_\mu; m_\pi = 139.57018 \text{ MeV}, \tau = 2.6033 \times 10^{-8} \text{ s}, (99.99\% \text{ branch}), c\tau = 7.8045 \text{ m}$$

and

$$K^\pm \to \mu^\pm + \nu_\mu; m_K = 493.677 \text{ MeV}, \tau = 1.2380 \times 10^{-8} \text{ s}, (63.55\% \text{ branch}), c\tau = 3.711 \text{ m}$$

The shielding of muons and the production of muons are covered in more detail in Section 4.7.3. Muon radiation fields are forward peaked and normally dominated by those from pion decay. Usually, Monte Carlo techniques are needed to accurately estimate muon intensities, since one needs to simultaneously

- Calculate the production of pions from the proton interactions
- Follow the pions until they decay or interact

TABLE 4.1

Summary of Removal Mean Free Paths for High-Energy Protons

Material	Density	Removal Mean Free Path (g cm^{-2})	Removal Mean Free Path (cm)
Hydrogen gas @ STP	9.00×10^{-5}	43.3	4.81×10^5
Beryllium	1.85	55.8	30.16
Carbon	2.27	60.2	26.58
Aluminum	2.70	70.6	26.15
Iron	7.87	82.8	10.52
Copper	8.96	85.6	9.55
Lead	11.35	116.2	10.24
Uranium	18.95	117.0	6.17
Air @ STP	1.29×10^{-3}	62.0	4.81×10^4
Water	1.00	60.1	60.1
Concrete (typical)	2.50	67.4	26.96
Silicon dioxide (quartz)	2.64	66.5	25.19
Plastics (polyethylene)	0.93	56.9	61.29

Source: Adapted from Beringer, J. et al. (Particle Data Group). 2012. *Physical Review* D86:010001. This report is updated and periodically republished. Current tabulations are available online at http://pdg.lbl.gov/ (accessed October 17, 2018).

TABLE 4.2

Material-Dependent Parameters to Be Used in Equation 4.3

Target	A	B	C
H$_2$	0.55	−0.30	2.68
Be	0.68	−0.39	3.12
Fe	0.92	−0.75	2.90
Pb	1.14	−1.06	2.73

Source: Adapted from Ranft, J. 1967. *Nuclear Instruments and Methods* 48:133–140.

- Adequately account for the range-energy relation and range straggling
- Track the muons to the point of interest, for example, through magnetic fields

4.3 Primary Radiation Fields at Ion Accelerators

Section 4.2 discussed general considerations appropriate for the primary radiation fields generated by accelerated ions as well as protons. In this section special issues found in radiation fields produced by ions other than protons are described. A comprehensive reference on this topic is that of Nakamura and Heilbronn (2005).

4.3.1 Light Ions (Ion Mass Number A < 5)

For such ions there are exothermic reactions that should be treated as special cases. Noteworthy examples followed by their reaction Q-values Q_v in parentheses

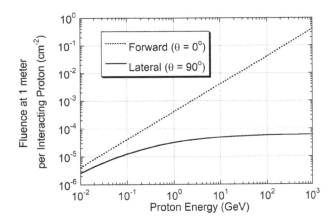

FIGURE 4.9
Plot of Equation 4.4 for two different values of θ according to Equation 4.4 The protons are interacting in a copper target.

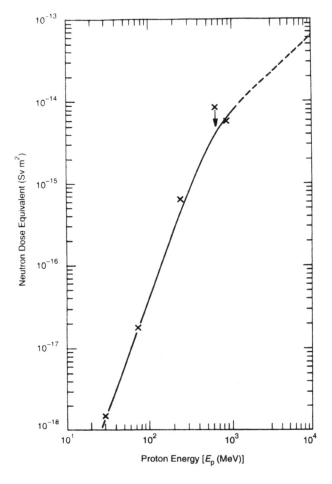

FIGURE 4.10
Dose equivalent per proton due to neutrons at $\theta = 90°$ with energies higher than 8.0 MeV at a distance of 1.0 m from a copper target. The curve is an interpolation between the experimental measurements denoted by the "x" symbols. (Reproduced with permission from Tesch, K. 1985. *Radiation Protection and Dosimetry* 11:165–172.)

(NNDC 2018) are D(d,n)^{3}He ($Q_v = 3.268$ MeV), ^{9}Be(α,n)^{12}C ($Q_v = 5.702$ MeV), and ^{3}H(d,n)^{4}He ($Q_v = 17.589$ MeV).

In some cases monoenergetic beams of neutrons can be produced using these or the following slightly endothermic reactions: ^{12}C(d,n)^{13}N ($Q_v = -0.281$ MeV), ^{3}H(p,n)^{3}He ($Q_v = -0.764$ MeV), and ^{7}Li(p,n)^{7}Be ($Q_v = -1.644$ MeV). The energies of such neutrons can range from 0 to 24 MeV for bombarding energies up to 10 MeV. In general, deuteron stripping and breakup reactions (d,n) have the highest yields, because the binding energy of the deuteron is only 2.225 MeV. In effect, one gets an easily liberated neutron carrying half of the kinetic energy of the deuteron essentially "for free." Furthermore, the neutrons due to deuteron stripping reactions typically have a kinetic energy of about half that of the incident deuteron if the latter has a kinetic energy that is large compared with the binding energy of the target nucleus. This phenomenon is especially pronounced at the lower energies. In the low-energy region, and especially with light ions, one should carefully consider all possible reactions given the materials present in conjunction with the ions that are being accelerated. Patterson and Thomas (1973) have summarized total neutron yields for light ions. In general, the yields for the various light ions behave similarly to those due to protons. That is, the yield is within, typically, a factor of three of that expected for proton beams. A good measurement of neutron yields due to 40 MeV α-particles has been provided by Shin et al. (1995). Higher-energy neutron production data for 640 and 710 MeV α-particles has been provided by Cecil et al. (1980). The NNDC has comprehensive tabulations (NNDC 2018).

4.3.2 Heavy Ions (Ions with A > 4)

At higher energies and especially at higher masses, neutron yield and dose data and calculations are somewhat sparse. These data are usually normalized in terms of kinetic energy per atomic mass unit, the *specific energy*, expressed in units of energy nucleon^{-1} (e.g., MeV/nucleon) because of some scaling of reaction parameters with that normalization. Commonly one simply divides the kinetic energy by the atomic mass number as an *integer*, ignoring the small, but sometimes important differences between the integer mass number and the actual atomic weight that reflects nuclear binding energies. In some papers, the specific energy is stated in units of MeV amu^{-1}, an equivalent terminology. For ions up to 20 MeV amu^{-1}, Ohnesorge et al. (1980) have measured dose equivalent rates at 1.0 m and $\theta = 90°$ from thick targets of iron, nickel, or copper bombarded by ^{4}He, ^{12}C, ^{14}N, ^{16}O, and ^{20}Ne beams. The dose was found to be essentially independent of ion type as a function of specific energy. At a specific energy of 10 MeV amu^{-1}, a value of 6.3×10^{-18} Sv (incident ion)$^{-1}$ was measured, while at 20 MeV amu^{-1}, a value of 3.6×10^{-17} Sv (incident ion)$^{-1}$ was found. Other data relevant to this general energy region are exemplified by those of Hubbard et al. (1960), Aleinikov et al. (1985), and especially Nakamura (1985).

Tuyn et al. (1984) reported studies done with 86 MeV/nucleon ^{12}C ions incident on Fe targets slightly thicker (longer) than an interaction length. The results are shown in Figure 4.11.

In these measurements, activation detectors, as explained in Section 9.5.3, with the following sensitive regions in neutron energy E_n were used: moderated indium foils ("Indium") ($0.4 < E_n < 107$ eV), ^{32}S(n,p)^{32}P ($E_n > 3.0$ MeV), ^{27}Al(n,α)^{24}Na ($E_n > 7.0$ MeV), and ^{12}C(n,2n)^{11}C ($E_n > 20$ MeV).

At a specific energy of 155 MeV/nucleon, Britvitch et al. (1999, 2001) have measured energy spectra and total neutron yields and angular distributions for ^{4}He, ^{12}C, and ^{16}O ions

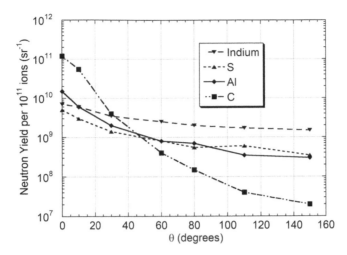

FIGURE 4.11

Measured neutron yields for 86 MeV/nucleon ^{12}C ions incident on an iron target. The lines are intended to guide the eye. (Adapted from the results reported by Tuyn, J. W. N. et al. 1984. *Some radiation aspects of heavy ion acceleration.* European Organization for Nuclear Research: CERN Report TIS-RP/125/CF. Geneva, Switzerland.)

stopping in a thick target of an alloy of tungsten, nickel, and copper commonly known as "Hevimet." The differential yields $dY/d\Omega$ were fit by the form

$$\frac{dY}{d\Omega} = C\exp(-\beta\theta) \tag{4.7}$$

with the total yields being found by performing the integration,

$$Y_{total} = 2\pi \int\limits_{0}^{\pi} d\theta \sin\theta \frac{dY(\theta)}{d\Omega} = 2\pi C \frac{\left[\exp(-\beta\pi)+1\right]}{(\beta^2+1)} \tag{4.8}$$

The results are presented in Figure 4.12. The total neutron yields for ^{4}He, ^{12}C, and ^{16}O were found to be 4.90, 1.56, and 1.74 neutrons per incident ion, respectively.

Clapier and Zaidins (1983) surveyed a sample of ion data from 3 to 86 MeV/nucleon and offered approximations to the total neutron yields and angular distributions over that domain. They found the total yield per ion Y to be given by

$$Y(W,Z) = C(Z)W^{\eta(Z)}\ (\text{neutrons ion}^{-1})\ \text{with} \tag{4.9}$$

$$\eta(Z) = 1.22\sqrt{Z}\ \text{and} \tag{4.10}$$

$$C(Z) = \frac{1.96\times10^{-4}}{Z^{2.75}}\exp\{-0.475\,(\ln Z)^2\} \tag{4.11}$$

where Z is the atomic number of the projectile, and W is the specific energy (MeV/nucleon). They found essentially no dependence on the atomic number of the target material

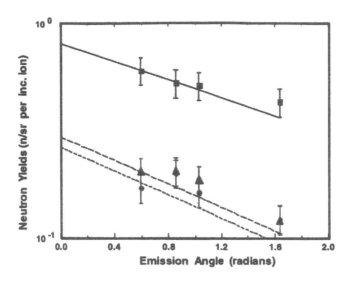

FIGURE 4.12
Neutron yields per incident ion for 155 MeV/nucleon ions reported by Britvich et al. (1999). The filled square boxes ■ are measurements for ^{4}He which were fitted by parameters (C{neutrons (incident ion)$^{-1}$}and β{radian^{-1}}) of (0.8, 0.49) as defined by Equations 4.7 and 4.8. The results for ^{12}C are denoted by filled circles ● and were fit by (C,β) values of (0.26, 0.51). The results for ^{16}O are denoted by filled ▲ triangles and were fit by (C,β) values of (0.29, 0.51). (Reprinted with permission from Britvich, G. I. et al. 1999. Measurements of thick target neutron yields and shielding studies using beams of ^{4}He, ^{12}C, and ^{16}O at 155 MeV/nucleon from the K1200 cyclotron at the National Superconducting Cyclotron Laboratory. *Review of Scientific Instruments* 70:2314–2324. Copyright 1999, American Institute of Physics.)

and assert that an average neutron energy of 6–7 MeV is appropriate. To fit the angular distribution $dY/d\Omega$, the *form factor* $F(\theta,\xi)$ defined as follows was found to be useful:

$$F(\theta,\xi) = \frac{1}{4\pi}\left(\frac{1}{\ln\{1+1/\xi\}}\right)\left(\frac{1}{\xi+\sin^2(\theta/2)}\right) \quad (4.12)$$

where θ is defined as usual, and the fitting parameter ξ is related to the ratio of fluences Φ at $\theta = 0$ and 90°, and thus related to "forward peakedness." In a subsequent paper, Aleinikov et al. (1985) developed the following parameterization for ξ:

$$\xi = \frac{\Phi(90°)}{\Phi(0°)-\Phi(90°)} = \frac{1}{\Phi(0°)/\Phi(90°)-1} \quad (4.13)$$

In this scheme,

$$\frac{dY(\theta)}{d\Omega} = Y(W,Z)F(\theta,\xi) \quad (4.14)$$

Values of the parameters $C(Z)$ and $\eta(Z)$ for specific circumstances are given in Table 4.3. Aleinikov et al. (1985) further give a few examples of the values of their parameter ξ; 0.07 for uranium incident on uranium at 9 MeV/nucleon, 0.025 for neutrons of energy $E_n < 20$ MeV produced by ^{12}C ions at 86 MeV/nucleon incident on iron, and 3×10^{-4} for neutrons of energy $E_n > 20$ MeV produced these same ions, based on an analysis of the data presented in Figure 4.11. In principle one could use values given in Table 4.3 or the direct

TABLE 4.3

Values of the Parameters $\eta(Z)$ and $C(Z)$ as Used in Equations 4.9 through 4.11

Atomic Number	Element	$\eta(Z)$	$C(Z)$
1	Hydrogen	1.5	1.7×10^{-4}
2	Helium	2.6	3.9×10^{-6}
6	Carbon	1.7	2.5×10^{-6}
8	Oxygen	3.6	3.6×10^{-7}
10	Neon	7.0	2.7×10^{-10}
18	Argon	7.0	5.1×10^{-11}
36	Krypton	7.9	6.0×10^{-12}
82	Lead	11.0	1.7×10^{-13}

Source: Adapted from Clapier, F., and C. S. Zaidins. 1983. *Nuclear Instruments and Methods* 217:489–494.

calculation using Equations 4.9 and 4.11 and obtain some idea of the uncertainties inherent in this fit of such a broad range of data. However, the uncertainties in this type of fit are quite large due to the functional forms that were used.

McCaslin et al. (1985) measured the angular distribution yields of 670 MeV/nucleon Ne and Si ions stopped by ionization in a copper target. The distributions for the two different projectiles were similar with fits to the data provided for the ^{20}Ne ions. For these, including all neutrons above 6.5 MeV at a radius of 1.0 m, McCaslin et al. found

$$\Phi(\theta) = 372\frac{1}{\theta}\text{neutrons m}^{-2} \text{ per ion for } 2° < \theta < 180°, \theta \text{ in degrees} \tag{4.15}$$

$$\Phi(\theta) = 248e^{-0.2\theta}\text{neutrons m}^{-2} \text{ per ion for } 0° < \theta < 20°, \theta \text{ in degrees, and} \tag{4.16}$$

$$\Phi(\theta) = 10e^{-0.038\theta}\text{neutrons m}^{-2} \text{ per ion for } 20° < \theta < 120°, \theta \text{ in degrees.} \tag{4.17}$$

The neutron yields at this high specific energy for heavy ions turn out to be quite large. By integrating these properties over all angles, one finds a total yield of 74.1 neutrons per incident ion for $E_n > 6.5$ MeV for ^{20}Ne incident ions.

A more complete picture of heavy ion neutron yields is clearly desirable for intermediate to high energies. The results of Kurosawa et al. (1999), Heilbronn et al. (1999), and Kurosawa et al. (2000) spanning the periodic table represent a major advance. Along with a good parameterization of measured neutron angular distributions not discussed in detail here, Kurosawa et al. (2000) presents a useful simple formula based on geometrical considerations that describe total yield Y of neutrons per incident ion having energies above 5.0 MeV emitted into the hemisphere $0 < \theta < 90°$. This *heavy ion neutron yield formula* is

$$Y = \frac{1.5 \times 10^{-6}}{N_T^{1/3}} W_P^2 \left(A_P^{1/3} + A_T^{1/3}\right)^2 N_P \frac{A_P}{Z_P^2} \text{ (neutrons particle}^{-1}) \tag{4.18}$$

where the subscripts P and T denote the projectile ion and the target, respectively; and Z, N, and A have their usual meanings of atomic number, neutron number, and mass number. W_P is the projectile specific kinetic energy in MeV/nucleon. This formula describes data generally within factors of two or three for ions from He to Xe incident ions fully stopped in

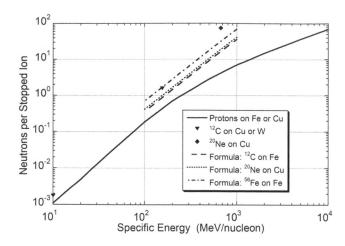

FIGURE 4.13
Neutron yields as a function of specific energy for selected projectiles and targets as reported by the cited references along with results obtained using Equation 4.18.

targets ranging from C to Pb over the specific energy domain $100 < E_P < 800$ MeV/nucleon. Figure 4.13 shows representative total neutron yields for heavy ions from measurements and calculations using Equation 4.18 compared with the yields found for protons at the same specific energy.

It should be noted that measurements of neutron yields using heavy ions are rather difficult. Compared with protons or light ions, target thickness effects are more important due to the shorter ionization ranges in materials for heavy ions. Results are sometimes inconsistent. For example, Equation 4.18 gives a yield of only 18.6 neutrons ion^{-1} for the situation studied by McCaslin et al. (1985). The factor of four discrepancy could be due to a collection of variables such as target sizes and detector efficiencies. Escape of energy via mesons can be important at the higher energies.

4.4 Hadron (Neutron) Shielding for Low-Energy Incident Protons ($E_o < 15$ MeV)

Neutron shielding in this region is complicated by significant nuclear structure effects. There are many resonances associated with compound nucleus that can be excited. There are also nuclear reaction channels leading to a large number of nuclear excited states with a wide variety of nuclear structure quantum numbers and very narrow widths in energy. Clark (1971) has expressed some general rules of attack on the neutron shielding problem. *Clark's principles* are as follows:

- "The shield must be sufficiently thick and the neutrons so distributed in energy that only a narrow band of the most penetrating source neutrons give any appreciable ultimate contribution to the dose outside the shield."

- "There must be sufficient hydrogen in the shield-intimately mixed or in the final shield region-[*sic*] to assure a very short characteristic transport length from about one MeV to absorption at or near thermal energy."

- "The source energy distribution and shield material (non-hydrogenous) properties must be such as to assure a short transport distance for slowing down from the most penetrating energies to one MeV."

An elementary method used to calculate shielding thicknesses in this energy domain is *removal cross section theory*. It has been found that the dose equivalent H as a function of shield thickness t is approximately given for these neutrons by

$$H(t) = \Phi_o PG \exp(-\Sigma_r t) \tag{4.19}$$

where Φ_o is the fluence before the shielding as measured or perhaps calculated from neutron yield information, P is the "average" dose equivalent per fluence factor, G is a "geometry factor," and t (cm) is the thickness of the shield. For parallel beams, $G = 1.0$, while for an isotropic point source, $G = 1/r^2$. Σ_r is the *macroscopic removal cross section*:

$$\Sigma_r = \frac{0.602\sigma_r \rho}{A} \ (\text{cm}^{-1}) \tag{4.20}$$

where σ_r is the microscope removal cross section (barns), and ρ (g cm^{-3}) is the density of the material. For a shielding material that is a mixture of constituents, the overall microscopic removal cross section can be obtained by summing over the partial densities of the constituents. For $A > 8$, a reasonable approximate is

$$\sigma_r = 0.21A^{-0.58} \ (\text{barns}) \tag{4.21}$$

for neutrons of approximately 8 MeV.

Figure 4.14 shows the values of σ_r as a function of mass number at this energy.

Table 4.4 gives representative values for σ_r for sample energies where this approach is applicable.

The use of removal cross sections describes attenuation data rather effectively despite the fact that as more shielding is penetrated, neutrons of lower energy tend to dominate the spectrum over those found in the few MeV region.

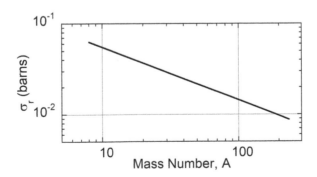

FIGURE 4.14

Removal cross sections per unit atomic mass number for fission neutrons as a function of mass number at a neutron energy of 8.0 MeV as calculated by Equation 4.21. Over the range $8 \leq A \leq 240$, the values are well fit by this equation.

TABLE 4.4

Removal Cross Sections, σ_r (Barns), for Low-Energy Neutrons

Element	1.0 MeV	Fission Spectrum	2.9 MeV	4.0 MeV	6.7 MeV	14.9 MeV
Carbon		0.9	1.58	1.05	0.83	0.50
Aluminum		1.31				
Iron	1.1	1.96	1.94	1.98	2.26	1.60
Copper		2.04				
Lead		3.28	2.70	3.44	3.77	2.95

Note: The typical accuracy is said to be ±5%. (Adapted from Patterson, H. W., and R. H. Thomas. 1973. *Accelerator health physics.* New York, NY: Academic Press.)

4.5 Limiting Attenuation at High Energy

The most important feature of neutron shielding at higher-energy accelerators is the fact that the attenuation length becomes an approximate constant at high energy. Perhaps first noticed by Lindenbaum (1961), as the energy increases the neutron inelastic cross sections increase rapidly until about 25 MeV where they generally level off and then fall rapidly with energy in the region $25 < E_n < 100$ MeV to a value that becomes independent of energy, aside from a slight, gradual *increase* at the very highest energies available to modern particle accelerators. The result is that high-energy neutrons attenuate approximately exponentially with an attenuation length λ_{atten} that is rather insensitive to energy. Thus, in units of length,

$$\lambda_{atten} = \frac{1}{N\sigma_{in}} \ (\text{cm}) \tag{4.22}$$

where σ_{in} is the inelastic cross section, roughly equivalent to the "absorption cross section," and N, as before, is the number of absorber nuclei per unit volume. σ_{in} specifically does not include elastic scattering and thus is always *smaller* than the total cross section that is larger because it includes the elastic scattering process. In a simple-minded approach, σ_{in} can be taken approximately to be the *geometrical cross section* with the nucleon radius taken to be 1.2×10^{-13} cm. It follows that in the high-energy limit, one can multiply by the density to get

$$\rho\lambda_{atten} = 36.7 A^{1/3} \ (\text{g cm}^{-2}) \tag{4.23}$$

For our present purposes, σ_{in} is essentially constant at higher energies having the values listed in Table 1.2. At lower kinetic energies, those roughly bounded by 150–200 MeV, the values of σ_{in} are larger by a factor of as much as approximately three as the cross sections are enhanced by numerous possible nuclear reactions leading to excited nuclear states (Lindenbaum 1961).

The high-energy asymptotes were first verified by historic cosmic ray data and are well-represented by

$$\sigma_{in} = 43 A^{0.69} \ (\text{mb}) \tag{4.24}$$

In the high-energy limit, the *interaction length* λ_{inel} is thus given by

$$\lambda_{inel} = \frac{\rho}{N\sigma_{in}} = 38.5 A^{0.31} \; (\mathrm{g\,cm^{-2}}) \tag{4.25}$$

Thus, this *geometric approximation* is reasonably accurate. Figure 4.15 shows the results for absorption cross sections based on these values.

Fassò et al. (1990) have provided extensive tabulations of the value of σ_{in} (mb) for a variety of particles, energies, and materials in the high-energy region as functions of particle momenta up to 10 TeV/c.

The leveling off of the attenuation length as a function of energy for ordinary concrete, the most common shielding material, is essentially equivalent to that of earth shielding if proper consideration of the density of the material is observed (see Chapter 6). This is of special importance due to the widespread use of both concrete and earth in shielding at large accelerators. Figure 4.16 gives the results for both neutrons and protons.

An important feature of these results is the equivalence of the attenuation lengths for protons and neutrons at high energies. Due to the similarities of chemical composition, results for soil shielding in this energy regime can be taken to be the same as those for concrete when λ is expressed in units of areal density, for example, in g cm^{-2}, when one adjusts for the difference in material density.

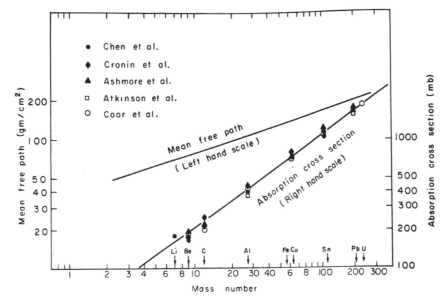

FIGURE 4.15
Inelastic mean free path and cross section as a function of mass number *A*. In this figure, the experimental results of Chen et al. (1955), Cronin et al. (1957), Ashmore et al. (1960), Atkinson et al. (1961), and Coor et al. (1955) were used as indicated by the first author's name for each of the cited references. (Reprinted from *Accelerator health physics, Patterson*, H. W., and R. H. Thomas. New York, NY: Academic Press. Copyright 1973, with permission from Elsevier.)

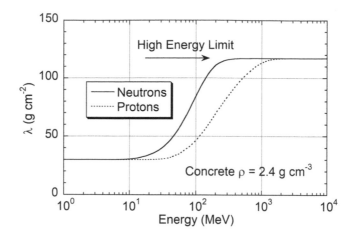

FIGURE 4.16
The attenuation length λ for a broad beam of monoenergetic neutrons and protons in concrete shielding as a function of particle kinetic energy. The high-energy limit is 117 g cm^{-2}. (Adapted from Thomas, R. H., and G. R. Stevenson. 1988. *Radiological safety aspects of the operation of proton accelerators*. International Atomic Energy Agency: IAEA Technical Report No. 283. Vienna, Austria. Used with permission.)

4.6 Intermediate- and High-Energy Shielding: Hadronic Cascade

4.6.1 Hadronic Cascade from a Conceptual Standpoint

The *hadronic cascade* is initiated at proton accelerators when the beam interacts with targets, beam absorbers, and accelerator components to produce neutrons and other particles. Such cascades can also arise at electron accelerators since, as was discussed in Chapter 3, high-energy secondary hadrons can also result from electromagnetic interactions. The collision of a high-energy nucleon with a nucleus produces a large number of particles: pions, kaons, and other nucleons as well as fragments of the target nucleus. According to Thomas and Stevenson (1988), above about 1.0 GeV and at forward angles, the pions, protons, and neutrons can be nearly equal in number. The neutrons may be classified as either evaporation neutrons or cascade neutrons (see Section 4.2.2.2). The spectrum extends in energy up to the incident energy with diminishing probability and follows an energy dependence roughly proportional to $1/E$.

As the proton kinetic energy increases, other particles, notably $\pi^{\pm}$, π^{o}, and K$^{\pm}$, play roles in the cascade when their production becomes energetically possible. They are absorbed with absorption lengths comparable in magnitude to, but not identical with, those of protons. These particles also decay into muons. Because of their long ionization ranges and lack of nuclear interactions, muons provide a pathway for energy to escape the cascade. Hadrons, principally neutrons with $E_n > 150$ MeV, propagate the cascade. This is clear from the attenuation lengths shown in Figures 4.15 and 4.16. Nucleons in the range $20 < E_n < 150$ MeV also deposit their energy predominantly by nuclear interactions, but their energy gets distributed over many particles of all types energetically possible. The charged particles produced in such cascades are generally "ranged-out" by ionization in material or create yet other particles in the cascade. The role played by the energy of approximately 150 MeV for hadronic cascade propagation is *qualitatively* analogous, in some respects, to that of the critical energy for electromagnetic cascades.

Neutral pions (π^o) are produced when the kinetic energy of the incident proton significantly exceeds the π^o rest energy of 134.98 MeV. The π^o mean-life of 8.52×10^{-17} s is very short so that for the π^o, $c\tau = 25.5$ nanometers (Beringer et al. Particle Data Group 2012). Hence, π^o's travel only, figuratively speaking, "microscopic" distances before decaying. The principal decay (98.8% branching ratio) is into a pair of photons emitted in opposite directions in the center of mass reference frame of the moving π^o. An energetic π^o thus appears as two forward-peaked photons each with half of the π^o's total energy. The decay photons from π^o decay readily initiate an electromagnetic cascade along with the hadronic one. It is possible for the electromagnetic channel to feed back into the hadronic cascade because it, too, produces high-energy hadrons. However, this effect is generally of little importance and, for most shielding calculations, the electromagnetic component of a hadronic cascade can be ignored. The principal exception involves energy deposition calculations at forward angles, that is, for small values of θ. In fact, at hundreds of GeV, electromagnetic cascades dominate the energy deposition at very forward angles. This can have important ramifications for radiation damage to equipment, the heat load on cryogenic systems, and the ability of targets to survive bombardment.

In general, the neutrons are the principal drivers of the cascade because of the ionization energy loss for pions and for protons below 450 MeV, the energy at which the ionization range becomes roughly equal to the interaction length. Also, any magnetic fields that are present which can deflect and disperse charged particles present will not, of course, affect the neutrons. Furthermore, neutrons are produced in large quantities at large values of θ compared with the forward-peaked pions. These phenomena, in general, apply also to ions heavier than the proton with suitable corrections (especially at low energies) for nuclear structure effects. Scaling of proton results for heavier ions will, in general, roughly be according to the specific energy (MeV/amu). Figure 4.17 is a schematic flowchart of the hadronic cascade process (International Commission on Radiation Units and Measurements [ICRU] 1978).

4.6.2 Simple One-Dimensional Cascade Model

A simple one-dimensional model of the hadronic cascade was first proposed by Lindenbaum (1961) and sometimes known as the *Lindenbaum model*. This approach provides some intuition into the nature of the hadronic cascade. Figure 4.18 defines the geometry of this model.

Suppose one initially has N_o incident high-energy nucleons. After an individual collision, one of them continues in its original direction at a reduced energy but with the same attenuation length λ or will generate one or more secondary particles also with the same value of λ. The value of λ is approximately constant due to the limiting attenuation at high energy. This process continues until a number of collisions n have occurred that are sufficient to degrade the particle energies to approximately 150 MeV. Below 150 MeV the inelastic cross sections greatly increase with decreasing particle energy, as discussed earlier. This additional absorption at the lower energies can be said to "remove" the particle. For the present discussion, it is assumed that n is an integer, an approximation since in reality n has a statistical distribution. Following these assumptions, the number ν_1 that reach z having made no additional collisions that produce secondary particles is

$$\nu_1 = N_o \exp(-z/\lambda) \tag{4.26}$$

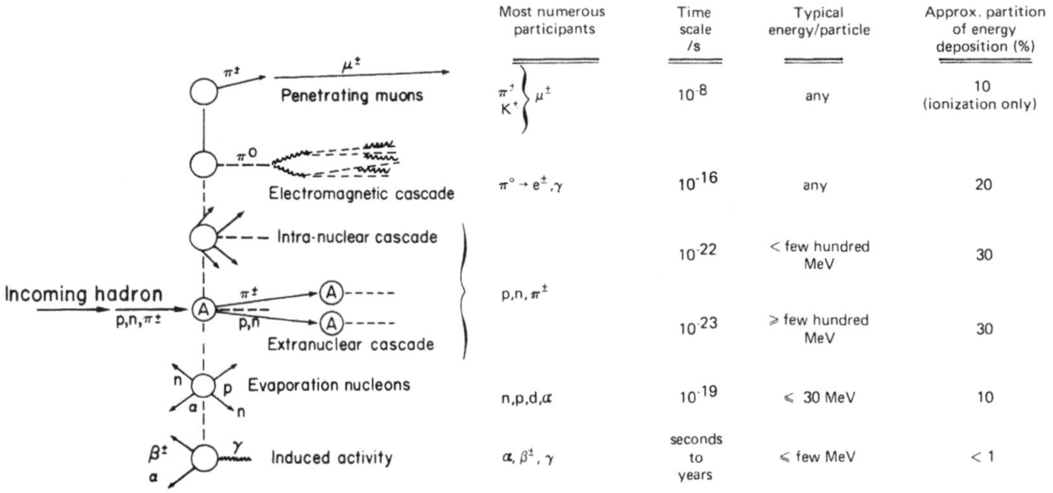

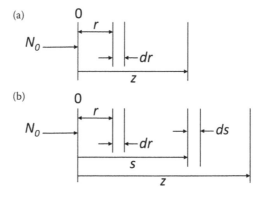

	Most numerous participants	Time scale /s	Typical energy/particle	Approx. partition of energy deposition (%)
Penetrating muons	$\left.\begin{array}{c}\pi^{\pm}\\K^{\pm}\end{array}\right\}\mu^{\pm}$	10^{-8}	any	10 (ionization only)
Electromagnetic cascade	$\pi^0 \to e^{\pm}, \gamma$	10^{-16}	any	20
Intra-nuclear cascade		10^{-22}	< few hundred MeV	30
Extranuclear cascade	$p, n, \pi^{\pm}$	10^{-23}	⩾ few hundred MeV	30
Evaporation nucleons	n, p, d, α	10^{-19}	⩽ 30 MeV	10
Induced activity	$\alpha, \beta^{\pm}, \gamma$	seconds to years	⩽ few MeV	< 1

FIGURE 4.17
Development of the hadronic cascade and the major participants in any given path. The approximate time scales, the typical energies, and the fraction of the total energy deposition due to these participants are also shown. The symbols, "$\pi^{\pm}$" are representative also of other charged hadrons that can be produced in such a cascade. (Reprinted with permission from International Commission on Radiation Units and Measurements (ICRU). 1978. *Basic aspects of high-energy particle interactions and radiation dosimetry.* ICRU Report No. 28. Washington, DC, http://ICRU.org).)

Now suppose that there is one additional collision between 0 and z that produces secondary particles. The number ν_2 of additional particles that reach z is given by the product of the number that reach elemental coordinate dr; the probability of subsequently reaching z; the probability of interacting in dr, dr/λ; and the *multiplicity* m_1 of particles produced in the interaction as illustrated in Figure 4.18a. Integrating over dr,

$$\int_0^z dr \frac{m_1}{\lambda}[N_0 \exp(-r/\lambda)][\exp\{-(z-r)/\lambda\}] = \left(N_0 m_1 \frac{z}{\lambda}\right)\exp(-z/\lambda) = \nu_2 \qquad (4.27)$$

FIGURE 4.18
(a) Single-collision geometry for the Lindenbaum model. (b) Two-collision geometry for the Lindenbaum one-dimensional model. (Adapted from Thomas, R. H., and G. R. Stevenson. 1988. *Radiological safety aspects of the operation of proton accelerators.* International Atomic Energy Agency: IAEA Technical Report No. 283. Vienna, Austria. Used with permission.)

Continuing, now suppose two additional collisions occur that produce secondary particles as in Figure 4.18b. The number ν_3 of additional particles that reach z is the product of those that reach s having made one particle-producing collision; the probability of subsequently reaching z; the multiplicity in the second interaction m_2; and the probability of interacting in ds:

$$\int_0^z ds \frac{m_2}{\lambda} \left[N_0 m_1 \frac{s}{\lambda} \exp(-s/\lambda) \right] [\exp\{-(z-s)/\lambda\}]$$

$$= \left\{ N_0 m_1 m_2 \frac{z}{\lambda^2} \exp(-z/\lambda) \right\} \int_0^z s\, ds = \left\{ N_0 m_1 m_2 \frac{z^2}{2\lambda^2} \exp(-z/\lambda) \right\} = \nu_3 \tag{4.28}$$

Therefore, with n defined as shown, one can write for the total number of particles that reach z:

$$N_n(z) = N_0 \beta_n (z/\lambda) \exp(-z/\lambda) \tag{4.29}$$

where β is a buildup factor:
for $n = 1$; $N_1 = \nu_1$ and $\beta_1 = 1$,
for $n = 2$; $N_2 = \nu_1 + \nu_2$ and $\beta_2 = 1 + (m_1 z/\lambda)$, while
for $n = 3$; $N_2 = \nu_1 + \nu_2 + \nu_3$ and $\beta_3 = 1 + (m_1 z/\lambda) + (m_1 m_2 z^2/2\lambda^2)$.

For arbitrary n,

$$\beta_n = 1 + \frac{m_1 z}{\lambda} + \frac{m_1 m_2 z^2}{2\lambda^2} + \ldots + \frac{1}{(n-1)!} \left(\frac{z^{n-1}}{\lambda^{n-1}} \right) \prod_{i=1}^{i=n-1} m_i \tag{4.30}$$

(In Equation 4.30, the *product* operator $\prod_{i=1}^{i=n-1}$ is used to indicate the taking of the mathematical *product* of a sequence of *indexed* terms, here denoted m_i, analogous to the way that the more commonly encountered summation operator, e.g., $\sum_{i=1}^{i=n-1}$, is used to indicate the taking of the *sum* of such *indexed* terms.) Thus, this buildup factor is a monotonically increasing function of z. If one makes the assumptions that the multiplicity stays the same for all interactions, $m1 = m_2 = \ldots = m$, and that n is large, comparison with the series expansion of the exponential function reveals that β_n approximates an exponential dependence on z. The condition on n implies that the shield must be quite thick. The general result is that the *attenuation length of the cascade* λ_{cas} is somewhat larger than the value of the interaction length λ for a single interaction. Figure 4.19 is a plot of the results of this model for several values of m and n as a function of z/λ.

Figure 4.20 shows a plot for a specific case ($n = 3$ and $m = 2$). The "exponential" region is not completely achieved until $z/\lambda \approx 5$. In Figure 4.20, this calculation is also compared with the results of measurements of tracks in photographic emulsions in the experiment of Citron et al. (1965) that used 19.2 GeV/c protons incident on an iron slab. Measurements of such tracks are sensitive to individual nuclear interactions.

4.6.3 Semiempirical Method: Moyer Model for a Point Source

A number of references (Routti and Thomas 1969; Patterson and Thomas 1973; Cossairt et al. 1982; Stevenson et al. 1982; Tesch 1983; Thomas and Thomas 1984; Cossairt et al. 1985a;

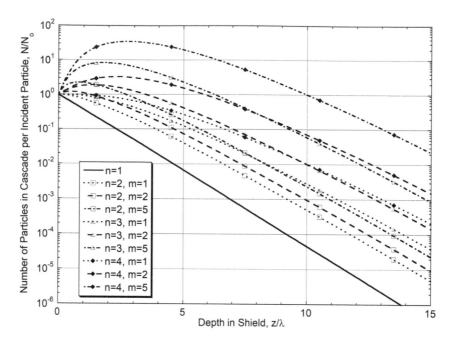

FIGURE 4.19
Development of a one-dimensional cascade in the Lindenbaum model for $n = 1, 2, 3,$ and 4 and for $m = 1, 2,$ and 5.

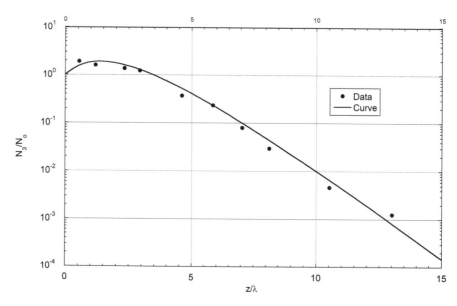

FIGURE 4.20
The Lindenbaum model approximation plotted with $n = 3$ and $m = 2$ labeled "Curve" compared with the laterally integrated track density in nuclear emulsions produced by a 19.2 GeV/c proton beam incident on an iron slab measured by Citron et al. (1965), which is labeled "Data." The value of λ used here was that of the nuclear interaction length of iron listed in Table 1.2. The normalization of the data was adjusted by an approximate factor of two to facilitate the comparison with the Lindenbaum model. (Adapted from the data reported by Citron, A. et al. 1965. *Nuclear Instruments and Methods* 32:48–52.)

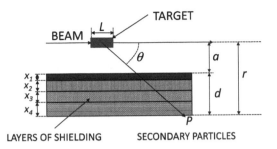

FIGURE 4.21
Sketch of the geometry for the semiempirical Moyer model. A beam of N_p protons impinges on the target of length L on the axis of a cylindrical shield consisting of several layers of shielding materials. From inside (near the beam) to outside, the shield materials represented by the layers x_i thick, could for example be iron, heavy concrete, regular concrete, and earth, respectively. a is the inner radius of the tunnel. The observer at location **P** is situated at a radial thickness of d equal to the sum of the thicknesses of the layers of shielding, here shown as four in number, and is at a distance $r' = r \csc \theta$ from the beam-target interaction point.

Tesch 1985; McCaslin et al. 1987b) bear on the development of the *Moyer model* predominantly based on an exponential approximation with constants fitted to measurements spanning proton beam energies from 7.4 to 800 GeV. The summary of this method presented here is largely taken from Patterson and Thomas (1973) and Fassò et al. (1990).

This model was first developed by Burton J. Moyer to solve particular shielding problems related to the 6 GeV Bevatron at the Lawrence Radiation Laboratory (now the Lawrence Berkeley National Laboratory). The model predates the development of large, fast computers and advanced Monte Carlo techniques but remains useful as a means of checking more sophisticated calculations. Consider the situation shown in Figure 4.21 for a "point" target source surrounded by a cylindrical shield consisting of several layers of possibly different materials.

In the Moyer model the number of neutrons dN/dE that are emitted into a given element of solid angle $d\Omega$ at angle θ relative to a target struck by Np protons in an energy interval $E + dE$ is given for a single shield material of thickness d by

$$\frac{dN}{dE} = N_p B(E) \left(\frac{d^2Y}{dEd\Omega} \right) d\Omega \exp\left(-\frac{d \csc \theta}{\lambda(E)} \right) \tag{4.31}$$

where $B(E)$ is a buildup factor that approximates the development of a multiplicity of particles in the hadronic cascade, and the exponential function accounts for the attenuation of the radiation field by shielding of thickness d at the angle θ. The energy-dependent interaction length is denoted by $\lambda(E)$. $\lambda(E)$ and d must be in the same units, usually of length in meters or centimeters. The role of the double differential of the yield is obvious. In Equation 4.31, the flux density at coordinates (r,θ) can be obtained at distance $r' = r \csc\theta$ from the target by including the factor

$$\frac{d\Omega}{dA} = \frac{1}{(a+d)^2 \csc^2 \theta} = \frac{1}{r^2 \csc^2 \theta} = \frac{1}{r'^2} \tag{4.32}$$

The total fluence Φ at the point where the ray emerges from the shield is given by

$$\Phi = \frac{N_p}{r'^2} \int_{E_{min}}^{E_{max}} dE \, \frac{d^2Y}{dEd\Omega} B(E) \exp\left(-\frac{d \csc \theta}{\lambda(E)} \right) \tag{4.33}$$

The following are Moyer's simplifying assumptions:

A. $\lambda(E) = \lambda = $ constant for $E \geq 150$ MeV and $\lambda(E) = 0$ for $E < 150$ MeV. This approximates the leveling-off of the inelastic cross section at high energy. Thus,

$$\Phi(E_n > 150 \, \text{MeV}) = \frac{N_p}{r'^2} \exp\left(-\frac{d \csc \theta}{\lambda}\right) \int_{150 \, \text{MeV}}^{E_{\text{max}}} dE \frac{d^2 Y}{dE d\Omega} B(E) \tag{4.34}$$

B. The neutrons emitted at angle θ can be represented by a simple function $f(\theta)$ multiplied by a multiplicity factor $M(E_{\text{max}})$ that depends only on the incident energy; thus,

$$\Phi(E_n > 150 \, \text{MeV}) = \frac{N_p}{r'^2} \exp\left(-\frac{d \csc \theta}{\lambda}\right) M(E_{\text{max}}) f(\theta)$$

$$= \frac{N_p}{r'^2} \exp\left(-\frac{d \csc \theta}{\lambda}\right) g(E_{\text{max}}, \theta) \tag{4.35}$$

where $g(E_{\text{max}}, \theta)$ is an angular distribution function that is constant for a given value of E_{max} and for a particular target.

C. The dose equivalent per fluence P for neutrons is not strongly dependent on energy over a rather wide energy range near $E \approx 150$ MeV (see Figure 1.5).

D. Thus, the dose equivalent just outside of the shield due to neutrons with $E > 150$ MeV can be taken to be $H_{150} \approx P_{150} \Phi (E_n > 150 \, \text{MeV})$, where P_{150} is the value of this conversion factor at 150 MeV.

The total dose equivalent H_{equiv} is then given by

$$H_{equiv} = k H_{150} \tag{4.36}$$

In order to include the dose equivalent from neutrons with $E < 150$ MeV, $k \geq 1$. This implicitly assumes that the low-energy neutrons are in equilibrium with those having $E > 150$ MeV so that the spectrum no longer changes with depth. This is a valid assumption for a shield more than a few mean free paths thick. These assumptions lead to

$$H_{equiv} = \frac{k P_{150} N_p g(E_{\text{max}}, \theta)}{(a+d)^2 \csc^2 \theta} \exp\left(-\frac{d \csc \theta}{\lambda}\right) \tag{4.37}$$

Figure 4.21 shows multiple materials in the shield, a typical condition found in actual practice. The parameter ζ, which replaces the ratio d/λ in the argument of the exponential function in Equation 4.37, is introduced to account for n different cylindrical shielding layers, shown as four in the example of Figure 4.21):

$$\zeta = \sum_{l=1}^{n} \frac{x_i}{\lambda_i} \tag{4.38}$$

where the sum is over the n layers of shielding.

Moyer model parameters have been determined by experiment. Stevenson et al. (1982) and Thomas and Thomas (1984) have determined from global fits to data over a wide domain of energy that $f(\theta)$ is well described by

$$f(\theta) = \exp(-\beta\theta) \tag{4.39}$$

where θ is in radians, β is in radians^{-1}. Thus,

$$H_{equiv} = \frac{H_{o,equiv}(E_p)\exp(-\beta\theta)\exp(-\zeta\csc\theta)}{(r\csc\theta)^2} \tag{4.40}$$

with

$$r = a + \sum_{i=1}^{n} x_i \tag{4.41}$$

For proton kinetic energies well above $E_n = 150$ MeV, it was determined that $\beta \approx 2.3$ rad^{-1}. The value of $H_{o,equiv}(E_p)\exp(-\beta\theta)$ is determined from the yield data and empirical measurements. $H_{o,equiv}(E_p)$ is best fit as a power law of form $H_{o,equiv}(E_p) = kE^{\eta}$. From such results, per incident proton, with statistical errors in the fitting parameters reported by Thomas and Thomas (1984):

$$H_{o,equiv}(E_p) = [(2.84 \pm 0.14)\times 10^{-13}]E_p^{(0.80\pm0.10)} \text{ (Sv m}^2) \tag{4.42}$$
$$= 2.84\times 10^{-8}E_p^{0.8} \text{ (mrem m}^2) = 2.84\times 10^{-4}E_p^{0.8} \text{ (mrem cm}^2)$$

with E_p in GeV. Using the 1990 System, Cossairt has determined that to obtain effective dose H_{eff} with the Moyer model, the constant 2.84 ± 0.14 should be replaced with 3.98 ± 0.20, with all other parameters remaining the same (Cossairt 2013a). These results are derived for relatively "thick" targets (like accelerator magnets) in tunnel configurations. Fassò et al. (1990), based on Monte Carlo results, gave values for "thin" targets of $k = 2.0 \times 10^{-14}$ (Sv m^2) and $n = 0.5$ for determining dose equivalent (1973 System). A thin-walled beam pipe would be an example of a "thin" target. The variations thus reflect buildup in the shower. For thick lateral shields near the beam where the cascade immediately becomes fully developed and self-shielding arises, $k = (6.9 \pm 0.1) \times 10^{-15}$ (Sv m^2) and $n = 0.8$ independent of target material (Stevenson 1987; Fassò et al. 1990) (1973 System). The value of $n = 0.8$ for thick shields has also been rigorously discussed by Gabriel et al. (1994) and verified by Torres (1996).

Similarly, within the context of the Moyer model, recommended, but likely conservatively large, semi-empirical values of λ for concrete and other materials as a function of mass number A are

concrete: 1170 ± 20 kg m^{-2} = 117 g cm^{-2}

other materials: $428A^{1/3}$ kg m^{-2} = $42.8A^{1/3}$ g cm^{-2}

These values of λ are 15%–30% larger than the high-energy nuclear interaction lengths listed in Table 1.2, a result qualitatively consistent with the discussion of buildup in connection with the results of Lindenbaum's one-dimensional approximation. These semiempirical values of λ are also larger than those found in realistic Monte Carlo calculations (see examples later in this chapter and in the Appendix), since they are based on a point source model, a useful construct, but one that is never actually achieved with thick targets.

If one sets the partial derivative $\partial H/\partial\theta$ equal to zero, one can derive an equation for determining the value of $\theta = \theta'$ at which the maximum dose equivalent occurs:

$$\zeta\cos\theta' - \beta\sin^2\theta' + 2\cos\theta'\sin\theta' = 0 \tag{4.43}$$

Generally this equation can be solved by successive approximation methods. One can substitute the resulting value of θ into Equation 4.40 to get the maximum dose equivalent at a given radial depth. According to McCaslin et al. (1987b), with r in meters and over a wide range of values of ζ, the following holds for dose equivalent:

$$H_{\text{max}} = 1.66 \times 10^{-14} E_p^{0.8} \exp(-\zeta)\frac{\zeta^{-0.245}}{r^2} \text{ (Sv per incident proton)} \tag{4.44}$$

For values of $\zeta > 2.0$, the following is an equally accurate approximation for the maximum dose equivalent:

$$H_{\text{max}} = 1.26 \times 10^{-14} E_p^{0.8}\frac{\exp(-1.023\zeta)}{r^2} \text{ (Sv per incident proton)} \tag{4.45}$$

4.6.4 Moyer Model for a Line Source

The Moyer model can be extended to address a uniform line source placed on the axis of a cylindrical shield consisting of several layers of possibly different materials. Assume a uniform source of one proton interacting per unit length. Then, the doses from the individual increments along the line source contribute to the total at any given point P external to the shield. Figure 4.22 shows the integration variables.

One can integrate contributions of the elements dl of a line source at given perpendicular distance r as follows. Making the change of variable of integration from the line integral to an integral over angle θ ($dl = r\csc^2\theta\,d\theta$):

$$
\begin{aligned}
H &= H_0(E_p)\int_{-\infty}^{\infty} d\ell\,\frac{\exp(-\beta\theta)\exp(-\zeta\csc\theta)}{r^2\csc^2\theta} \\
&= H_0(E_p)\int_0^{\pi} d\theta\, r\csc^2\theta\,\frac{\exp(-\beta\theta)\exp(-\zeta\csc\theta)}{r^2\csc^2\theta} \\
&= \frac{H_0(E_p)}{r}\int_0^{\pi} d\theta\,\exp(-\beta\theta)\exp(-\zeta\csc\theta) = \frac{H_0(E_p)}{r}M(\beta,\zeta)
\end{aligned}
\tag{4.46}
$$

(per interacting proton per unit length)

The integral in the previous equation denoted $M(\beta,\zeta)$ is known as the *Moyer integral*. The values of this integral have been tabulated by Routti and Thomas (1969) and have been approximated by Cossairt (2013a). In view of the results found empirically for point sources,

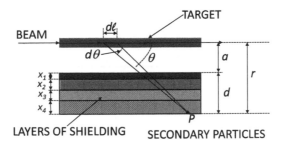

FIGURE 4.22
Variables of integration of Moyer point source result needed to obtain Moyer line source results. As in Figure 4.21, the shielding of thickness d could be composed of multiple layers, here shown as four in number, of thickness ζ mean free paths.

$M(2.3,\zeta)$ has obvious special significance and is tabulated extensively by, among others, Fassò et al. (1990). Tesch (1983) made an important contribution in that he determined an approximation to this integral that has become known as the *Tesch approximation*:

$$M_T(2.3,\zeta) = 0.065\exp(-1.09\zeta) \qquad (4.47)$$

For "intermediate" values of ζ, $M_T(2.3,\zeta)$ can be used instead of $M(2.3,\zeta)$ to simplify calculations. Table 4.5 gives the ratio $M_T(2.3,\ \zeta)/M(2.3,\ \zeta)$ as a function of ζ.

Of course, few "line sources" are actually *infinite* in length. Thus, the integration can be limited to a finite angular range. Likewise, only a limited angular range (and hence length) contributes significantly to the Moyer integral.

Tables 4.6 and 4.7 give angular integration limits that define the region contributing 90% of the value of $M(2.3,\zeta)$ as a function of ζ (Table 4.6) and the distances along the z-axis that likewise contribute to 90% of the value of $M(2.3,\zeta)$ as a function of the radial distance and ζ (Table 4.7).

These calculations were done for concrete shields.

TABLE 4.5

Values of the Ratio $M_T(2.3,\zeta)/M(2.3,\zeta)$ as a Function of ζ

ζ	$M_T(2.3,\zeta)/M(2.3,\zeta)$	ζ	$M_T(2.3,\zeta)/M(2.3,\zeta)$
0.2	0.27	11	1.02
1	0.53	12	0.99
2	0.75	13	0.95
3	0.90	14	0.91
4	1.00	15	0.86
5	1.06	16	0.82
6	1.09	17	0.78
7	1.10	18	0.73
8	1.10	19	0.69
9	1.08	20	0.65
10	1.06		

Source: Adapted from Fassò, A. et al. 1990. *Landolt-Börnstein numerical data and functional relationships in science and technology new series; Group I: Nuclear and particle physics Volume II: Shielding against high energy radiation,* ed. H. Schopper. Berlin, Germany: Springer-Verlag.

TABLE 4.6

Angular Integration Limits in θ (Degrees) which Contain 90% of the Moyer Integral $M(2.3,\zeta)$

ζ	Lower Limit	Upper Limit	ζ	Lower Limit	Upper Limit
2.5	31.52	106.58	12	57.25	106.29
3	24.35	107.15	13	58.45	106.04
4	39.00	107.64	14	59.74	105.78
5	42.67	107.73	15	60.66	105.54
6	45.77	107.66	16	61.49	105.29
7	48.51	107.48	17	62.34	105.04
8	50.69	107.28	18	63.22	104.80
9	52.7	107.04	19	64.08	104.54
10	54.34	106.79	20	64.63	104.30
11	56.07	106.54			

Source: Adapted from Fassò, A. et al. 1990. *Landolt-Börnstein numerical data and functional relationships in science and technology new series; Group I: Nuclear and particle physics Volume II: Shielding against high energy radiation,* ed. H. Schopper. Berlin, Germany: Springer-Verlag.

TABLE 4.7

Distances Corresponding to 90% Limits in Moyer Integrals

Radial Distance (meters)	Thickness (concrete) (meters)	Thickness (concrete) ζ	Upstream Limit, z_1 (meters)	Downstream Limit, z_2 (meters)	Total Length z_2-z_1 (meters)
1.5	0.5	1.0	−4.2	0.3	4.5
2.0	1.0	2.0	−3.7	0.6	4.3
3.5	2.5	5.0	−3.8	1.1	4.9
6.0	5.0	10.0	−4.3	1.8	6.1
8.5	7.5	15.0	−4.8	2.4	7.2
11.0	10.0	20.0	−5.2	2.8	8.0

Source: Adapted from Fassò, A. et al. 1990. *Landolt-Börnstein numerical data and functional relationships in science and technology new series; Group I: Nuclear and particle physics Volume II: Shielding against high energy radia,* ed. H. Schopper. Berlin, Germany: Springer-Verlag.

McCaslin et al. (1985) demonstrated that the Moyer model approach is also effective for moderately energetic heavy ions. It has also been found that the Moyer model approach works well even in the intermediate-energy region of $200 < E_o < 1000$ MeV. This may be interpreted as due to the relatively smooth dependence of neutron yield on incident proton kinetic energy. The Moyer model generally does *not* provide sufficiently accurate results at forward angles. For these situations, the Boltzmann equation must be solved usually with Monte Carlo calculations.

4.7 Use of Monte Carlo Shielding Codes for Hadronic Cascades

4.7.1 Examples of Results of Monte Carlo Calculations

It should be obvious that validity of the Moyer model approach is limited to simple shielding configurations. Furthermore, one cannot include magnetic fields. Also, the model is not valid at forward angles and for kinetic energies lower than a few hundred megaelectron volts (MeV). It is also incapable of handling the production of other types of particles aside from neutrons that can often be copiously produced at forward angles. Labyrinth penetrations and residual radioactivity considerations are unaddressed. Thus, the Monte Carlo technique is the primary one to use in such work.

The Appendix describes several Monte Carlo programs that have been developed at various laboratories for a variety of purposes. In this section, methods of using results from such computations are reviewed.

The code HETC remains an important benchmark in the history of development of such codes. A simple example of the results of a calculation performed using this code is shown in Figure 4.23 taken from results (Alsmiller et al. 1975) for 200 MeV protons incident on "thin" and "thick" aluminum targets. It is a plot of r^2H as a function of angle for several intervals of θ in a *spherical* concrete shield with the beam incident on a target at the center of the sphere.

For higher energies, CASIM and FLUKA have also served the role as benchmark programs, while MARS is undoubtedly at the time of this writing the most versatile. FLUKA and MARS model a multitude of physical effects in detail. Representative results for solid iron and concrete cylinders bombarded by protons of various energies are provided in Figures 4.24 through 4.26.

These values allow one to estimate the dose equivalent per incident proton at various locations and for various proton beam energies. They are also useful for obtaining a quick understanding of the effects of a beam absorber. Detailed calculations should be performed to assure adequately accurate designs.

4.7.2 General Comments on Monte Carlo Star-to-Dose Conversions

Several of these codes calculate the *star density* as their most basic output quantity. This quantity, generally denoted by S, is more correctly called the *density of inelastic interactions* (stars cm^{-3}) and is relatively easy to tabulate as the calculation proceeds, since only a simple counting process is involved. The term "star" comes from historic cosmic ray experiments in which the high-energy interaction events, with their large multiplicities, appeared as tracks originating from a point. In a shield composed of more than one material, the star density may change dramatically from one material boundary to the other, reflective of differing material densities and atomic numbers. A related quantity is the *star fluence*, denoted by Φ_S. Star fluence is the product of the star density and the nuclear interaction length. The star fluence roughly corresponds to the fluence of hadrons having energies above that where the cross section "levels off." It is also reflective of any "artificial" thresholds in the calculation. In contrast with star density, due to the property of continuity, the star fluence is conserved across material boundaries.

The *dose equivalent per star density* is a rather important parameter of radiation protection calculations. Perhaps the best results have been provided by Stevenson (1986). While this conversion factor is somewhat dependent on the position in the shield, after a shield thickness sufficient to establish "equilibrium" spectra, a constant value may be used for high-energy protons, loosely defined as protons exceeding a few hundred MeV of kinetic energy. and other hadrons, within a given material. In other words, the energy and spatial dependences are rather weak. Values for these quantities, as well as the related *dose equivalent per star fluence conversion factors*, are given in Table 4.8. A star density is transformed into the corresponding star fluence by the relation

$$\Phi_S = S\lambda \tag{4.48}$$

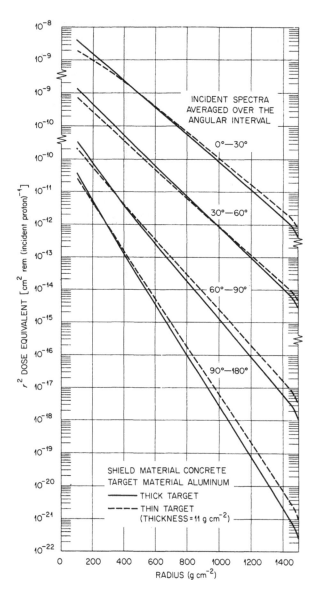

FIGURE 4.23
HETC calculations of r^2H as a function of **CONCRETE** shield thickness d averaged over several intervals of θ for 200 MeV protons incident on an aluminum target centered in a spherical shield. The "thin" target was 11 g cm^{-2} thick, about 1/10 of a mean free path. The "thick" target was just sufficient to stop the incident protons by ionization. (Alsmiller, Jr., R. G. et al. 1975. *Particle Accelerators* 7:1–7. Reprinted with permission from Taylor and Francis Ltd. http://www.tandfonline.com.)

where λ is the nuclear interaction length. As discussed in detail in Section 6.3.5, iron shielding presents a unique problem due to the copious emission of low-energy neutrons in shields of modest thickness. The values reported here are for relatively thin iron shields of only one or two mean free paths. If a thick iron shield is encountered that is not "finished" with at least 50 cm, or so, of concrete as the outermost layer, one should multiply these conversion factors by a factor of approximately 5. Table 4.8 implies use of the material density values of Table 1.2.

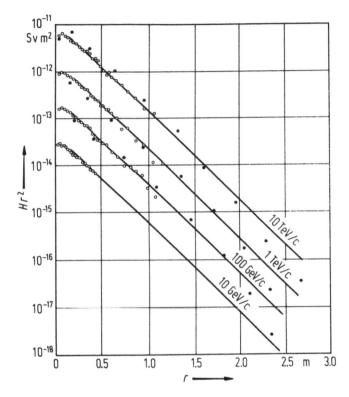

FIGURE 4.24
Variation of the dose equivalent per proton at the position of the longitudinal maximum multiplied by the square of the radius Hr^2 versus radius r for proton-induced cascades in **IRON** of density 7.2 g cm^{-3}. The coordinate r is defined as in Figure 4.21. Open circles are results of FLUKA calculations while full circles are results of CASIM calculations. The solid lines are derived using an empirical parametrization. (With kind permission from Springer Science+Business Media: *Landolt-Börnstein numerical data and functional relationships in science and technology new series; Group I: Nuclear and particle physics Volume II: Shielding against high energy radiation*, 1990, ed. H. Schopper. Fassò, A. et al. Berlin, Germany: Springer-Verlag. Copyright 1990.)

Compilations of such calculations have been given by Van Ginneken and Awschalom (1975), and Van Ginneken et al. (1987), Cossairt (1982), and Reitzner (2012). Fassò et al. (1990) have also compiled a comprehensive set of Monte Carlo results. A convenient way to display these results is to provide contour plots of star density as functions of longitudinal coordinate Z and radial coordinate R assuming cylindrical symmetry. The Appendix provides examples of results of hadronic Monte Carlo calculations that are meant to illustrate a number of situations commonly encountered. One of the salient advantages of the Monte Carlo method is the ability to handle configurations of arbitrary complexity and results for both solid cylinders and more complicated commonly encountered configurations for which examples are included in the Appendix.

4.7.3 Shielding against Muons at Proton Accelerators

The production of muons was discussed in Section 4.2.4. At higher energies, there are significant complications in that muon creation mechanisms, in addition to the production of pions and kaons and their subsequent decays, are possible. However, the muons from pion decay and kaon decay generally, but not universally, represent the most important

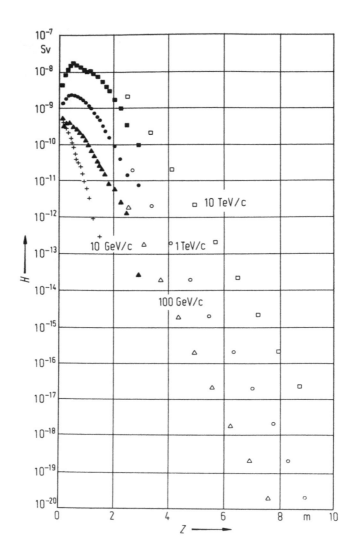

FIGURE 4.25

Dose equivalent per proton H on the longitudinal axis Z as a function of depth Z in the shield for proton-induced cascades in **IRON** of density 7.2 g cm^{-3}. The solid symbols represent FLUKA calculations for incidents protons of momenta 10 GeV/c, 100 GeV/c, 1 TeV/c, and 10 TeV/c. The open symbols correspond to CASIM calculations at the marked proton momenta. (With kind permission from Springer Science+Business Media: *Landolt-Börnstein numerical data and functional relationships in science and technology new series; Group I: Nuclear and particle physics Volume II: Shielding against high energy radiation*, 1990, ed. H. Schopper. Fassò, A. et al. Berlin, Germany: Springer-Verlag. Copyright 1990.)

consideration in practical shielding calculations. In Monte Carlo calculations, it is straightforward to mathematically "create" muons and follow them through the shielding medium. Muon transport in material is well understood (e.g., Cossairt et al. 1989a,b).

Particle energy downgrades quickly in hadronic showers so the most penetrating muons must originate in the first few generations of the cascade process. These energetic muons are not distributed over a large volume of space as are the neutrons. However, geometric effects such as collimation, or deflections by magnetic fields encountered near the point of production, can affect the muon fluence at large distances. Thus, the presence of large "empty" spaces, that is, decay paths (vacuum or air), near the point of interaction provide opportunity for the pions or kaons to decay into muons before they can be removed

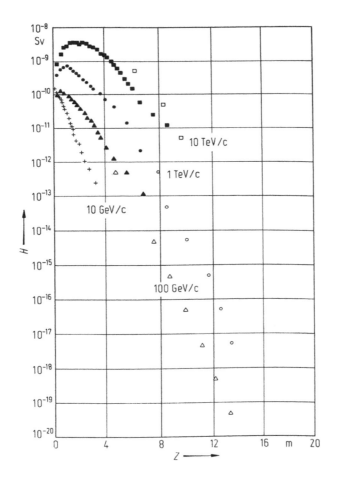

FIGURE 4.26
Dose equivalent per proton H on the longitudinal axis Z as a function of depth Z in the shield for proton-induced cascades in **CONCRETE** of density 2.4 g cm^{-3}. The solid symbols represent FLUKA calculations for incidents protons of momenta 10 GeV/c, 100 GeV/c, 1 TeV/c, and 10 TeV/c. The open symbols correspond to CASIM calculations at the marked proton momenta. (With kind permission from Springer Science+Business Media: *Landolt-Börnstein numerical data and functional relationships in science and technology new series; Group I: Nuclear and particle physics Volume II: Shielding against high energy radiation*, 1990, ed. H. Schopper. Fassò, A. et al. Berlin, Germany: Springer-Verlag. Copyright 1990.)

by nuclear interactions in solid materials. This is particularly important for the typical situation of a target used to produce secondary beams followed (downstream) by an air or vacuum gap (the space for decay into muons) and then a beam absorber. If magnetic fields are present, the muon fluence generally peaks in the bend plane. Multiple Coulomb scattering from nuclei is an important effect in muon transport.

Generally the dominant sources of muons are those due to the decay of pions and kaons. There are several important facts about such muons that are summarized as follows:

A. The decay lengths (mean length for $\pi^{\pm}$ or K$^{\pm}$ to decay), Λ, are given, respectively, by the following:

$\Lambda_{\pi} = 44.9p_{\pi}$ (meters), where p_{π} is the pion momentum in GeV/c, and

$\Lambda_K = 7.51p_K$ (meters), where p_K is the kaon momentum in GeV/c.

TABLE 4.8

Coefficients to Convert Star Densities S and Star Fluence Φ_S into Dose Equivalent

Proton Energy (GeV)	Absorber Material	Dose Equivalent/Star Density (Sv cm³ star⁻¹) (×10⁻⁸)	λ (cm)	Dose Equivalent/Star Fluence (Sv cm² star⁻¹)(×10⁻⁹)
10	Iron	2.04 ± 0.06	17.1	1.19 ± 0.04
100	Iron	2.15 ± 0.08	17.8	1.21 ± 0.05
1000	Iron	2.12 ± 0.08	17.2	1.23 ± 0.05
Mean	Iron	2.10 ± 0.04		
100	Aluminum	4.62 ± 0.17	38.6	1.20 ± 0.04
100	Tungsten	1.19 ± 0.05	9.25	1.29 ± 0.05
	Concrete	4.9	40.0	1.22
Mean	All			1.22 ± 0.02

Source: Adapted from Stevenson, G. R. 1986. *Dose equivalent per star in hadron cascade calculations.* European Organization for Nuclear Research: CERN Report TIS-RP/173. Geneva, Switzerland.

The available decay path in conjunction with the decay length can be used to estimate the total number of muons present. For example, a beam of 10^7 such pions at 20 GeV/c over a distance of 50 m will decay into $10^7 \times [50 \text{ m}]/[(56 \times 20)$ meters decay length$] = 4.5 \times 10^5$ muons. This uses the fact that the decay path (50 m) is small compared with the mean decay length of 1120 meters. If the decay path x were comparable to the decay length Λ_π, the final intensity would need to be multiplied by the exponential factor $\{1 - \exp(x/\Lambda_\pi)\}$.

B. If $\beta \approx 1$, relativistic kinematics determines that the ratio k_i of the *minimum* momentum of the daughter muon $(p_{\mu,min})$ to the momentum of the parent pion or kaon (p_i) is given by

$$k_i = p_{\mu,min}/p_{parent} = (m_\mu/m_{parent})^2 \tag{4.49}$$

The result is that k_i has a value of 0.57 for muons with pion parents and 0.046 for muons with kaon parents. Thus if, say, a beam transport system restricts the momentum of pions to some minimum value, then the momentum of the decay muons has a minimum value given by the previous equation.

C. Since in the frame of reference of the kaon or pion parent the decay is isotropic, and there is a one-to-one relationship between the muon momentum p_μ, and the angle of emission, for $p_\mu \gg m_{parent}$ (in units where $c = 1$) the momentum spectrum of the muons can be expressed as

$$\frac{dN}{dp_\mu} = \frac{1}{p_{parent}(1-k_i)} \tag{4.50}$$

This means that the spectrum of daughter muons uniformly extends from the momentum of the parent down to the minimum established in Equation 4.49.

D. Relativistic kinematics also gives the result that in the laboratory frame of reference the maximum angle θ_{max} between the momentum vector of the muon and that of the parent particle is given by

$$\tan\theta_{max} = \frac{m^2_{parent} - m^2_\mu}{2p_{parent}m_\mu} \qquad (4.51)$$

For muons originating from pion decay, θ_{max} is at most several milliradians. However, for muons originating from, say, the decay of 5.0 GeV kaons, θ_{max} is a relatively large 0.21 radians (12°). Thus $\pi \rightarrow \mu$ decays can be assumed to be approximately collinear, while $K \rightarrow \mu$ decays have significant divergence at the lower energies.

Monte Carlo calculations are needed to adequately describe the production and transport of muons because of the sensitivity to details of the geometry that determine the pion and kaon flight paths and influence the muon populations. Fassò et al. (1990) has presented some useful information about the yield of muons that one can use to make approximate estimates by giving calculated values of angular distributions of muon spectra with an absolute normalization from pion and kaon decays for 1.0 m decay paths. Neither the effects of absorbers nor magnetic fields are included in these results. For other decay paths that are short compared with the decay length, one can simply scale by the length of the actual decay path in meters. The results are displayed in Figure 4.27.

Decays of other particles can be important sources of muons at higher energies, especially those found in hadron-hadron collisions at high-energy colliders. Notable are those from

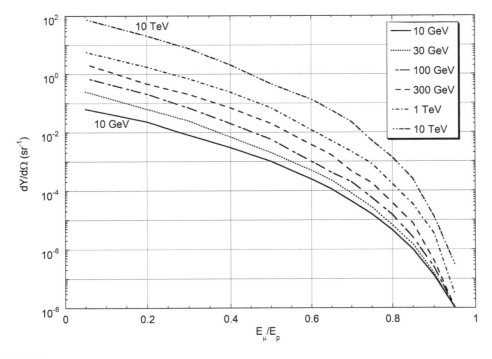

FIGURE 4.27
Yield of muons from the decay of pions and kaons of both charges produced in proton-Fe collisions at several energies of the incident proton at $\theta = 0$. The distance available for decay (the decay path) is taken to be 1.0 m. The abscissa E_μ/E_p is the muon energy expressed as a fraction of the incident proton energy. The ordinate $dY/d\Omega$ is the number of muons per unit solid angle (sr^{-1}) per incident proton having an energy greater than E_μ. (Data from results presented by Fassò, A. et al. 1990. *Landolt-Börnstein numerical data and functional relationships in science and technology new series; Group I: Nuclear and particle physics Volume II: Shielding against high energy radiation*, ed. H. Schopper. Berlin, Germany: Springer-Verlag.)

charm (D) and bottom (B) meson decays (Fassò et al. 1990). The muons from these sources are often called *direct muons* due to the short lifetimes and decay lengths involved. The masses of these parent particles and their mean-lives τ are as follows (Beringer et al. Particle Data Group 2012):

$$m(D^{\pm}) = 1869.62 \pm 0.15 \text{ MeV}, \tau = (1.040 \pm 0.007) \times 10^{-12} \text{ s}, c\tau = 311.8 \text{ μm and}$$

$$m(B^{\pm}) = 5279.25 \pm 0.17 \text{ MeV}, \tau = (1.641 \pm 0.008) \times 10^{-12} \text{ s}, c\tau = 492.0 \text{ μm}$$

Figures 4.28 and 4.29 give results for muons originating from these decays in the same format as used in Figure 4.27. The length of the decay path is irrelevant for these small values of $c\tau$.

Sullivan's approximation for muons is a method of estimating muon flux densities at proton accelerators based upon a semiempirical fit to existing muon production data (Sullivan 1992). Equation 4.52 gives Sullivan's result for the flux density of muons per meter of decay path as a function of shield thickness found along the proton beam axis (i.e., on the straight-ahead maximum of the muons):

$$\Phi = 0.085 \frac{Ez}{Z^2} \exp\left\{-\frac{\alpha t}{E}\right\} \tag{4.52}$$

where Φ is the fluence (muons m^{-2}) per interacting proton; E is the proton beam energy (GeV); Z is the distance of the point of concern to the point of production of the pions and kaons (meters); z is the average path length (i.e., the decay path) of the pions and kaons in air, gases, or vacuum prior to their absorption by solids or liquids; and α is an effective average

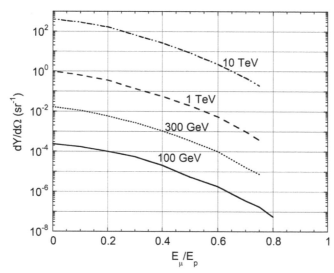

FIGURE 4.28

Yield of muons from the decay of D-mesons produced in proton-proton collisions at four incident proton energies and at $\theta = 0$. The abscissa $E\mu/E_p$ is the muon energy expressed as a fraction of the incident proton energy. The ordinate $dY/d\Omega$ is the number of muons per unit solid angle (sr^{-1}) per incident proton having an energy greater than E_μ. (Data from results presented by Fassò, A. et al. 1990. *Landolt-Börnstein numerical data and functional relationships in science and technology new series; Group I: Nuclear and particle physics Volume II: Shielding against high energy radiation*, ed. H. Schopper. Berlin, Germany: Springer-Verlag.)

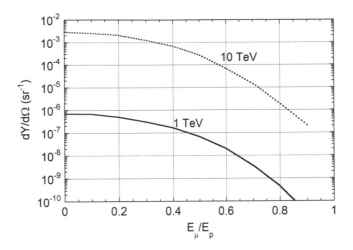

FIGURE 4.29
Yield of muons from the decay of B-mesons produced in proton-proton collisions at various energies of the incident proton and at $\theta = 0$. The abscissa $E\mu/E_p$ is the muon energy expressed as a fraction of the incident proton energy. The ordinate $dY/d\Omega$ is the number of muons per unit solid angle (sr^{-1}) per incident proton having an energy greater than E_μ. (Data from results presented by Fassò, A. et al. 1990. *Landolt-Börnstein numerical data and functional relationships in science and technology new series; Group I: Nuclear and particle physics Volume II: Shielding against high energy radiation*, ed. H. Schopper. Berlin, Germany: Springer-Verlag.)

energy loss rate (GeV meter^{-1}) for the muons in a shield of thickness t (meters). Values of α for typical shielding materials are provided in Table 4.9. The value of z can be taken to be the actual physical length of the decay path, or according to Sullivan, for a solid beam absorber, z can reasonably be taken to be 1.8 times the hadron nuclear interaction mean free path for the material comprising the beam absorber. It is obvious that the argument of the exponential in Equation 4.52 can be expanded as the sum over the materials comprising a composite shield. Sullivan has also given a prescription for calculating the full width at half maximum (FWHM) of the muon distribution at the boundary of such a shield:

$$\text{FWHM} = 4.6 \frac{z}{\sqrt{\alpha E t}} \qquad (4.53)$$

TABLE 4.9

Values of α for Typical Shielding Materials for Use in Equations 4.52 and 4.53

Material	α (GeV m^{-1})	Density, ρ(g cm^{-3})
Concrete[a]	9.0	2.35
Water	4.0	1.0
Iron	23.0	7.4
Lead	29.0	11.3

Source: Adapted from Sullivan, A. H. 1992. *A guide to radiation and radioactivity levels near high energy particle accelerators.* Ashford, Kent, UK: Nuclear Technology.

[a] The value for concrete can be used for earth if one adjusts it to the correct density.

PROBLEMS

1. One can use measurement results to check Sullivan's formula, Equation 4.4, for hadron fluence above 40 MeV for high-energy proton interactions. Check the agreement for the 22 and 225 GeV/c data in Figures 4.7 and 4.8 for three representative angles at 1.0 m. (Ignore the fact that the formula is for hadrons >40 MeV while the only data provided are for hadrons with energies >35 MeV and >50 MeV, but do *not* ignore the difference between normalizing to *incident* versus *interacting* protons.) Comment on the quality of the agreement.

2. Calculations can also be used to check the Tesch curve for dose equivalent at $\theta = 90°$ (Figure 4.10). Use the 200 MeV calculations in Figure 4.4 to do this by crudely numerically integrating the $60° < \theta < 90°$ yields to determine the average energy of the neutrons and the total fluence at $\theta = 90°$ and at 1.0 m. Use the results along with dose equivalent per fluence curves to obtain the dose equivalent per proton to compare with Tesch's result. Iron is considered equivalent to copper for this problem.

3. A copper target at an accelerator is struck by 1.0 μA of 100 MeV protons.

   a. Use Tesch's curve in Figure 4.10 to calculate the dose equivalent rate at 2.0 m and $\theta = 90°$ relative to this target.

   b. Compare this result with the neutron dose equivalent rate calculated in Chapter 3, Problem 5 for an *electron* accelerator having the same intensity and beam energy and discuss. (Scale the relevant result of Chapter 3, Problem 5 by the appropriate yield for copper versus tungsten.)

4. It is often necessary to work from fragmentary data to determine other quantities.

   a. Use McCaslin's results, Equations 4.15 through 4.17, and the appropriate dose equivalent fluence^{-1} factors to calculate the dose equivalent rate at 1.0 m and at $\theta = 30°$ for a target struck by 10^8 670 MeV/amu ^{20}Ne ions per second. Compare this with the effective dose rate. (Hint: Use all available spectra information.)

   b. Use McCaslin's results to obtain the total yield of neutrons per ion with $E_n > 6.5$ MeV. Assuming the target to be iron or copper, how does this yield correspond to that due to 700 MeV protons? Do this for both $E_n > 6.5$ MeV and $E_n > 20$ MeV to understand the overall composition. Hint: Integrate over the unit sphere (i.e., perform a double integral over spherical coordinates θ and ϕ) and convert all quantities associated with angles from degrees to radians. The following indefinite integrals are needed:

$$\int \frac{dx \sin x}{x} = x - \frac{x^3}{3 \times 3!} + \frac{x^5}{5 \times 5!} - \frac{x^7}{7 \times 7!} + \cdots \quad \text{and}$$

$$\int dx e^{ax} \sin bx = \frac{e^{ax}\left[a \sin bx - b \cos bx\right]}{a^2 + b^2}$$

The elemental area on the sphere of radius r is $dA = r^2 \sin \theta \, d\theta d\phi$, where ϕ is the standard azimuthal coordinate in a spherical coordinate system.

5. It is asserted that if the assumption is made that the limiting attenuation is simply geometric, with the nucleon radius equal to 1.2×10^{-13} cm, then $\rho\lambda_{atten} = 36.7A^{1/3}$ (g cm^{-2}). Show this to be the case using the volume of a nucleus and nucleons along with the cross section.

6. This problem explores the usage of the Moyer Model.

    a. Use the Moyer Model to calculate the dose equivalent rate (mrem h^{-1}) lateral ($\theta = 90°$) to a magnet centered in a 1.5 m radius tunnel. The magnet is struck by 1012 protons at 100 GeV per second. The tunnel walls consist of 0.333 m concrete followed by soil having the same composition but of lower density; ρ(concrete) = 2.5 g cm^{-3}, ρ(soil) = 2.0 g cm^{-3}]. Perform the same calculation for several thicknesses of soil out to 6.0 m of soil radially. Do this for increments of 1.0 m from 1.0 m to 6.0 m of soil.

    b. Calculate the result if the same beam loss occurs uniformly over a string of such magnets 100 m long in the same tunnel at the same soil thicknesses as above. Use the Tesch approximation. In this configuration, outside of 6.0 m of lateral soil shield, approximately how many how many linear meters of such beam loss are required to be the source of 90% of the calculated dose equivalent rate at 6.0 m of lateral soil shield?

    c. For the point loss in part (a), at what value of θ does the maximum dose equivalent rate occur, and what is its magnitude outside of 6.0 m of soil shield? (Use successive approximations to solve.)

7. a. An accelerator delivers 10^{12} 120 GeV protons per second head-on on the inner edge of a magnet located in a cylindrical tunnel centered on the beam axis. Use the results of the most relevant MARS calculation found in the Appendix to determine the approximate effective dose rate at $R = 400$ cm and compare with a result using the Moyer equation for a point loss. Both calculations should be at the location of the maximum effective dose. Assume the geometrical parameters to be identical to those in the Appendix calculation, ρ(concrete) = 2.4 g cm^{-3}, and ρ(soil) = 2.24 g cm^{-3}. Ignore the difference between effective dose and dose equivalent for this comparison. What might explain the difference between the two results?

    This problem explores the usage of results of Monte Carlo calculations.

    b. As a further comparison, consider an accelerator that delivers 10^{12} 1.0 TeV protons per second head-on on the inner edge of a magnet in a tunnel. Use the CASIM calculations found in the Appendix to determine the approximate dose equivalent rate at $R = 400$ cm and compare with a result using the Moyer equation for point loss. Both calculations should be at the location of the maximum dose equivalent. Once again assume the geometrical parameters to be identical to those in Appendix calculation and ρ (concrete) = 2.4 g cm^{-3} and ρ (soil) = 2.25 g cm^{-3}. What might explain the apparent disagreement between the two results?

    c. Discuss any general observations from the results of these two calculations.

8. Using the results of Monte Carlo hadron calculations (FLUKA/CASIM) calculate, for solid shields of iron (cylinders), what longitudinal thickness of iron is needed to achieve the same *hadron* dose equivalent per proton on the beam axis as found at $R = 50$ cm at 10 GeV/c, 100 GeV/c, 1000 GeV/c, and 10 TeV/c. Use the maximum value of H ($r = 50$ cm).

9. In Figure 4.4, we have calculations of neutron energy spectra for 200 MeV protons incident on an aluminum target. In Figure 4.23, calculations of dose equivalent values for spherical concrete shielding surrounding aluminum targets at $E_p = 200$ MeV are given. At shielding thicknesses approaching zero and at forward angles, are the

two results "sensible" (i.e., approximate, agreement)? (Hint: "Integrate" crudely over the forward spectrum to obtain the fluence/proton and convert this fluence to dose equivalent.)

a. Make the comparison for zero shield thickness and in the angular range $0 < \theta < 30°$.

b. Now use the shielding calculations to obtain the dose equivalent rate (rem h^{-1}) due to a 1.0 μA beam incident at 200 MeV on such a thick target at a distance of 4 m from the target with 0, 1, 2, and 3 m of intervening concrete shielding ($\rho = 2.5$ g cm^{-3}) for $\theta = 15°$ and $\theta = 75°$. (Hint: Use the center of the angular bins.)

10. Assume that a target is struck by 100 GeV protons and that a 10 m long decay path exists for $\pi^{\pm}$ and K$^{\pm}$ decay. Use the curves in Figure 4.27 to crudely estimate the muon flux density and dose equivalent rates (mrem h^{-1}) at 1.0 km away and at $\theta = 0°$ if 10^{12} protons/second are targeted in this manner with the following additional assumptions:

a. Assume that there is no shielding present and neglect air scattering and in-scattering from the ground. (Hint: The muon yield for this decay path will scale with the length of the decay path.)

b. Assume there are 100 m of intervening shielding of earth ($\rho = 2$ g cm^{-3}). (Hint: Use Figure 1.12 range-energy curves to determine the mean energy of muons that will penetrate this much shielding). Neglect multiple scattering and range-straggling.

c. If the beam operates for 4000 h yr^{-1}, is 100 mrem yr^{-1} exceeded? Will multiple scattering increase or decrease this dose equivalent? (Answer both questions for the soil-shielded case only.)

d. Repeat part (b) of the same calculation using Sullivan's semiempirical approach. If the disagreement between the results obtained using the two methods is large, suggest an explanation of a possible cause of the difference.

5

Unique Low-Energy Prompt Radiation Phenomena

5.1 Introduction

In this chapter two phenomena that involve low-energy particles, most importantly neutrons, are discussed: the transmission of photons and neutrons through penetrations and the control of neutron "skyshine." They must be understood at most accelerators, and clearly at all those that operate above the energy thresholds for producing neutrons. Both phenomena exhibit behavior qualitatively independent of incident particle type and energy.

5.2 Transmission of Photons and Neutrons through Penetrations

All neutron-producing accelerators need to control the transmission of neutrons, and also photons, through penetrations since all have accessways to permit entry of personnel and equipment. Penetrations for cables, cooling water piping, radiofrequency (RF) waveguides, and other equipment are also present. Personnel access penetrations will typically have cross-sectional dimensions of about 1.0 m by 2.0 m (i.e., door-sized), while utility ducts will generally be much smaller. Often the utility penetrations are partially filled with cables and other items as well as pipes delivering cooling water.

Penetrations are commonly called *labyrinths*, *mazes*, or even *chicanes*. Labyrinth is the term of choice in this book. Two general rules are advised for all penetrations of accelerator shielding:

1. A particle or photon beam should not be aimed directly toward a penetration. This assures that the penetrations are transmitting primarily scattered particles.

2. For any labyrinth, the sum of the wall thickness between the source and the "outside" should be equivalent to that which would be required if the labyrinth were not present. The "void" presented by the passageway does not provide any shielding laterally.

5.2.1 Albedo Coefficients

Before describing the details of penetration design, one should review some simple parameterizations of the scattering of photons and neutrons. These reflections can be treated by means of *reflection* or *albedo coefficients*. Such coefficients provide the fraction of incident particles at a given angle of incidence θ_i that "reflect" at a particular angle

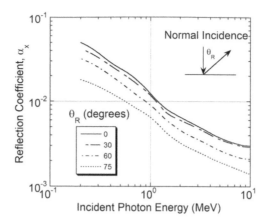

FIGURE 5.1

Reflection coefficients α_x for monoenergetic PHOTONS with normal (perpendicular) incidence on ordinary concrete as a function of incident photon energy for several angles of reflection θ_R. For photon energies higher than 10 MeV, the use of the 10 MeV values of α_x is expected to be a conservative, but useful choice. (Reprinted with permission from NCRP. 2003. *Radiation protection for particle accelerator facilities*. NCRP Report No. 144. Bethesda, MD. https://NCRPonline.org.)

of reflection θ_R. This is analogous to the diffuse reflection of visible light by various kinds of surfaces more complex than are simple mirrors. They take into account the appropriate *microscopic* scattering cross sections in a *macroscopic* way. These have general applications beyond the design of penetrations. Figures 5.1 through 5.4 give the albedo coefficients α_x and α_n for monoenergetic photons and neutrons, respectively, incident on flat surfaces of infinite dimensions of concrete plotted as functions of energy for various conditions of incidence. Clearly, the albedo of neutrons is typically larger and somewhat less strongly dependent on energy than is that of photons. Chilton et al. (1984) have given more detailed results for concrete and for other materials. A good summary of the results

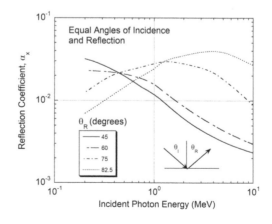

FIGURE 5.2

Reflection coefficients α_x for monoenergetic PHOTONS incident on ordinary concrete as a function of incident photon energy for equal angles of incidence and reflection $\theta_I = \theta_R$. For photon energies higher than 10 MeV, the use of the 10 MeV values of α_x is expected to be a conservative, but useful choice. (Reprinted with permission from NCRP. 2003. *Radiation protection for particle accelerator facilities*. NCRP Report No. 144. Bethesda, MD. https://NCRPonline.org.)

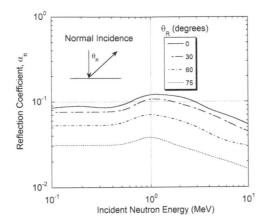

FIGURE 5.3
Reflection coefficients α_n for monoenergetic NEUTRONS with normal (perpendicular) incidence on ordinary concrete as a function of incident photon energy for several angles of reflection θ_R. (Reprinted with permission from NCRP. 2003. *Radiation protection for particle accelerator facilities.* NCRP Report No. 144. Bethesda, MD. https:// NCRPonline.org.)

has been provided by the National Council on Radiation Protection and Measurements (NCRP 2003).

The curves in Figures 5.1 through 5.4 were calculated for irradiated surface areas of 1.0 m² and at a distance of 1.0 m and include absorption within the scattering material (NCRP 2003). The effects of such absorption are most evident in Figures 5.1 and 5.3 where the values of α_x and α_n at a given energy decrease almost uniformly as θ_R increases, because the scattered photons or neutrons must penetrate more absorber material to emerge at these larger angles. In Figures 5.2 and 5.4, the dependence on the value of θ_R is slightly more pronounced for the photons because of the importance of Compton scattering. In general, the albedo of neutrons is less energy dependent and angle dependent than that of photons scattered under comparable scenarios.

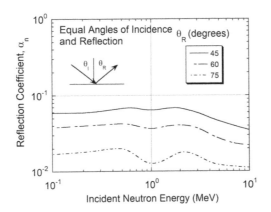

FIGURE 5.4
Reflection coefficients α_n for monoenergetic NEUTRONS incident on ordinary concrete as a function of incident photon energy for equal angles of incidence and reflection $\theta_I = \theta_R$. (Reprinted with permission from NCRP. 2003. *Radiation protection for particle accelerator facilities.* NCRP Report No. 144. Bethesda, MD. https:// NCRPonline.org.)

5.2.1.1 Usage of Photon Albedo Coefficients

A particular application of these coefficients in the design of labyrinths is given here as an illustration. Figure 5.5 shows an example of a labyrinth providing access to a collimated photon source of some known dose equivalent or effective dose (or dose rate, with inclusion of units of inverse time) H_o determined at some reference distance d_o.

To use these coefficients correctly, knowledge of the photon energy spectrum at this location is also needed. For example, such a de facto photon "beam" can arise from the forward-peaked photons due to the targeting of a beam from an electron accelerator. With the reflection coefficients α_x, one can use the following formula to obtain a conservative estimate of the dose equivalent or effective dose (or dose rate) H_{r_j} after j sections (not counting the initial path length to the wall d_i) of the labyrinth:

$$H_{r_j} = H_0 \left(\frac{d_0^2}{d_i^2} \right) \left(\frac{\alpha_1 A_1}{d_{r_1}^2} \right) \prod_{k=2}^{k=j} \left(\frac{\alpha_k A_k}{d_{r_k}^2} \right), j \geq 2 \tag{5.1}$$

In this formula the coefficient α_1 is selected to be representative of that expected at the *initial* photon energy, while A_1 estimates the cross-sectional area of the wall struck by the initial

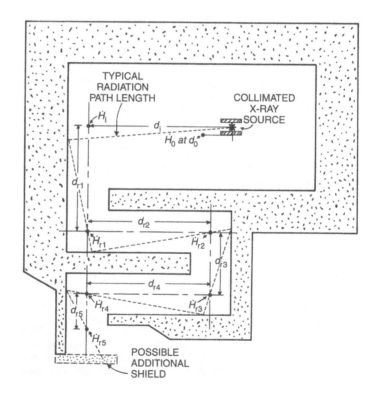

FIGURE 5.5

Generalized labyrinth design illustrating successive reflections of photons from a collimated source through the maze. The source could just as well originate from an electron beam originating from the right side of the figure incident on a target located at the point in space labeled "collimated x-ray source." The various path lengths can be approximated by a sequence of centerline distances, as shown in the diagram. (Reprinted with permission from NCRP. 2003. *Radiation protection for particle accelerator facilities*. NCRP Report No. 144. Bethesda, MD. https://NCRPonline.org.)

photons evaluated by projecting the beam profile to the wall. For $k \geq 2$, A_k is the cross-sectional area of the kth leg of the maze. From the left, the first parenthetical factor is just inverse square propagation of the beam to the wall, the second factor models "reflection" into the first leg, and the product factor accounts for reflection into the remaining legs. For right-angle, straight labyrinths such as this one, it is reasonable to use the values plotted for normal incidence ($\theta_I = 90^0$) and $\theta_R = 75°$. For successive legs after the first, taking the value of α_k to be that for 0.511 MeV photons is often considered to be a conservative approach. This is a result of the fact that if E_o is the initial photon energy in MeV, the energy of the scattered photon E_{scatt} (MeV), following Compton scattering is given by

$$E_{scatt} = \frac{E_0}{1 + (E_0/0.511)(1 - \cos\theta)} \tag{5.2}$$

Thus, E_{scatt} has a maximum value of 0.511 MeV after a scatter of 90° for $E_o \gg 0.511$ MeV, the rest energy of the scattered electron. If the maze is of uniform cross section A and has j legs, then the product in the numerator is simply αA raised to the $(j - 1)$th power, $(\alpha A)^{j-1}$, where $\alpha = \alpha_k$ for all legs after the first. In the denominator, the distances are just those defined in Figure 5.5 and, of course, represent the inverse-square law dependence. This formula is "conservative" for photon energies exceeding 10 MeV, but at the higher energies the uncertainties are larger. This approach is probably most accurate if the ratios $d_{rk}/(A_k)^{1/2}$ lie between 2.0 and 6.0 (NCRP 2003).

5.2.2 Neutron Attenuation in Labyrinths: General Considerations

For neutrons, the more complex physics of their transport discourages the use of a formula similar to that employed previously using photon albedo coefficients. First, the radiation source, or potential radiation source for situations of concern from the standpoint of accidental beam losses, should be evaluated according to the methods described previously. Typical methods for addressing the attenuation of radiation by penetrations involve the use of the results of calculations performed using Monte Carlo codes. These can be used for both straight and curved labyrinths, but most practical experience is with the straight variety. In this section, the results of such work will be presented to give the reader useful information in the evaluation of such penetrations.

An overwhelming conclusion drawn from the existing body of data is that the bombarding particle energy, or even particle type, has very little effect on the attenuation by a labyrinth viewing a source of beam loss other than the fact that the total yield of "source" neutrons increases as a function of incident energy, or specific energy for ions. One can thus estimate the absorbed dose, dose equivalent, effective dose, or neutron fluence at the exit of a labyrinth by using attenuation estimates in the *legs* multiplied by an estimate of the neutron fluence or dose found at the entrance, or *mouth*, of the penetration into the beam enclosure. The validity of this *factorization approximation* allows attenuation measurements and calculations obtained at proton accelerators to be of general utility.

A typical straight line, that is, "rectilinear," personnel access labyrinth is shown in Figure 5.6.

5.2.3 Attenuation in the First Legs of Straight Penetrations

For penetrations exposed to targets struck by hadrons, we first consider the straight penetration studied by Gilbert et al. (1968) that reports the measurement of the transmission

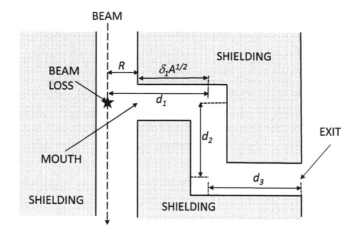

FIGURE 5.6
Plan view of a typical rectilinear personnel access labyrinth of three "legs" at an accelerator that defines the coordinate system and terminology associated with labyrinth calculations. The star symbol (✶) denotes the location of a localized loss of beam adjacent to the "mouth" of the labyrinth. The lengths of all legs after the first are measured between the *centers* of turns.

of neutrons in a very long, straight tunnel of dimensions 2.8 m high by 1.8 m wide and 100 m long. (This may be the longest tunnel in which such a labyrinth measurement has ever been made.) The 14 GeV protons were incident on a target providing a good "point source" 3.2 m from the tunnel mouth. The use of a set of activation detectors having different energy thresholds made it possible to obtain some information about the neutron energy spectrum as well. This measurement technique is discussed in somewhat more detail in Section 9.5.3. An absolute normalization to beam loss was not reported. Table 5.1 gives the thresholds, or approximate sensitive domains, of nuclear reactions used in this particular measurement. The dosimeters used to detect photons were sensitive to gamma rays produced by the capture of neutrons in the air and in the tunnel walls.

The results of the measurements are presented in Figure 5.7. The "fits" to the relative response R as a function of depth in the penetration d shown in this figure were arbitrarily matched to the measurements at a depth of 20 m in the tunnel and fit by an exponential attenuation multiplied by an inverse square-law dependence;

$$R(d) = R_{20} \left(\frac{20}{d_1}\right)^2 \exp\left[-\frac{d_1}{\lambda(E)}\right],$$ (5.3)

TABLE 5.1

Activation Detectors and Their Characteristics as Used in the Measurements Summarized in Figure 5.7 with Approximate Sensitive Energy Regions

Detector	Nuclear Reaction	Energy Range (MeV)
$\beta\gamma$ Dosimeters	Photons and charged particles	All
Gold (Au)	$^{197}Au(n,\gamma)^{198}Au$	Thermal energies
Aluminum (Al)	$^{27}Al(n,\alpha)^{24}Na$	E > 6 MeV
Carbon (C)	$^{12}C(n,2n)^{11}C$	E > 20 MeV

Source: Adapted from Gilbert, W. S. et al. 1968. *1966 CERN-LRL-RHEL shielding experiment at the CERN proton synchrotron.* University of California Radiation Laboratory: Report UCRL 17941. Berkeley, CA.

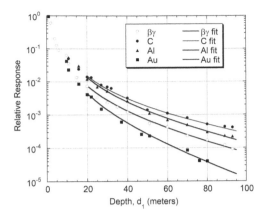

FIGURE 5.7
The relative transmission of neutron flux density and gamma dose rate along a large straight tunnel described in the text. The measurement results are shown as the symbols, while the solid lines represent the fits described in the text matched to the data at a depth $d_1 = 20$ m. (Adapted from Gilbert, W. S. et al. 1968. *1966 CERN-LRL-RHEL shielding experiment at the CERN proton synchrotron*. University of California Radiation Laboratory: Report UCRL 17941. Berkeley, CA.)

where R_{20} is the response measured at $d_1 = 20$ m and $\lambda(E)$ is an energy-dependent attenuation length.

The responses as a function of depth d_1 are quite revealing. At short depths into the tunnel (<20 m) the "attenuation" of the fast neutrons is almost entirely accounted for by inverse-square law considerations. Further into the tunnel, the responses clearly illustrate that neutrons of lower energy illustrated by the "Au" curve attenuate more rapidly by air and wall scattering than do the higher-energy neutrons corresponding to the "Al" and "C" curves. Factoring out the inverse-square dependence for this long tunnel, the remaining attenuation is well described by exponential absorption with effective mean free paths λ corresponding to effective removal cross sections for the unique neutron spectral regions "seen" by the individual detectors. The λ values determined by fitting these data are given in Table 5.2.

The effective removal cross sections determined by this measurement are about a factor of 1.5–2.0 *smaller* than those that would be inferred from the known absorption cross sections of the constituents of air. This is evidence of "in-scattering" by the concrete walls since *more* neutrons than expected were observed at the larger distances into the tunnel.

An important principle is the *labyrinth scaling rule*. The attenuation of neutrons in the legs of labyrinths generally scale with a *unit length* equal to the square root of its cross-sectional

TABLE 5.2

Mean Free Paths and Removal Cross Sections for Tunnel Transmission as Exhibited by the Measurements Summarized in Figure 5.7

Detector	λ, Mean Free Path (meters)	Inferred Removal Cross Section (barns)
βγ Dosimeters	55	3.3
Gold (Au)	30	6.2
Aluminum (Al)	60	3.2
Carbon (C)	100	1.9

Source: Adapted from Gilbert, W. S. et al. 1968. *1966 CERN-LRL-RHEL shielding experiment at the CERN proton synchrotron*. University of California Radiation Laboratory: Report UCRL 17941. Berkeley, CA.

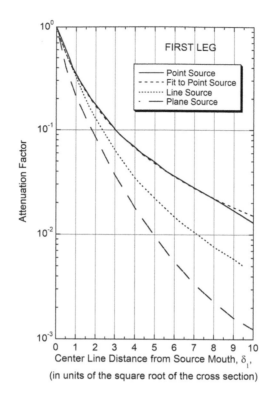

FIGURE 5.8

Universal transmission curves for the first leg of a labyrinth as a function of normalized distance δ_1 from the mouth. δ_1 is the distance from the source mouth expressed in units of the square root of the cross-sectional area of the passageway (see Figure 5.6). The fit for the point source curve represented by Equation 5.4 is also included. The curve for a plane source is also suitable to use with an off-axis point source. (Adapted from Goebel, K. et al. 1975. *Evaluating dose rates due to neutron leakage through the access tunnels of the SPS*. European Organization for Nuclear Research: CERN Report LABII-RA/Note/75-10. Geneva, Switzerland; Cossairt, J. D. 2013b. *Approximate technique for estimating labyrinth attenuation of accelerator-produced neutrons*. Fermi National Accelerator Laboratory: Fermilab Radiation Physics Note No. 118. Batavia, IL.)

area, provided that the height-to-width ratio does not vary greatly outside of the range 0.5–2.0 (Thomas and Stevenson 1988). Some details of the source geometry are very important in such a straight penetration. Goebel et al. (1975) has calculated universal attenuation curves for "first" legs of labyrinths (i.e., those sections first encountered as one moves outward from the beam). The three situations of point source, line source, and plane or point source off-axis for a straight tunnel displayed as *universal dose attenuation curves* as calculated by Goebel et al. (1975) are given in Figure 5.8.

Goebel et al. (1975) obtained their results from the codes SAM-CE (Cohen et al. 1973), AMC (Maerker and Cain 1967), and ZEUS (d'Hombres et al. 1968; deSereville and Tardy-Joubert 1971). Gollon and Awschalom (1971) have generated similar curves using the ZEUS code for a variety of geometries. An *off-axis point source* is one not centered in front of the labyrinth mouth. For neutron radiation fields, the attenuation of a labyrinth is essentially the same for all of the "dose" quantities discussed in Chapter 1: H_{equiv}, H_{eff}, D, etc. Clearly, neutrons from a plane or off-axis sources are more readily attenuated since the tunnel aperture provides less solid angle for acceptance.

Cossairt (2013b) found that the point source dependence found by Goebel et al. in a tunnel of cross-sectional area A is well approximated by the following expression, where $\delta_1 = (d_1 - R)/A^{1/2}$ and r_o is a fitting parameter:

$$H_1(\delta_1) = \left[\frac{r_o}{\delta_1 + r_o}\right]^2 H_o(R) \text{ with } r_o = 1.4 \qquad (5.4)$$

$H_1(\delta_1)$ is the dose at distance δ_1 in the first leg as measured from the mouth of the passageway in "units" of the square root of the cross-sectional area of the first leg (see Figure 5.6). $H_o(R)$ is the dose at the mouth, the determination of which is discussed later in this chapter.

Over the domain of $0 < \delta_1 < 9.0$, the expression fits the Goebel curve within $\pm 10\%$, sufficiently accurate for most radiation protection purposes. The domain in δ_1 is an appropriate one given the fact that most "personnel" labyrinths are of cross-sectional area of about 1.0×2.0 m^2 with a typical unit length of about 1.4 m. A 10 "unit" long first leg is thus about 14 m, a rather long passageway segment.

Tesch (1982) has developed a very simple approach to the problem of dose attenuation by multilegged labyrinths at proton accelerators typical of personnel passageways of approximately 2.0 m^2 cross section. For the first leg the expression is an inverse-square law dependence augmented by a factor of 2.0 to approximate "in-scattering":

$$H_1(d_1) = 2H_o(R)\left(\frac{R}{d_1}\right)^2 \qquad (5.5)$$

In Equation 5.5, the distance into the labyrinth d_1 (defined as in Figure 5.6) is in meters and is *not* scaled by the cross-sectional area of the passageway. Equation 5.5 is valid only for personnel tunnels of approximately 2 m^2 cross-sectional area.

5.2.4 Attenuation in Second and Successive Legs of Straight Penetrations

Stevenson and Squier (1973) reported the results of measurements in a two-legged penetration at the NIMROD synchrotron. This penetration was of cross section 2.3×2.3 m^2, and the walls were made of concrete. The target at the mouth of the labyrinth was bombarded by 7.0 GeV protons. Figure 5.9 is a plot of the transmission of particle flux density along this tunnel using different nuclear reactions, again employed because of their unique nuclear reaction thresholds.

In Figure 5.9, the indicated nuclear reactions used as neutron detectors by means of radioactivation have the following approximate sensitive ranges for detecting neutrons: ^{12}C(n,2n)^{11}C ($E_n > 20$ MeV), ^{19}F(n,2n)^{18}F ($E_n > 11$ MeV), ^{27}Al(n,α)^{24}Na ($E_n > 3.2$ MeV), and ^{197}Au(n,γ)^{198}Au (thermal).

One can see that, proceeding from the target outward in the passageway, beyond the abrupt jump that arises as the corner "hides" the target from view, the fast neutron components are attenuated more readily than is the thermal one. Second and successive legs of such straight penetrations thus change the situation dramatically, principally by modifying the spectrum of the transmitted neutrons. As described shortly, the phenomenon associated with "turning the corner" was also verified by Cossairt et al. (1985b). Goebel et al. (1975) calculated a universal curve for second and succeeding legs that can serve as a companion to that given for the first leg in Figure 5.8. These results are plotted in Figure 5.10.

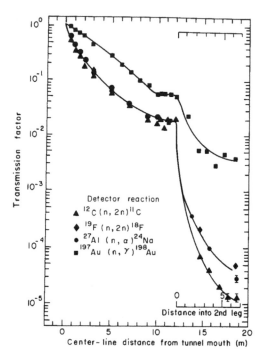

FIGURE 5.9

Relative transmission of particle flux density along a two-legged labyrinth using threshold detectors. (Reprinted with permission from Stevenson, G. R., and D. M. Squier. 1973. *Health Physics* 24:87–93. https://journals.lww.com/health-physics/pages/default.aspx.)

Working outward from the radiation source, the distance from the center of the preceding turn down the subsequent labyrinth leg is, as was done with first legs, normalized to the square root of the cross-sectional area of the ith leg A_i in unit lengths; $\delta_i = d_i/A^{1/2}$, for second and succeeding legs. Cossairt (2013b) found the following recursive expression to describe this curve, where δ_i is the distance in the ith leg measured in "units" of the square root of the cross-sectional area of the ith leg:

$$H_i(\delta_i) = \left\{ \frac{\exp(-\delta_i/a) + A\exp(-\delta_i/b) + B\exp(-\delta_i/c)}{1 + A + B} \right\} H_{i-1}(\delta_{i-1}) \; i\text{th leg } (i > 1) \qquad (5.6)$$

with the fitting parameters $a = 0.17$, $A = 0.21$, $b = 1.17$, $B = 0.00147$, and $c = 5.25$. Figure 5.10 displays these results.

Tesch (1982) also has developed the following formula for the transmission of the second and successive legs:

$$H_i(d_i) = \left\{ \frac{\exp(-d_i/0.45) + 0.022A_i^{1.3}\exp(-d_i/2.35)}{1 + 0.022A_i^{1.3}} \right\} H_{i-1} \; (i\text{th leg}, i > 1) \qquad (5.7)$$

Here A_i is the cross-sectional area of the ith leg in units of square meters. As was the case with respect to Equation 5.5, this formula uses the distances d_i (meters) along the labyrinth (not in units of the square root of the cross-sectional area). As it was for Equation 5.5, the results are valid for "door-sized" labyrinths with cross-sectional areas of about 2.0 m².

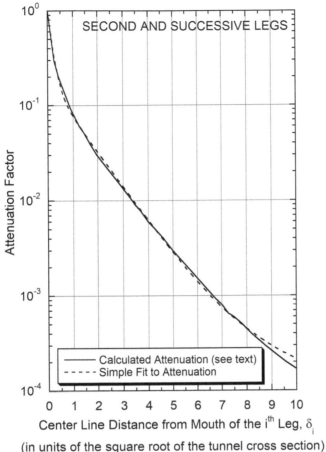

Attenuation Factor

Center Line Distance from Mouth of the ith Leg, δ_i

(in units of the square root of the tunnel cross section)

—— Calculated Attenuation (see text)
----- Simple Fit to Attenuation

SECOND AND SUCCESSIVE LEGS

FIGURE 5.10
Universal transmission curve for the second and subsequent legs of labyrinths as a function of normalized distance from the center of the previous turn δ_i. The calculated attenuation (solid curve) is that reported by Goebel et al. (1975) using the code AMC. The dashed curve is the fit provided by Equation 5.6. (Adapted from Goebel, K. et al. 1975. *Evaluating dose rates due to neutron leakage through the access tunnels of the SPS.* European Organization for Nuclear Research: CERN Report LABII-RA/Note/75-10. Geneva, Switzerland; Cossairt, J. D. 2013b. *Approximate technique for estimating labyrinth attenuation of accelerator-produced neutrons.* Fermi National Accelerator Laboratory: Fermilab Radiation Physics Note No. 118. Batavia, IL.)

Figure 5.11 shows a four-legged labyrinth providing entrance to a tunnel above a target struck by 400 GeV protons accelerated by the Tevatron at Fermilab studied in a set of measurements conducted by Cossairt et al. (1985b).

Figure 5.12 compares experimental measurements (Cossairt et al. 1985b) of absorbed dose as a function of position throughout this labyrinth with several methods of calculation.

As one can see, all methods of calculating the attenuation discussed here are approximately valid even for this four-legged labyrinth. Though the first leg does not have a truly "open view" of the target, it is well described by these methods as a "first" leg. Further, these results show that the behavior in legs beyond the second one is that of successive "second" legs.

It should be noted that for the labyrinth shown in Figure 5.11, the shielding blocks were, in fact, aligned precisely as indicated. This is, indeed, contrary to good design practice as the blocks should have been overlapped to prevent "streaming" of neutrons between the

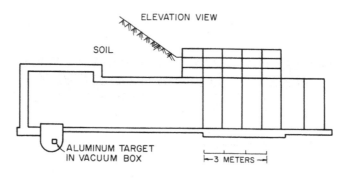

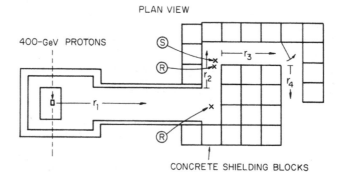

FIGURE 5.11

Labyrinth enclosure in which 400 GeV protons interacted with an aluminum target located beneath the floor as shown. The neutron energy spectrum was measured at the location denoted **S** and the quality factor of the radiation field (1973 System) was measured at the locations denoted **R**. (Reprinted with permission from Cossairt, J. D. et al. 1985b. *Health Physics* 49:907–917. https://journals.lww.com/health-physics/pages/default.aspx.)

blocks. This deficiency of design may explain the excess of measurement over calculation at the end of the third leg, where there well might have been some enhanced "streaming" of neutrons through the undesirably aligned cracks in the blocks from the first leg into the third.

For the labyrinth shown in Figure 5.11, a recombination chamber technique (see Section 9.5.7) was used to measure the neutron quality factor Q at two locations, one at the end of the first leg and one in the middle of the short second leg denoted by R. The values of Q were determined in terms of the 1973 System. The results were $Q = 5.5 \pm 0.6$ (first leg) and $Q = 3.4 \pm 0.1$ (second leg). This indicates a reduction of the average neutron energy in the second leg which was further verified by a measurement of the neutron energy spectrum (see Figure 6.14) at the location denoted by **S** in Figure 5.11 using a multisphere technique (see Section 9.5.2.1) that resulted in $Q = 3.1 \pm 0.7$. This neutron energy spectrum was measured at the location denoted by S in Figure 5.11. The spectrum measured in the second leg exhibited domination by thermal, or near-thermal, neutrons. It is clear that several approaches to the design of labyrinths are equally effective for practical radiation protection work.

5.2.5 Attenuation in Curved Tunnels

Curved tunnels are principally used to provide access for large, especially lengthy, equipment items that cannot negotiate right-angle bends. These have not been treated in

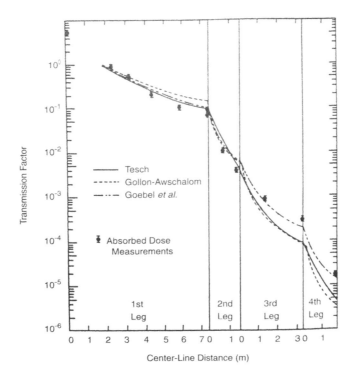

FIGURE 5.12

Measurements and predictions of transmission in a tunnel at Fermilab. The results of Tesch (1982), Goebel et al. (1975), and Gollon and Awschalom (1971) are compared with measurements of absorbed dose conducted at the position shown in the four-legged labyrinth displayed in Figure 5.11. Fortuitously, the "Transmission Factor" plotted as the ordinate is also the absolute scale of the absorbed dose measurement in units of mrad per 10^{10} incident 400 GeV protons (fGy proton^{-1}). (Reprinted with permission from Cossairt, J. D. et al. 1985b. *Health Physics* 49:907–917. https://journals.lww.com/health-physics/pages/default.aspx.)

nearly the same detail as have the rectilinear passageways. It appears that the attenuation is described by an exponential function having an attenuation length λ that is only a function of the radius R of the tunnel. Patterson and Thomas (1973) determined that

$$\lambda = 0.7\sqrt{R} \tag{5.8}$$

where R is in meters and $4 < R < 40$ m. Thus, the dose $H(r)$ or fluence at any circumferential distance through the tunnel x is given by

$$H(x) = H_o \exp\left(-\frac{x}{\lambda}\right) \tag{5.9}$$

where x and λ are expressed in mutually consistent units.

5.2.6 Attenuation beyond the Exit

A question that often arises in discussion of labyrinths is how to determine the dose rate outside of the exit of the passageway. Qualitative experience is that beyond the exit, the neutrons "disappear" rather rapidly. This phenomenon is plausibly a result of the fact, as

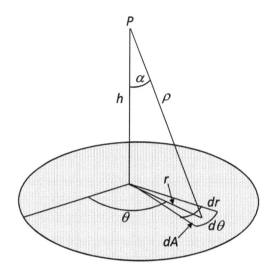

FIGURE 5.13
Diagram of labyrinth exit neutron calculation. The coordinates are explained in the text.

indicated previously, that the neutron energy spectrum is heavily dominated by thermal and near-thermal neutrons in all legs after the first. Such neutrons, therefore, having suffered many scatters, are not being collimated in any particular direction, being a thermalized "gas." Elwyn (1991, private communication to J. D. Cossairt) has studied this by assuming that the exit of the labyrinth is a circular disk of area A, equivalent in area to that of the actual exit. Further, it is assumed that the neutrons emerge from this disk uniformly and at all random directions with source fluence S_A. Figure 5.13 illustrates the geometry.

It is further assumed that only emission into the 2π steradian hemisphere outside the exit matters beyond the exit. The differential flux density at P on the axis of the disk is

$$d\phi = \frac{S_A \cos \alpha \, dA}{2\pi\rho^2} \tag{5.10}$$

where $dA = rdrd\theta$, $\rho^2 = h^2 + r^2$, and angle α is defined in Figure 5.13 ($\cos \alpha = h/\rho$). The $\cos \alpha$ factor is present to account for the solid angle of the source elemental area subtended at point P.

Thus,

$$d\phi = \frac{S_A hr}{2\pi\rho^3} drd\theta \tag{5.11}$$

Integrating,

$$\phi(h) = \frac{S_A h}{2\pi} \int_0^R dr \int_0^{2\pi} d\theta \frac{r}{(r^2 + h^2)^{3/2}} = S_A h \int_0^R dr \frac{r}{(r^2 + h^2)^{3/2}}$$

$$= S_A h \left[\frac{-1}{\sqrt{r^2 + h^2}} \right]_0^R = S_A \left[1 - \left\{ 1 + \left(\frac{R}{h} \right)^2 \right\}^{-(1/2)} \right] \tag{5.12}$$

TABLE 5.3

Estimates of Relative Neutron Fluence or Dose
Equivalent as a Function of Scaled Distance
from the Exit of a Labyrinth in Equation 5.12

h/R	$\phi(h)/S_A$
0.5	0.55
1.0	0.29
2.0	0.11
4.0	0.03
10.0	0.005

where attenuation by the air is neglected. One uses this result by approximating the area of the exit opening by the area of a disk having an equivalent area. At large distances, one can apply a "point source" approximation due to the fact that

$$\phi(h) \approx \frac{S_A}{2}\left(\frac{R}{h}\right)^2 \quad \text{for } h \gg R \tag{5.13}$$

For $h = 0$, $\phi(0) = S_A$ as expected. The dramatic decrease of fluence with distance is illustrated by the tabulation of a few values in Table 5.3.

5.2.7 Determination of the Source Factor

Generally, the dose at the mouth of a labyrinth can be obtained using Monte Carlo techniques or by directly using the information about neutron yields. For protons, approximations that use Moyer model parameters discussed in Chapter 4 are likely to overestimate the dose at the entrance. This is because the Moyer parameters implicitly assume development of the shower (intrinsically a "buildup" mechanism, as seen in Chapter 4) in the enclosure shielding. This buildup does not happen in the passageway.

For high-energy proton accelerators, a rule of thumb for the source term found to be sufficiently accurate for most personnel protection purposes has been developed by Ruffin and Moore (1976). It was improved by inclusion of Moyer energy scaling by Rameika (1991, Labyrinths and penetration methodology, version 1.3, private communication to J. D. Cossairt). In this model, about one fast neutron proton^{-1} GeV^{-1} of proton beam energy is taken to be produced with an isotropic distribution. These could be considered to be evaporation neutrons, not the cascade neutrons produced with much higher multiplicity in the forward direction. These isotropically produced neutrons dominate the spectrum and determine the dose equivalent (1973 System) or effective dose (1990 System) of those that enter the labyrinth and have kinetic energies between 1.0 and 10 MeV. From the dose equivalent per fluence factors $P(E)$ of Figure 1.5, 1.0 rem of 1–10 MeV neutrons represents a fluence of approximately 3×10^7 cm^{-2}. Thus, at distance R (cm) from the source,

$$H_{equiv} = \frac{E_0^{0.8} N_p}{4\pi R^2 (3\times10^7)} = 2.65\times10^{-9}\,\frac{E_0^{0.8} N_p}{R^2}\,\text{(rem)} \tag{5.14}$$

where E_o is in GeV, and N_p is the number of incident protons. The constant, 2.65×10^{-9} (rem cm^2), turns out to be approximately one-third the value obtained by using the Moyer source parameter along with high-energy value of the Moyer angular factor at $\theta = \pi/2$ as indicated by Equations 4.39 and 4.42:

$$(2.8 \times 10^{-7}\ \text{rem cm}^2)\exp\left(-2.3\frac{\pi}{2}\right) = 7.6 \times 10^{-9}\ \text{rem cm}^2 \qquad (5.15)$$

For effective dose, H_{eff}, the constant 2.65×10^{-9} should be replaced with 3.71×10^{-9} (Cossairt 2013b).

To obtain the source factor for neutrons produced by electrons, the neutron yields discussed in Section 3.2.3 can be utilized.

5.3 Skyshine

Thin roof shielding represents a serious problem that has plagued a number of accelerators such as the Cosmotron at Brookhaven National Laboratory, the Bevatron at the Lawrence Berkeley National Laboratory, the Fermilab experimental areas, and likely elsewhere. The phenomenon of interest, known as *skyshine*, is the situation in which the roof of some portion of the accelerator or an associated experimental facility is shielded more thinly than are the sides of the same enclosure that directly view the radiation source. The first widely documented calculation of the skyshine radiation field was that of Lindenbaum (1961). Fassò et al. (1990) give a rather complete description of the phenomena, while Patterson and Thomas (1973), Rindi and Thomas (1975), Stevenson and Thomas (1984), and Cossairt and Coulson (1985) present refinements and specific results. Neutron skyshine continues to arise as a problem due to failure to identify it as an issue during design or, subsequently, the identification of the need to repurpose an existing facility for energies and/or beam powers exceeding those within the original design scope. Commonly, other constraints must be met, notably the minimization of the weight of shielding borne by the roofs of large experimental halls, that exclude the obvious, simple solution of just adding more shielding.

While neutron skyshine is the topic of this section, it is noted that under some conditions *photon* skyshine can also be of importance at accelerators. In particular, the phenomenon can be significant when very highly activated items must be repaired or otherwise handled with limited shielding present. Photon skyshine has been discussed extensively by the NCRP (2003).

5.3.1 Simple Parameterizations of Neutron Skyshine

Figure 5.14 shows the geometry of a typical skyshine problem. In the model considered in this section, azimuthal isotropicity is assumed.

When addressing the skyshine question, it is often customary to plot the neutron fluence, dose equivalent, or effective dose as a function of distance from the source by multiplying it by the square of the distance from the source, that is, as $r^2\Phi(r)$.

Stevenson and Thomas (1984) included plots of several measurements of neutron skyshine obtained at proton accelerators producing protons of energies ranging from 30 MeV to 30 GeV, and also at high-energy electron accelerators having energies of 7.5 and 12 GeV. In general, the quantity $r^2\Phi(r)$ as a function of r is characterized by a buildup region followed by exponential falloff. Most skyshine distributions are azimuthally isotropic at ground level

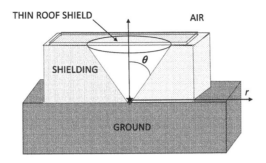

FIGURE 5.14
Geometry of a typical skyshine scenario. The emission of the neutrons is presumed to be centered on the location of the star symbol (★). The roof shielding is taken to be thinner than the wall shielding, or perhaps nonexistent. The emission is approximated by a cone defined by the semivertical angle θ constrained by the shielding geometry. The emitted neutrons thus "escape" vertically through the much more thinly shielded roof to scatter back to the ground at distance r on the ground surface, usually taken to be flat for purposes of simplicity.

(i.e., independent direction with respect to the beam axis). In typical skyshine data λ, the effective attenuation length measured at ground level, has been found to vary between a minimum value of about 200 m and much larger values that approach 1.0 km. This quantity is dependent on the energy spectrum of the neutron radiation field that is the source of the skyshine. Patterson and Thomas (1973) give a formula that describes such behavior for r (meters) greater than about 20 m:

$$\Phi(r) = \frac{aQ}{4\pi r^2}(1 - e^{-r/\mu})e^{-r/\lambda} \tag{5.16}$$

In this equation, a value of $a = 2.8$ fits data well and represents an empirical buildup factor, while μ is the corresponding buildup relaxation length. Nearly all existing measurements are well described by taking μ to be 56 m. Q is the source strength and dimensionally must be consistent with $\Phi(r)$. Thus, for the usual meaning of $\Phi(r)$ as the fluence, Q is the number of neutrons emitted by the source needed to result in that fluence at r. A plot of $r^2\Phi(r)$ for a variety of choices of the value of the scaling length λ in Equation 5.16 is given in Figure 5.15.

λ is determined by the neutron energy spectrum present at the thinly shielded location. A value of 830 m (100 g cm^{-2} of air at standard temperature and pressure) corresponds to the interaction length in air of the neutrons of approximately 100 MeV likely to control the propagation of hadronic cascades. Values of $\lambda > 830$ m are possible if very high–energy neutrons ($E \gg 150$ MeV) are present, or if multiple sources or a large area source are present. In such circumstances, the radiation field may also not necessarily be azimuthally isotropic. Cossairt and Coulson (1985) described an example of a nonisotropic, complex skyshine source involving high energies, an extended source, and a very thin shield that resulted in an extraordinary value of $\lambda \approx 1200$ m.

The procedure for using Equation 5.16 is to do the following:

1. Estimate the total emission rate of neutrons from the source. This can be done by using information about the neutron spectrum at the source to choose an "average" energy and intensity. The dose per fluence factor at that energy can then be used in conjunction with a dose rate survey over the thinly shielded region to determine the total neutron emission rate Q by numerically integrating over the area of the top of the shield.

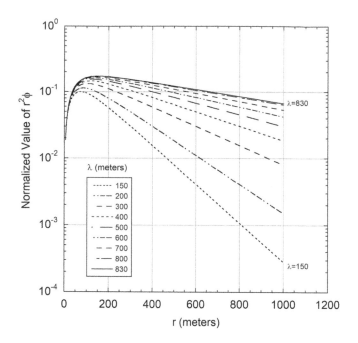

FIGURE 5.15

Plot of skyshine distributions according to Equation 5.16 for a variety of values of λ. The ordinate is the quantity $r^2\phi\,(r)$ in that equation for a value of $Q = 1$.

2. Estimate the value of λ from the available neutron energy spectrum information.

3. Apply Equation 5.16 to determine the radial dependence.

5.3.2 A More Rigorous Treatment

A more rigorous treatment has been reported by Stevenson and Thomas (1984) based on the work of Alsmiller et al. (1981) and Nakamura and Kosako (1981). The latter two groups have independently performed extensive calculations of the neutrons emitted into cones of small vertex angle. Alsmiller et al. (1981) used the Discrete Ordinates Transport Code DOT, while Nakamura and Kosako (1981) used the Monte Carlo code MORSE.

For selected distances from the skyshine source, these workers have calculated the dose equivalent (1973 System) as a function of both the source neutron energy and the emission cone's semivertical angle (i.e., the half-angle the rotation of which defines the cone into which the neutrons are emitted). The authors define this quantity, the *neutron importance*, as the dose equivalent per emitted neutron as a function of the energy of the emitted neutron and of the distance from the source. This somewhat erudeitely named quantity is a measure of how "important" a given emitted neutron is in delivering radiation dose equivalent to a point on the ground located at a given distance from the skyshine source. The results of the Alsmiller calculations for a selected semivertical angle of 37°, a value that is "sensible" for many practical situations, are plotted in Figure 5.16.

Numerical tabulations of these neutron importance functions determined by Alsmiller et al. (1981) for a variety of domains of semivertical angle are given by the NCRP (2003) and by Thomas and Stevenson (1988). The results of Nakamura and Kosako (1981) are essentially identical to but less detailed than those of Alsmiller et al. (1981).

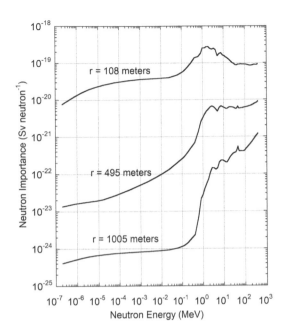

FIGURE 5.16
Neutron skyshine importance functions for a semivertical cone angle of 37° at three different values of the distance from a point source calculated by Alsmiller et al. (1981). (Adapted from tabulations published by the NCRP. 2003. *Radiation protection for particle accelerator facilities*. NCRP Report No. 144. Bethesda, MD.)

Stevenson and Thomas (1984) were able to derive an alternative "recipe" for skyshine neutron calculations to that expressed in Equation 5.16 by making two assumptions:

A. The neutron energy spectrum is proportional to $1/E$ up to the proton energy or some other limiting *upper energy* and zero at higher energies. This likely overestimates the contribution of the higher-energy neutrons and, as we see in Chapter 6, represents some oversimplification of neutron spectra.

B. The neutrons are solely emitted into a cone whose semivertical angle is about 35°–40°, values common for typical structures and thus approximately 37°. This may overestimate the skyshine radiation at a given location by a factor of three for sources of smaller semivertical angles.

Stevenson and Thomas (1984) parameterized the skyshine phenomena as follows:

$$\Phi(r) = \frac{Q'}{4\pi r^2} e^{-r/\lambda} \tag{5.17}$$

In this equation, the buildup exponential factor has been suppressed so the formula is valid only at large distances (i.e., $r \gg 56$ m).

The Alsmiller importance functions were used to estimate the values of λ as a function of the high-energy upper limit of the $1/E$ spectrum, called here the *upper energy*. Figure 5.17 displays the results of doing this by applying the importance functions to several choices of upper energies of $1/E$ spectra by connecting the values at the same three values of r for which results are shown in Figure 5.16 with straight lines in a plot in which the

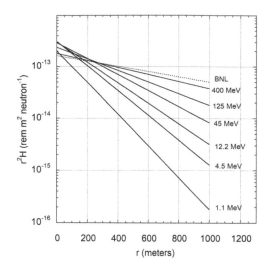

FIGURE 5.17
Variation of dose equivalent with distance r for $1/E$ neutron energy spectra with different upper energies, as labeled on the right side of the figure. The ordinate is the dose equivalent H multiplied by r^2. The curve labeled "BNL" is the result of a measurement at the Brookhaven National Laboratory Alternating Gradient Synchrotron, a 28 GeV proton accelerator, for which very energetic neutrons were present, equivalent to an upper energy >500 MeV, and is provided for comparison. (Reprinted with permission from Stevenson, G. R., and R. H. Thomas. 1984. *Health Physics* 46:115–122. https://journals.lww.com/health-physics/pages/default.aspx.)

inverse-square dependence is suppressed. A comparison with experimental data from a measurement conducted at Brookhaven National Laboratory is also provided.

The slopes of the lines in Figure 5.17 were used to obtain theoretical values of λ as a function of upper energy that are plotted in Figure 5.18.

To determine the source term for use with Equation 5.17, the straight lines in Figure 5.17 (on the semilogarithmic plot) were extrapolated to $r = 0$, and the ordinate axis intercepts were found to range from 1.5×10^{-15} to 3×10^{-15} Sv m^2 neutron^{-1} (1.5×10^{-13} to 3×10^{-13} rem m^2

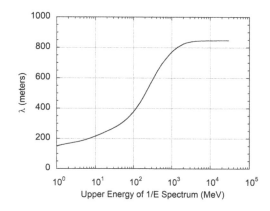

FIGURE 5.18
Effective absorption length λ as a function of upper neutron energy E for $1/E$ spectra. (Reprinted with permission from Stevenson, G. R., and R. H. Thomas. 1984. *Health Physics* 46:115–122. https://journals.lww.com/health-physics/pages/default.aspx.)

neutron^{-1}). Hence, conservatively, it was found that over a rather large range of incident proton energies, the spatial dependence of the dose equivalent $H(r)$ can be described by

$$H(r) = \frac{3 \times 10^{-13}}{r^2} e^{-r/\lambda} \text{(rem/emitted neutron, } r \text{ in meters)} \qquad (5.18)$$

To use Equation 5.18 one needs to determine the total number of neutrons emitted. This can be done as before by measuring the integral of dose equivalent times the area over the thinly shielded location and using the reciprocal of the dose equivalent per fluence conversion factor appropriate for the neutron energy spectrum at hand to get the total number of neutrons emitted. The use of Equation 5.18 will lead to an overestimate of neutrons for values of r less than approximately 100 m because the extrapolation ignores the observed exponential buildup of the skyshine explicitly included in Equation 5.16, but not in Equation 5.18.

Stevenson and Thomas (1984) give a convenient table, useful for general purposes, of dose equivalent per fluence conversion factors provided by the 1973 System integrated over such $1/E$ spectra in Table 5.4. The use of 1990 System quantities, for example, H_{eff} instead of H_{equiv} will require a small adjustment of the result of Equation 5.18 that is not provided in this chapter.

TABLE 5.4

Dose Equivalent per Fluence for $1/E$ Neutron Spectra of Different Upper Energies

Upper Energy (MeV)	Spectrum Averaged Dose Equivalent Conversion Factor (10^{-9} rem cm² neutron^{-1})
1.6	3.9
2.5	4.8
4.0	5.6
6.3	6.4
10	7.2
16	7.9
25	8.6
40	9.4
63	10.1
100	10.9
160	11.7
250	12.5
400	13.4
630	14.6
1000	16.2
1600	18.4
2500	21.2
4000	25.0
6300	30.0
10,000	36.5

Source: Adapted from Stevenson, G. R., and R. H. Thomas. 1984. *Health Physics* 46:115–122.

5.3.3 Examples of Experimental Verifications

Measurements at Fermilab (Cossairt and Coulson 1985) have confirmed the validity of these methods for a "source" involving the targetry of 400 GeV protons. Figure 5.19 shows two measured and fitted radial distributions made using Equation 5.16.

In Figure 5.19 "Survey 2" corresponds to a shielding configuration where the neutron energy spectrum was inferred to be of very high energy and with considerable evidence of not being isotropic and perhaps not well localized. "Survey 4" was likely to involve a much less energetic spectrum and was likely to be much more isotropic. Survey 4 was made for the same beam and target after the concrete shield thickness above the source was greatly increased compared with that present when Survey 2 was obtained. The normalization to "COUNTS-M^2 HR^{-1}" refers to an integration of an instrumental response over the surface area of the source and was approximately proportional to the emitted neutron fluence. The instrument calibration of "COUNTS-M^2 HR^{-1}" made possible an estimate of the dose equivalent at $r = 200$ m for the two surveys. Based on configuration details not described in detail here, one can estimate that the spectrum of emitted neutrons of Survey 2 had an upper energy of $\approx$1.0 GeV, while the spectrum of emitted neutrons of Survey 4 had an upper energy of $\approx$100 MeV.

Using the appropriate dose equivalent per fluence conversion factor, the value of Q for the Survey 2 conditions was determined experimentally to be 2.5×10^5 mrem m^2 hr^{-1}. This was obtained from numerically integrating the measured absorbed dose rate over the area of the source and determining a source strength Q of 5×10^4 mrad m^2 hr^{-1}, assumed a quality factor of 5.0. Similarly, for the Survey 4 conditions, Q was found to be 4.0×10^4 mrem m^2 hr^{-1}. For Survey 4, the measured absorbed dose surface integral was 8.1×10^3 mrad m^2 hr^{-1}, and the quality factor assumed to be 5.0. Table 5.5 makes a comparison with the prescription of Stevenson and Thomas (1984) for these data.

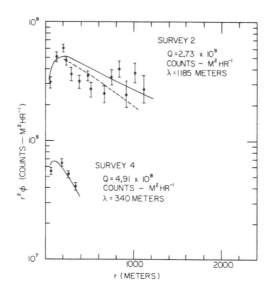

FIGURE 5.19

Skyshine data from two different surveys plotted as $r^2\phi$ as a function of distance from the source r. The solid curves are from the least squares fit of Equation 5.16 to the data points, while the dashed curve is the fit if λ is constrained to have a value of 830 m. Error bars represent one standard deviation counting statistics. (Reprinted with permission from Cossairt, J. D., and L. V. Coulson. 1985. *Health Physics* 48:171–181. https://journals.lww.com/health-physics/pages/default.aspx.)

TABLE 5.5

Comparisons of Fermilab Neutron Skyshine Data (Cossairt and Coulson 1985) with Results of Parameterizations of Surveys Shown in Figure 5.19, Assuming $1/E$ Spectra with Estimated Upper Energies E_{max} for a Measured Period of 1.0 Hour Duration

Survey	λ (meters)	E_{max} (Estimated) (MeV)	Dose Equivalent per Fluence (mrem cm²)	Q (Measured) (mrem m²)	H (200 m) (Calculated) (mrem)	H (200 m) (Measured) (mrem)
Survey 2	1200	1000	16.2×10^{-6}	2.5×10^5	1.0	1.6
Survey 4	340	100	10.9×10^{-6}	4.0×10^4	0.15	0.15

In this table, H is the dose equivalent in one hour at 200 m. The prescription of Stevenson and Thomas (1984) is used to calculate the dose equivalent in one hour at 200 m. The agreement is well within all uncertainties involved.

Another illustration is provided by Elwyn and Cossairt (1986) in connection with neutron radiation field emerging from an iron shield that is more fully described in Section 6.3.5. Figure 5.20 shows the measured radial dependence of neutron flux as a function of distance from that iron shield.

From other considerations pertaining to an iron shield discussed in Section 6.3.5, it is known that the radiation field is dominated by neutrons of energies near 847 keV. Using the measured data, normalized to 10^{12} incident protons the parameters $Q = 1.75 \times 10^{10}$ and

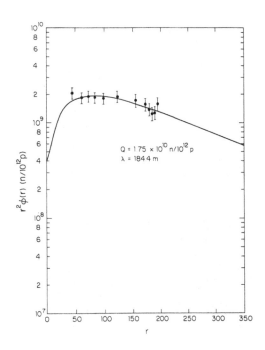

FIGURE 5.20

The product of r^2 and the neutron fluence $\phi(r)$ per 10^{12} protons incident on a target as a function of the distance from the source r. The source is that described in connection with Figure 6.15, in its final shielding condition with additional shield blocks added. The smooth curve is a fit to Equation 5.16 with parameters $\lambda = 184.4$ m and $Q = 1.74 \times 10^{10}$ neutrons per 10^{12} protons. (Reprinted with permission from Elwyn, A. J., and J. D. Cossairt. 1986. *Health Physics* 51:723–735. https://journals.lww.com/health-physics/pages/default.aspx.)

$\lambda = 184.4$ m were determined by fitting the skyshine data using Equation 5.16. Evaluating Φ at $r = 200$ m,

$$\Phi(200) = \frac{2.8(1.75 \times 10^{10})}{4\pi (200)^2}\left[1 - \exp\left(-\frac{200}{56}\right)\right]\exp\left(-\frac{200}{184.4}\right) = 3.20 \times 10^4 \text{ neutrons m}^{-2} \quad (5.19)$$

Thus, taking the measured neutron flux at $r = 200$ m and applying a dose equivalent per fluence value of 3.0×10^{-5} mrem cm^{-2} appropriate for 847 keV neutrons (see Figure 1.5) gives a dose equivalent per 10^{12} incident protons of 9.6×10^{-5} mrem at $r = 200$ m. The value of λ that fitted the skyshine data is also consistent with the neutron energy spectrum, known to be dominated by neutrons of about 1.0 MeV kinetic energy, as seen in Figure 5.18.

Elwyn and Cossairt (1986) also estimated the total neutron emission of the source to have a value of $(3.4 \pm 2.0) \times 10^{10}$ per 10^{12} incident protons by performing a numerical integration over the surface area of the source, separate from the result determined using the skyshine measurement. Applying the prescription of Stevenson and Thomas (1984) found in Equation 5.17,

$$H(200) = \frac{(3 \times 10^{-13})[(3.4 \pm 2.0) \times 10^{10}]}{(200)^2}\exp\left(-\frac{200}{184.4}\right) \quad (5.20)$$

$$= (8.6 \pm 5.1) \times 10^{-5} \text{ mrem per } 10^{12} \text{ protons}$$

at this same location. This result is quite consistent with the value of 9.6×10^{-5} mrem at $r = 200$ m already determined.

PROBLEMS

1. A 1.0 µA 100 MeV electron beam is incident on an "optimized bremsstrahlung" target in a shielding configuration and labyrinth like that in Figure 5.5. Using the facts given in Chapter 3 (Swanson's Rules of Thumb, etc.) about bremsstrahlung, calculate the dose equivalent rate at the exit of a labyrinth having two legs. Set all distances d_i, d_1, and d_2 to 3.0 m. If the goal is to get the dose equivalent rate at the exit to be <1.0 mrem h^{-1}, is this a sensible design? The legs are 1.0×2.0 m^2 in cross section. Since no other information is available, use $\alpha = 10^{-2}$ as a "conservative" value. For purposes of this problem, photons constitute the only component of radiation present. (Hint: One needs to calculate the projected diameter of the beam at the wall where the first scatter occurs. This can be done using Equation 3.13.)

2. A 500 GeV proton beam of 10^{11} protons/second strikes a magnet 2.0 m from the mouth of a three-legged labyrinth. Each of the three legs is 4.0 m long and 1.0×2.0 m^2 in cross section. The length of leg 1 is measured from the mouth of the labyrinth to the center of the first turn; all other lengths are measured between centers of turns. Assume the source is an on-axis "point source." Using Goebel's "universal" curves and Rameika's source term, what is the dose equivalent rate at the exit expressed in rem hr^{-1}? How far away from the exit does the value of dH/dt fall to 10 mrem hr^{-1}.

3. A high-energy accelerator has a section of beamline that was poorly designed. Beam losses and insufficient shielding have resulted in a region of roof 10 m wide and 50 m long where a neutron dose equivalent rate averaging 100 mrem h^{-1} (averaged over the surface of the weak shield) is found. An energy spectrum measurement indicates the spectrum shape to be approximately $1/E$ with an upper end point of approximately 500 MeV. Calculate the dose equivalent rate due to skyshine at distance $r = 50, 100, 200, 500,$ and 1000 m using both formulae presented here.

6

Shielding Materials and Neutron Energy Spectra

6.1 Introduction

In this chapter we discuss the relevant properties of the most common materials used in radiation shielding. Also, since many shielding problems are driven by the nature of the energy spectrum of the neutrons, such spectra are discussed here in some detail. Examples of neutron energy spectra measured external to shielding at various types of accelerator facilities are presented.

6.2 Discussion of Shielding Materials Commonly Used at Accelerators

Given the size of many modern accelerators, economic considerations commonly dominate shielding designs, requiring the use of relatively inexpensive, but not necessarily in all aspects optimum, shielding materials. In all situations good engineering practices concerning structural properties, appropriate floor loading strength, and fire protection must be taken into account to assure an acceptable level of occupational and public safety.

Briefly, low atomic number materials are best used for targets, collimators, and beam stops at electron accelerators to reduce photon production. High atomic number materials are preferred at proton and heavy ion accelerators for these components to reduce neutron production. The use of toxic or hazardous materials should be avoided or minimized to the extent possible. As discussed previously, at beam energies above about 5.0 MeV, neutrons are produced in most materials. Furthermore, some materials have superior heat transfer characteristics that enhance durability and reliability while reducing personnel exposures incurred in maintenance activities.

6.2.1 Earth

Earth, meaning the soil atop its crust, not the planet itself, has many admirable qualities as a shield material besides the obvious one, its low cost. Notably, the water it contains enhances the effectiveness of the neutron attenuation because the mass of a proton is essentially equal to that of a neutron. This facilitates the transfer of energy from the particle to the shielding medium. Due to conservation of energy and momentum, in an elastic collision the energy

ΔE that can be transferred from a neutron having kinetic energy E_o to a target nucleus as a function of scattering angle θ is given by

$$\frac{\Delta E}{E_o} = 4\frac{M}{m_n}(\cos^2\theta)\left(1+\frac{M}{m_n}\right)^{-2} \tag{6.1}$$

where M is the rest mass of the recoiling nucleus, and m_n is the rest mass of the incident neutron. Thus, at small scattering angles (i.e., $\theta \approx 0$), nearly all of the neutron kinetic energy can be transferred to the protons in the water. For comparison, a ^{12}C nucleus is capable of absorbing only a maximum of 28.4% of the incident neutron energy in a single collision. The proton energy will then be dissipated in the medium by means of ionization and nuclear interactions. Percentages of soil water content (% of dry weight) for different soil types are sand (0–10), sandy loam (5–20), loam (8–25), silty loam (10–30), dry loam (14–30), and clay (15–30). Earth includes sufficient constituent elements of higher atomic number to be generally effective against photons. It is generally a shield free of cracks and thus not prone to neutron leakage by *streaming* through such voids. The total density of earth including the water content varies widely, approximately from 1.70 to 2.25 g cm^{-3}, dependent on soil type and water content.

Points of confusion about the density of earth result from common practices of civil engineers to reference the dry density of earth not including the water content that is present. For shielding applications, the *wet* density including water content is more relevant and should be determined. In general, sandy soils have lower values of density than do heavy clays found in glacial deposits. Extrusive volcanic soils can have very low densities. In view of this variability, the detailed characteristics of the soil found at a particular accelerator site including water content may be needed for accurate shielding design. An example of an elemental composition of dry earth is given in Table 6.1.

TABLE 6.1

Example Elemental Composition, Dry Weight Percent, of a Representative Soil

Element	Global Average (%)
O	43.77
Si	28.1
Al	8.24
Fe	5.09
Mn	0.07 ± 0.06
Ti	0.45 ± 0.43
Ca	3.65
Mg	2.11
K	2.64
Na	2.84

Source: Adapted from Chilton, A. B. et al. 1984. *Principles of radiation shielding.* Englewood Cliffs, NJ: Prentice Hall.

6.2.2 Concrete

Concrete has obvious advantages in that it can either be poured in place permanently or cast into modular blocks in configurations having considerable structural strength. Typical steel reinforcement has essentially no effect on radiation shielding properties. Concrete blocks may be used to shield targets, beam stops, etc., in a manner that allows their ready access for maintenance. When concrete blocks are used, they generally should be overlapped to avoid streaming through the cracks. The concrete density, locally variable due to available ingredients in the aggregate, can be increased by adding a heavier material to the recipe to increase both the density and average atomic number. Table 6.2 (Chilton et al. 1984) gives examples of some partial densities of various concretes, some with additional additives designed to increase the density as well as the average atomic number.

These partial densities are locally variable due to their strong dependence on the aggregate material used in the concrete mix. When shielding neutrons, the concrete water content is quite important because it accounts for almost all of the hydrogen present. Under conditions of extremely low humidity and high temperatures such as found in desert geographical regions, the water content of concrete can decrease with time, to as little as 50% of the initial value over a 20-year period. Heating due to the energy deposition of the beam can also drive out the water (NCRP 2003).

6.2.3 Other Hydrogenous Materials

6.2.3.1 Polyethylene and Other Materials That Can Be Borated

Polyethylene $(CH_2)_n$ is a very effective neutron shield because of its hydrogen content (14% by weight) and its density (≈ 0.92 g cm^{-3}), because it can attenuate "fast" neutrons.

TABLE 6.2

Examples of Total and Elemental Partial Densities of Representative Concretes after Curing

Densities (g cm^{-3})	Ordinary	Magnetite (FeO, Fe$_2$O$_3$)	Barytes (BaSO$_4$)	Magnetite and Iron
H	0.013	0.011	0.012	0.011
O	1.165	1.168	1.043	0.638
Si	0.737	0.091	0.035	0.073
Ca	0.194	0.251	0.168	0.258
Na	0.040			
Mg	0.006	0.033	0.004	0.017
Al	0.107	0.083	0.014	0.048
S	0.003	0.005	0.361	
K	0.045		0.159	
Fe	0.029	1.676		3.512
Ti		0.192		0.074
Cr		0.006		
Mn		0.007		
V		0.011		0.003
Ba			1.551	
Total Density	2.34	3.53	3.35	4.64

Source: Adapted from Chilton, A. B. et al. 1984. *Principles of radiation shielding.* Englewood Cliffs, NJ: Prentice Hall. Also available from (NCRP). 2003. *Radiation protection for particle accelerator facilities.* NCRP Report No. 144. Bethesda, MD.

Thermal neutrons can be captured through the $^1H(n,\gamma)^2H$ reaction, a thermal neutron capture process having a cross section of 0.33 barn for neutrons in thermal equilibrium at room temperature ($E_n = 0.025$ eV). The emitted γ-ray of 2.2 MeV energy is sometimes a problem as an additional source of radiation that requires attention for optimization of the shielding. The fluence of these photons can be reduced by adding boron to the polyethylene. In such *borated polyethylene*, many of the thermal neutrons are captured with the $^{10}B(n,\alpha)^7Li$ reaction. The cross section of this reaction for room-temperature thermal neutrons is much larger than for hydrogen, 3840 barns (Knoll 2010). In 94% of these captures, the emitted α-particle is accompanied by a 0.48 MeV γ-ray. The α-particle is readily absorbed by ionization in the material, while the γ-ray has a somewhat shorter attenuation length than does the 2.2 MeV γ-ray (see Figure 3.11). Borated polyethylene, while somewhat costly, is commercially available with additives of boron (up to 32%), lithium (up to 10%), and lead (up to 80%) in various shapes; for example, planes, spheres, and cylinders. In addition to borated polyethylene, boron has been added to other materials including plastics, putties, clays, and glasses to accomplish specific shielding objectives. Some of these materials are available in powder form, for molding into a desired shape by the user. These materials can be useful if it is necessary to economize on space and also to accomplish shielding of photons and neutrons simultaneously.

Plastic materials like polyethylene can be subject to significant radiation damage at relatively low levels of integrated absorbed dose (Fassò et al. 1990), with possible effects on their structural integrity. Also, pure polyethylene is flammable, but some of the commercial products available contain self-extinguishing additives.

6.2.3.2 Water, Wood, and Paraffin

The high hydrogen content makes these materials seem attractive as a shielding material. However, *water* tends to rust out its containers with loss of shielding. Exposed to thermal neutrons, it also emits the 2.2 MeV capture γ-ray from the $^1H(n,\gamma)^2H$ thermal neutron capture reaction. Adding boron is difficult due to the relative insolubility of boron salts. An exception with better solubility is potassium tetraborate. *Wood* is approximately as effective per unit of linear thickness as is concrete for shielding intermediate energy neutrons. However, wood is flammable and prone to rot. Chemically treated wood that is nearly completely fireproof is available, but caution is warranted about the flammability of the material over time. Treated wood can also have reduced structural strength. *Paraffin*, a common composite of hydrocarbons, historically has been used for neutron shielding. Largely spurned in accelerator installations due to its high flammability, on occasion it has been used successfully packaged in metal containers. Various forms of paraffin have melting points in the range of 49°C–72°C, a potential problem for some applications.

6.2.4 Iron

With its relatively high density and low cost, iron is an attractive shielding material. The density of iron varies widely, from a low of 7.0 g cm^{-3} for some cast irons to a high of 7.8 g cm^{-3} for some steels. The "textbook" value of 7.87 g cm^{-3} in Table 1.2 is almost never attained in the bulk quantities of materials needed for radiation shielding of large important components at large accelerators such as target stations and beam absorbers. Because of its nonmagnetic properties and resistance to corrosion, stainless steel is often used in accelerator components. Due to concerns about accelerator-produced radioactivity (see Chapter 7), knowledge of the elemental composition of various alloys is useful. For example, long-lived ^{60}Co can be produced in stainless steel but not in pure iron. The use of shredded steel (i.e., "steel wool") to fill cracks in a large shield, a practice used sometimes by well-intended individuals to mitigate streaming through cracks in shielding at accelerators, is highly undesirable due to the removable radioactivity hazard presented of the resulting rust. Iron has a very important deficiency as a neutron shield discussed in Section 6.3.5.

6.2.5 High Atomic Number Materials: Lead, Tungsten, and Uranium

The materials in this category are valuable due to their high atomic number, especially when shielding photons.

Lead has a high density of 11.35 g cm^{-3} and has the benefit of being resistant to corrosion. Pure lead has major drawbacks due to its poor structural characteristics and low melting point (327.4°C). It is usually best used when laminated to some other, more structurally stable, material. Some alloys of lead also perform better structurally than does pure lead. Lead is available as an additive to other materials to improve the ability to shield photons. Fabric blankets containing shredded lead can be used to shield radioactivated components to minimize exposures associated with accelerator maintenance activities during operational shutdowns if the material is not allowed to become activated. Lead must be handled in accordance with good industrial hygiene practices due to its chemical toxicity. The chemical toxicity as well as any potential radioactive contamination renders the use of shredded lead (i.e., "lead wool") to fill cracks highly undesirable.

Bismuth, having a density of 9.75 g cm^{-3}, is sometimes used as a lower-toxicity substitute for lead for shielding against photons.

Tungsten is an excellent, but relatively expensive, shielding material for large-scale applications. Its high density (19.3 g cm^{-3}) and high melting temperature (3410°C) make it extremely useful as a component in photon shields, beam absorbers, and beam collimators. It is difficult to machine, so alloys such as Hevimet are commonly used. Hevimet consists of tungsten (90%), nickel (7.5%), and copper (2.5%), with a typical density ranging from 16.9 to 17.2 g cm^{-3} (Marion and Young 1968).

Uranium is superficially an attractive shielding material in its *depleted* form. In depleted uranium, the concentration of the readily fissionable isotopes ^{233}U and ^{235}U compared with the dominant ^{238}U is reduced from the natural value of 0.72%, usually down to $\leq$0.2%. Its high density (19.0 g cm^{-3}) and relatively high melting point (1133°C) can be useful. Uranium may not be a good choice in environments with a high neutron flux density due to its susceptibility to fission. Depleted uranium is relatively safe, but if it is combined with hydrogenous materials, nuclear fission criticality should be considered for the specific material and geometric arrangement to be employed. Even in the absence of hydrogen, the possibility of criticality can exist if the material is insufficiently depleted (Borak and Tuyn 1987).

Uranium has major deficiencies as a shielding material. It has a large anisotropic thermal expansion coefficient; at a given increase in temperature a plate or slab of uranium will expand more in one coordinate than another. Also, it readily oxidizes when exposed to air, especially under conditions of high humidity. The oxide is readily removable and presents a significant internal exposure hazard and can also represent a flammability hazard given that the material in small pieces is well known to be pyrophoric. Prevention of oxidation by sealing the material with epoxy or paint meets with only limited success due to eventual embrittlement and chipping caused by the intense β-radiation field at the material surface. Sealed containers filled with dry air or noble gases or liquefied noble gases such as argon appear to be the best storage solution to limit oxidation. The pure material in the form of small pieces such as "shavings," as is the case for some of the oxides, is *pyrophoric*, complicating machining-type processes by posing yet another safety hazard, that of spontaneous fire, that should be specifically analyzed and mitigated. Uranium, included depleted uranium, is classified as a "nuclear material" by the U.S. Department of Energy rendering it subject to stringent accountability requirements.

6.2.6 Miscellaneous Materials: Beryllium, Aluminum, and Zirconium

These three materials find considerable usage as accelerator components because of various properties. *Beryllium*, especially in the form of the oxide BeO with a melting point of 2530°C, is often used as a target material in intense beams because of its excellent heat transfer properties with a resultant ability to endure large values of energy deposition density. It has been used at high-energy accelerators in relatively large quantities as a "filter" to enrich one particle type at the expense of another, taking advantage of particle-specific variations in absorption cross sections. Of concern, especially when fabricating components, is the significant chemical toxicity of the metal and its compounds, and it must be handled in accordance with best industrial hygiene practices. *Aluminum* is used in accelerator components because of its nonmagnetic properties and its resistance to corrosion. It is a poor shield against neutrons. *Zirconium*, having good thermal conductivity properties and a high melting point (1852°C), has a very small thermal neutron capture cross section. It is not a good neutron absorber but has been found to be useful in beam-handling components.

6.3 Neutron Energy Spectra outside of Shields

As discussed previously, at most accelerators the shielding is largely designed to attenuate neutrons emitted in all directions. Exceptions are found at forward angles at higher-energy accelerators where energetic muons and in rare instances other, more exotic particles, need to be shielded. In this section examples of neutron energy spectra found at accelerators external to shielding are presented and discussed. These examples, not intended to be a comprehensive set, illustrate the general principles.

6.3.1 General Considerations

In the simplest approximation, outside of thick shields of soil or concrete that contain some hydrogen content (usually as water), accelerator neutron intensities are generally to first order proportional to inverse energy. Such 1/E *spectra* can span energies extending from those of thermal neutrons ($\langle E_n \rangle \approx 0.025$ eV) up to the energy of the accelerated particles. More commonly they are cut off at some upper energy of lower value. At this level of approximation, the spectrum is given by

$$\frac{d\phi(E)}{dE} \approx k \frac{1}{E} \qquad (6.2)$$

where *k* is a normalizing constant. Rohrig (1983) observed that it is often convenient to plot such spectra as flux per *logarithmic* energy interval by simply plotting $Ed\phi(E)/dE$:

$$\frac{d\phi(E)}{d\ln E} = E\frac{d\phi(E)}{dE} \tag{6.3}$$

In discussions of neutron physics, $E\phi(E)$ is sometimes called the fluence per unit "lethargy," and such a plot is called a *lethargy plot*. Such a plot suppresses the dominant $1/E$ dependence typically found in such spectra. Most of the example spectra discussed here are such lethargy plots.

6.3.2 Examples of Neutron Spectra due to Incident Electrons

Alsmiller and Barish (1973) calculated the neutron energy spectra that arise when 400 MeV electrons are incident on a thick copper target. Predictions of the neutron yields over several ranges of production angle θ resulted from these calculations. Four different shielding materials, soil, concrete, the mineral ilmenite ($FeTiO_3$), and iron, were included in these calculations. These results are for a radial depth in the shield of 7.0 mean free paths of the highest-energy neutrons found in this source spectrum within this angular range. Table 6.3 gives the densities and the neutron mean free paths used for the four materials.

The results of the calculations of neutron energy spectra are shown in Figure 6.1.

In Figure 6.1, the inverse square dependence was removed to eliminate the effect of "geometrical" attenuation within the shield. It should be clear that the neutron spectrum in the iron shield is markedly different from that found in the soil and concrete shields. The characteristics of the spectra found in the ilmenite are intermediate, perhaps related to the presence of iron in this material. This behavior associated with iron shielding is found at nearly all accelerators and is discussed in Section 6.3.5.

The fractional contributions to the total dose equivalent (1973 System) from neutrons with energies less than a given energy E for the same spectra for the angular region $0 < \theta < 30°$ are presented in Figure 6.2.

6.3.3 Examples of Neutron Spectra due to Low- and Intermediate-Energy Protons

Calculations and measurements of neutron energy spectra at various depths in several different shields due to 52 MeV protons have been reported by Uwamino et al. (1982) and are presented in Figures 6.3 through 6.5.

TABLE 6.3

Material Properties Used in the Calculations of Alsmiller and Barish

Material	Density (g cm^{-3})	Mean Free Path λ (g cm^{-2})
Soil	1.8	103.6
Concrete	2.3	105.3
Ilmenite	3.8	120.6
Iron	7.8	138.6

Source: Adapted from Alsmiller, Jr., R. G., and J. Barish. 1973. *Particle Accelerators* 5:155–159.

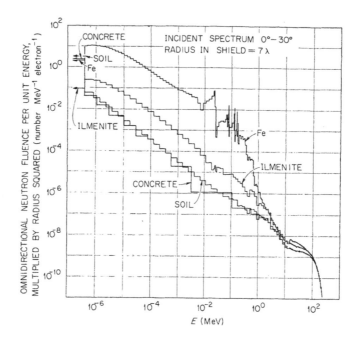

FIGURE 6.1

Calculated neutron energy spectra for 400 MeV electrons incident on a thick Cu target for the angular region $0 < \theta \leq 30°$. The plot shows the omnidirectional neutron fluence per unit energy multiplied by the square of the radial depth in the shield as a function of energy for the various shield materials studied by Alsmiller and Barish (1973). The calculations for all four materials were performed at a depth of 7λ, where λ is the neutron mean free path in the material at the highest energy under consideration. (Reprinted with permission from Alsmiller, Jr., R. G., and J. Barish. 1973. *Particle Accelerators* 5:155–159. Taylor and Francis Ltd. http://www.tandfonline.com.)

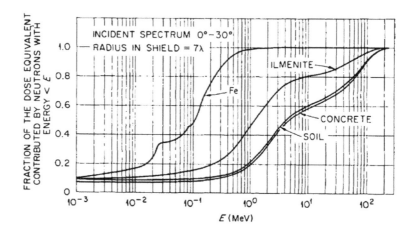

FIGURE 6.2

Fractional contribution to the total dose equivalent from neutrons with energies less than E as a function of E for the angular region $0 < \theta \leq 30°$ for the illustrated shielding materials for which representative neutron energy spectra were plotted in Figure 6.1. The calculations for all four materials were performed at a depth of 7λ, where λ is the neutron mean free path in the material at the highest energy under consideration. The values of λ for the four materials are given in Table 6.3. (Reprinted with permission from Alsmiller, Jr., R. G., and J. Barish. 1973. *Particle Accelerators* 5:155–159. Taylor and Francis Ltd. http://www.tandfonline.com.)

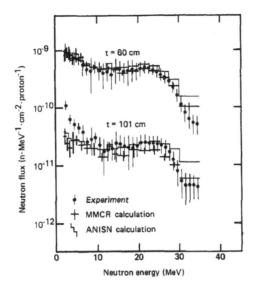

FIGURE 6.3
Forward neutron energy spectra and attenuations measured and calculated for 52 MeV protons incident through water assemblies at two different values of thickness t. Experimental data points are compared with calculations displayed as histograms for two different calculational tools, MMCR and ANISN, not discussed further here. (Reprinted with permission from Uwamino, Y. et al. 1982. *Nuclear Science and Engineering* 80:360–369. © American Nuclear Society, http://www.ans.org/, Taylor and Francis Ltd. http://www.tandfonline.com on behalf of American Nuclear Society.)

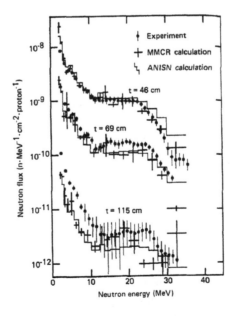

FIGURE 6.4
Forward neutron energy spectra and attenuations measured and calculated for 52 MeV protons incident on ordinary concrete at three different values of thickness t. Experimental data points are compared with calculations displayed as histograms for two different calculational tools, MMCR and ANISN, not discussed further here. (Reprinted with permission from Uwamino, Y. et al. 1982. *Nuclear Science and Engineering* 80:360–369. © American Nuclear Society, http://www.ans.org/, Taylor and Francis Ltd. http://www.tandfonline.com on behalf of American Nuclear Society.)

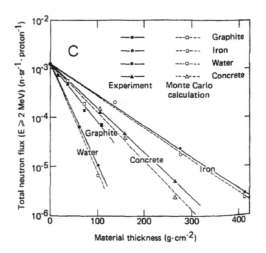

FIGURE 6.5
Measured and calculated attenuation profiles of the neutron fluences integrated above 2.0 MeV provides data on the attenuation profiles determined for four different materials. (Reprinted with permission from Uwamino, Y. et al. 1982. *Nuclear Science and Engineering* 80:360–369. © American Nuclear Society, http://www.ans.org/, Taylor and Francis Ltd. http://www.tandfonline.com on behalf of American Nuclear Society.)

Alsmiller et al. (1975) have provided predictions of neutron energy spectra averaged over specific angular intervals for 200 MeV protons stopped in a thick water target. The results are given for large angles and are presented in Figure 6.6.

6.3.4 Examples of Neutron Spectra due to High-Energy Protons

In the regime of proton energies well above 1.0 GeV, the details of the spectra are far more sensitive to geometrical considerations than they are dependent on the incident proton energy. O'Brien (1971) carried out a calculation of a generalized neutron spectrum to be found external laterally (i.e., at $\theta \approx 90°$) to a high-energy proton accelerator with a proton kinetic energy greater than 0.8 GeV. The results were compared with measurements and alternative "generic" spectrum calculations performed by Höfert and Stevenson (1984). The results are provided in Figure 6.7 for "forward" ($\theta \approx 0$) and Figure 6.8 "lateral" ($\theta \approx 90°$) angular regions, respectively, and include a comparison with the results of O'Brien. Those for forward angles also include the spectra of charged pions and protons. It is clear that at forward angles, the fluence of hadrons at high energies is likely to be a mixture of charged particles and neutrons.

Details of the geometry can produce peaks in the spectra. Examples have been reported by various workers (Patterson and Thomas 1973; Thomas and Stevenson 1988; Elwyn and Cossairt 1986; McCaslin et al. 1987b; Cossairt et al. 1988; Cossairt and Vaziri 2009). Such peaks are typically encountered in the few megaelectron volt (MeV) region. Figures 6.9 through 6.14 are drawings of the shielding geometry and plots of the corresponding neutron spectra energy spectra for three sets of measurements at Fermilab (McCaslin et al. 1987b; Cossairt and Vaziri 2009). These spectra were obtained (i.e., "unfolded") using the *Bonner sphere technique* discussed in Section 9.5.2.1 and are plotted as lethargy plots. In the drawings of the configurations involved, "spheres" denote the locations where the measurements were performed. The results are typical of the spectra found at high-energy proton accelerators.

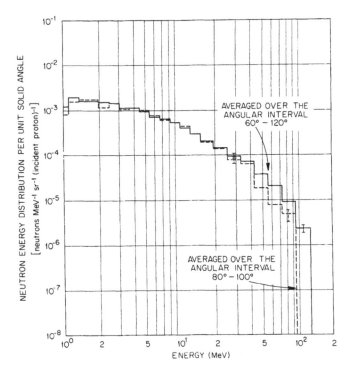

FIGURE 6.6
Energy distribution of neutrons averaged over particular angular intervals, produced when 200 MeV protons are stopped in a thick water target. The protons are incident at $\theta = 0°$. (Reprinted with permission from Alsmiller, Jr., R. G. et al. 1975. *Particle Accelerators* 7:1–7. Taylor and Francis Ltd. http://www.tandfonline.com.)

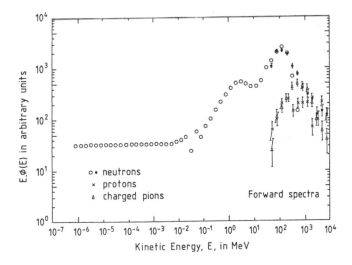

FIGURE 6.7
Hadron energy spectra outside of a concrete shield at a high-energy proton accelerator. The open circles represent the lateral ($\theta \approx 90°$) shielding calculations of O'Brien (OB71), while the other symbols represent the calculations at forward angles ($\theta \approx 0$) of Höfert and Stevenson. (Reprinted from Höfert, M., and G. R. Stevenson. 1984. *The assessment of dose equivalent in stray radiation fields around high-energy accelerators.* European Organization for Nuclear Research: CERN Report TIS-RP/131/CF. Geneva, Switzerland.)

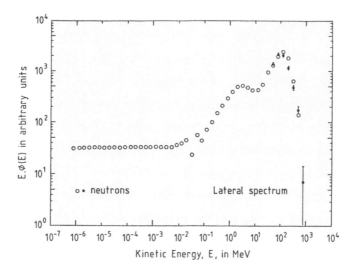

FIGURE 6.8

Hadron energy spectra outside of a concrete shield at a high-energy proton accelerator at large angles ($\theta \approx 90°$). The open circles represent the same calculation of O'Brien (OB71) shown in Figure 6.7, while the other symbols represent the calculations of Höfert and Stevenson. (Reprinted from Höfert, M., and G. R. Stevenson. 1984. *The assessment of dose equivalent in stray radiation fields around high-energy accelerators.* European Organization for Nuclear Research: CERN Report TIS-RP/131/CF. Geneva, Switzerland.)

Figure 6.10 ("Debuncher Ring") is rather typical of the spectra found external to earth and concrete shields lateral to high-energy proton accelerators. The neutron energy spectrum displayed in Figure 6.12 ("Tevatron Tunnel") is particularly interesting because its shape was demonstrated in the measurements to be essentially independent of proton energy over the range of 150–900 GeV (McCaslin et al. 1987a). Figure 6.14 ("Labyrinth") is typical of the results obtained in the second and succeeding sections ("legs") of a labyrinth penetration.

6.3.5 Leakage of Low-Energy Neutrons through Iron Shielding

One peak commonly found in such spectra is of particular importance. As observed by Alsmiller and Barish (1973), iron has a major deficiency as a shield for fast neutrons. The

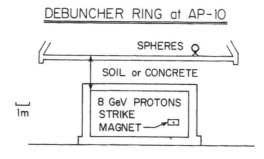

FIGURE 6.9

Cross-sectional view of the shielding measurement external to the Fermilab Tevatron-Era Debuncher Ring at Location AP-10 where 8 GeV protons struck the yoke of an iron magnet. The beam was directed into the page in this view. (Reprinted from Cossairt, J. D., and K. Vaziri. 2009. *Health Physics* 96:617–628. Rights to this figure are reserved by Fermi Research Alliance, LLC as manager and operator of the Fermi National Accelerator Laboratory.)

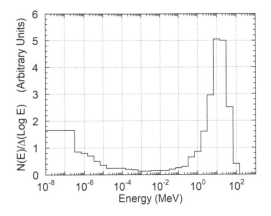

FIGURE 6.10
The neutron energy spectrum measured above the Fermilab Tevatron Era Debuncher Ring at Location AP-10 illustrated in Figure 6.9. The normalization of the spectrum is arbitrary. (Reprinted from Cossairt, J. D., and K. Vaziri. 2009. *Health Physics* 96:617–628. Rights to this figure are reserved by Fermi Research Alliance, LLC as manager and operator of the Fermi National Accelerator Laboratory.)

dominant mechanism by which fast neutrons lose energy is inelastic scattering. At energies below the first excited state of any nucleus, inelastic scattering becomes energetically impossible, and elastic scattering remains as the only removal process aside from nuclear reactions. As evident from Equation 6.1, elastic scattering is a very inefficient mechanism for energy transfer from neutrons scattering from a much more massive nucleus such as iron. Similarly, the scattering of neutrons by the "free" protons in hydrogenous materials transfers energy much better than does elastic scattering of neutrons from iron nuclei. The first excited state of ^{56}Fe, the dominant isotope in natural iron (92% natural abundance), is at 847 keV. Thus, neutrons having kinetic energies above 847 keV in a given spectrum will be slowed by inelastic scattering to $E_n \approx 847$ keV. Once below that energy, since elastic scattering of neutrons from iron is very inefficient, the process of slowing them further is inhibited. Thus, an accumulation of neutrons just below the energy of this first excited state will occur, since there are no longer any inelastic nuclear processes available to remove the energy. The effect is further enhanced since the radiation weighting or quality factor

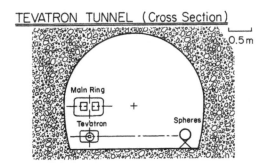

FIGURE 6.11
Cross-sectional view of the shielding measurement taken inside the Fermilab Tevatron tunnel in which 800 GeV protons interacted with a controlled leak of nitrogen (N$_2$) gas injected into the path of the beam during circulating beam conditions. The beam was directed into the page in this view. (Reprinted from McCaslin, J. B. et al. 1987a. *Radiation environment in the tunnel of a high-energy proton accelerator at energies near 1.0 TeV.* Lawrence Berkeley Laboratory: Report LBL-24640. Berkeley, CA; Cossairt, J. D., and K. Vaziri. 2009. *Health Physics* 96:617–628. Rights to this figure are reserved by Fermi Research Alliance, LLC as manager and operator of the Fermi National Accelerator Laboratory.)

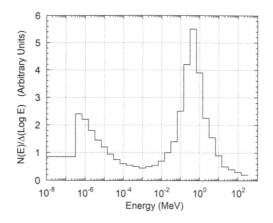

FIGURE 6.12

The neutron energy spectrum measured inside the Tevatron tunnel illustrated in Figure 6.11. The normalization of the spectrum is arbitrary. (Reprinted from McCaslin, J. B. et al. 1987a. *Radiation environment in the tunnel of a high-energy proton accelerator at energies near 1.0 TeV*. Lawrence Berkeley Laboratory: Report LBL-24640. Berkeley, CA; Cossairt, J. D., and K. Vaziri. 2009. *Health Physics* 96:617–628. Rights to this figure are reserved by Fermi Research Alliance, LLC as manager and operator of the Fermi National Accelerator Laboratory.)

for neutrons as a function of energy also has its maximum value very close to 845 keV, independent of whether the 1973 or 1990 System is used (see Figure 1.3).

This effect is clearly seen in the spectra of Figure 6.1. A perhaps clearer example of this phenomenon is provided by the geometry and spectra shown in Figures 6.15 through 6.17 that describe neutron energy spectra measurements made external to a thick iron shield at Fermilab (Elwyn and Cossairt 1986).

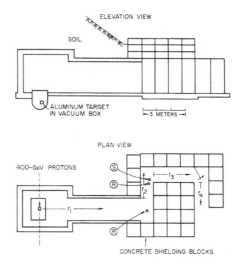

FIGURE 6.13

Cross-sectional (upper frame) and plan views of the labyrinth in which a neutron energy spectrum measurement was made at the Fermilab Tevatron in which 400 GeV protons interacted with an aluminum target located beneath the floor of the enclosure shown. The spectrum was measured in the second leg at the location denoted S. A radiation quality factor measurement was made at the locations denoted R (see Sections 5.2.4 and 9.5.7). (Reprinted from Cossairt, J. D. et al. 1985b. *Health Physics* 49:907–917; Cossairt, J. D., and K. Vaziri. 2009. *Health Physics* 96:617–628. Rights to this figure are reserved by Fermi Research Alliance, LLC as manager and operator of the Fermi National Accelerator Laboratory.)

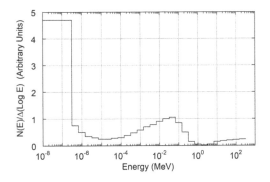

FIGURE 6.14
The neutron energy spectrum measured inside the labyrinth illustrated in Figure 6.13. The normalization of the spectrum is arbitrary. (Reprinted from Cossairt, J. D. et al. 1985b. *Health Physics* 49:907–917; Cossairt, J. D., and K. Vaziri. 2009. *Health Physics* 96:617–628. Rights to this figure are reserved by Fermi Research Alliance, LLC as manager and operator of the Fermi National Accelerator Laboratory.)

In the configuration shown in Figure 6.15, the beam axis was horizontal, 1.8 m above the floor, with the beam going from left to right in the figure. The target and primary proton beam absorber were far upstream (to the left) of the region shown here. Many hadrons produced by these interactions were intercepted by lead and polyethylene secondary particle absorber shown in the plan view. The shielding was composed of large, movable shielding blocks made of ordinary concrete. The five concrete shielding blocks that are

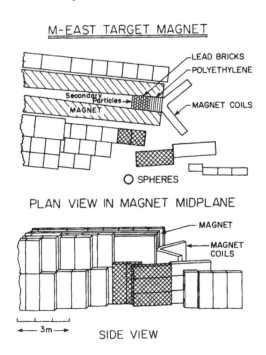

FIGURE 6.15
Shielding geometry for the measurements shown in Figures 6.16 and 6.17 shown in plan and elevation views. 800 GeV protons interacted with a production target and beam absorber within the gap of the magnet shown. (Reproduced with permission from Elwyn, A. J., and Cossairt, J. D. 1986. *Health Physics* 51:723–735. https://journals.lww.com/health-physics/pages/default.aspx.)

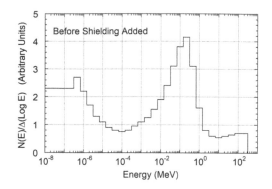

FIGURE 6.16

Neutron energy spectra obtained external to the shielding configuration shown *before* the addition of the cross-hatched concrete shielding blocks shown in Figure 6.15. The plot is a lethargy plot, and the normalization of the spectrum is arbitrary. (Reproduced with permission from Elwyn, A. J., and Cossairt, J. D. 1986. *Health Physics* 51:723–735. https://journals.lww.com/health-physics/pages/default.aspx.)

"cross-hatched" in Figure 6.15 were *not* present during initial operations of this target and beam absorber. The result was the measured spectrum shown in Figure 6.16.

Later, the cross-hatched blocks shown in Figure 6.15 were added, and the spectrum shown in Figure 6.17 was measured.

The neutron energy spectra shown in Figures 6.16 and 6.17 were measured about 1.0 m above the floor and 4.0 m horizontal distance from the face of the magnet return yoke at the longitudinal position of the front face of the set of lead bricks struck by the secondary particles. The five "cross-hatched" blocks were square in cross section, 0.914 m thick, and 2.29 m long, and were placed between the neutron detectors and the beam absorber up to a height of about 0.5 m above the beam line.

For the bare iron situation associated with Figure 6.16, the dose equivalent rate normalized to the proton beam intensity external to the shield was over 40 times that measured after the concrete was installed and as associated with Figure 6.17. This factor far exceeds the approximate factor of 10 expected from simple attenuation of the equilibrium cascade

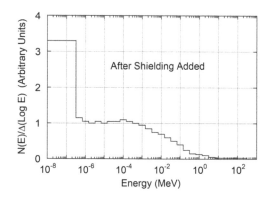

FIGURE 6.17

Neutron energy spectra obtained external to the shielding configuration shown *after* the addition of the cross-hatched concrete shielding blocks shown in Figure 6.15. The plot is a lethargy plot, and the normalization of the spectrum is arbitrary. (Reproduced with permission from Elwyn, A. J., and Cossairt, J. D. 1986. *Health Physics* 51:723–735. https://journals.lww.com/health-physics/pages/default.aspx.)

neutron spectrum, indicative of both the importance of the leakage neutrons and the maximization of their quality or radiation weighting factor.

In general, an iron shield "capped" or "backed" by such a concrete shield will be an efficient use of space. Evidently, about 60 cm of concrete is the most efficient thickness to use for this purpose (Yurista and Cossairt 1983; Zazula 1987). Such "capping" must obviously be done at the outside of the shield. Shielding properties of other elements with atomic mass numbers much larger than unity and roughly similar first excited state energies will be comparable. It is unlikely that this effect is important for "reinforced" concrete because the iron reinforcement materials represent a relatively small amount of the mass or volume of the material and are distributed in a matrix within it.

6.3.6 Neutron Spectra due to Ions

Measurements of neutron energy spectra due to ions remain somewhat rare. Britvich et al. (1999, 2001) report results for ^{12}C ions incident on a Hevimet target at 155 MeV/nucleon. The spectrum was measured at $\theta = 94°$ without shielding at a distance of 121 cm from the target as shown in Figure 6.18. Qualitatively similar spectra were obtained at this location with ^4He and ^{16}O ions, also at 155 MeV/nucleon.

The spectrum of neutrons due to the ^{12}C ions was also measured at $\theta = 94°$ at a distance of 403 cm from the target shielded by 308 g cm^{-2} of ordinary concrete and shown in Figure 6.19.

Intuitively, especially for ions of high atomic number, one might expect more copious production of neutrons given the increasing excess of neutron number over proton (atomic) number in heavy nuclei. This matter was studied in measurements conducted by Aroua et al. (1997) for lead ions having a specific energy of 160 GeV/nucleon. Comparisons of neutron energy spectra on top of a concrete shield surrounding a lead target were made with those obtained with 205 GeV protons, an available beam of roughly the same specific energy. The results are given in Figure 6.20.

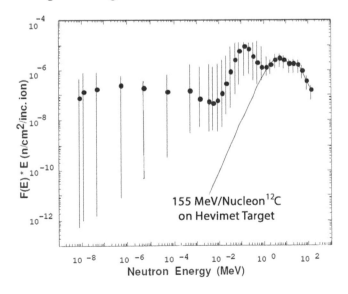

FIGURE 6.18
Lethargy plot of neutron energy spectrum $F(E)xE$, [i.e., $E\Phi(E)$] at 121 cm and $\theta = 94°$ from a thick Hevimet target bombarded by ^{12}C ions at 155 MeV/nucleon where F denotes the normalized neutron fluence elsewhere in this book usually denoted Φ. (Reprinted with permission from AIP Publishing from Britvich, G. I. et al. 1999. *Review of Scientific Instruments* 70:2314–2324.)

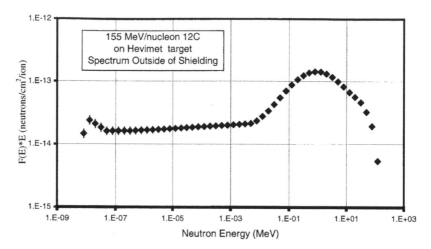

FIGURE 6.19

Lethargy plot of neutron energy spectrum $F(E)xE$ [i.e., $E\phi(E)$] at 403 cm and $\theta = 94°$, and external to 128.3 cm (308 g cm^{-2}) of concrete shielding, from a thick Hevimet target bombarded by ^{12}C ions at 155 MeV/nucleon. F denotes the normalized neutron fluence elsewhere usually denoted Φ. (Reprinted with permission from AIP Publishing from Britvich, G. I. et al. 2001. *Review of Scientific Instruments* 72:1600.)

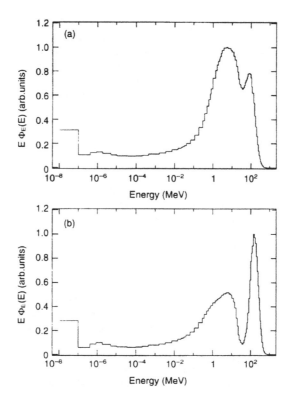

FIGURE 6.20

Lethargy plots of neutron energy spectra around a lead ion beam of 160 GeV/nucleon shielded by concrete (frame a) compared with that obtained with of a 205 GeV proton beam (frame b). The units used for the ordinate are arbitrary. (Reproduced with permission from Aroua, A. et al. 1997. *Radiation Protection Dosimetry* 70:437–440.)

TABLE 6.4

Percent Fluence in Specific Energy Bins for Selected Neutron Energy Spectra

Energy Range	Figure 6.10 Debuncher	Figure 6.12 Tevatron	Figure 6.14 Labyrinth	Figure 6.16 Iron (Bare)	Figure 6.17 Iron (Shielded)
<1.5 eV	31.5	19.5	71	28	55
0.0015–100 keV	12.5	36	24	46	43
0.1–2 MeV	8.5	36	2.0	17.5	2.0
2–25 MeV	40.5	7.0	1.0	4.5	0.1
>25 MeV	7.0	1.5	1.5	4.0	0.0

Source: Adapted from Cossairt, J. D. et al. 1988. Measurement of neutrons in enclosures and outside of shielding at the Tevatron. Fermi National Accelerator Laboratory: Fermilab Report FERMILAB-CONF-88/106. Batavia, IL.

6.3.7 Neutron Fluence and Dosimetry

The distribution of neutrons in different regions of the energy spectra can be quite important both in the potential to produce induced radioactivity (see Chapters 7 and 8) and with respect to the dose equivalent or effective dose that might be received by personnel present in the prompt radiation field. As examples, Table 6.4 gives the percent fluence determined to be in specific energy bins for the five spectra displayed in Figures 6.10, 6.12, 6.14, 6.16, and 6.17.

Table 6.5 gives these percentages for dose equivalent and average quality factor $\langle Q \rangle$ for these five spectra in accordance with the 1973 System.

Use of the 1990 System gives different values (Cossairt and Vaziri 2009). Table 6.6 gives the percentages of effective dose in these same energy bins along with average effective quality factors (i.e., radiation weighting factors) using the International Commission on Radiological Protection (ICRP) Report 60 (ICRP 1991) and ICRP Report 103 (ICRP 2007) prescriptions.

Figure 6.21 is a plot of cumulative values of similar quantities calculated for 1000 GeV protons incident on the face of a thick cylindrical concrete shield. As determined by Van Ginneken and Awschalom (1975), the dependence on incident proton energy of the distributions of fluence and dose equivalent is slight.

TABLE 6.5

Percent of Dose Equivalent H (1973 System) in Specific Energy Bins for Sample Spectra along with Average Quality Factor $\langle Q \rangle$

Energy Range	Figure 6.10 Debuncher	Figure 6.12 Tevatron	Figure 6.14 Labyrinth	Figure 6.16 Iron (Bare)	Figure 6.17 Iron (Shielded)
<1.5 eV	1.5	2.0	32	4.0	41.5
0.0015–100 keV	0.5	6.0	16	11.5	37
0.1–2 MeV	9.0	58.5	9.0	35	17
2–25 MeV	75	26	13	24	3.5
>25 MeV	14	7.5	30	25	1.0
$\langle Q \rangle$ (NCRP 1971)	5.8	6.9	3.1	5.4	2.5

Source: Adapted from Cossairt, J. D. et al. 1988. Measurement of neutrons in enclosures and outside of shielding at the Tevatron. Fermi National Accelerator Laboratory: Fermilab Report FERMILAB-CONF-88/106. Batavia, IL.

TABLE 6.6

Percent of Effective Dose H_{eff} (1990 System) in Specific Energy Bins for Sample Spectra along with Average Quality Factor $\langle Q \rangle$

Energy Range	Figure 6.10 Debuncher	Figure 6.12 Tevatron	Figure 6.14 Labyrinth	Figure 6.16 Iron (Bare)	Figure 6.17 Iron (Shielded)
<1.5 eV	1.0	1.8	22	3.9	31
0.0015–100 keV	0.6	4.6	12	9.1	30
0.1–2 MeV	5.1	45	9.0	31	16
2–25 MeV	63	35	15	22	12
>25 MeV	30.3	13.6	42	34	11
$\langle Q \rangle$ (ICRP 1991)	7.7	11	6.1	8.2	6.4
$\langle Q \rangle$ (ICRP 2007)	7.5	9.5	4.2	6.6	3.6

Source: Adapted from Cossairt, J. D., and K. Vaziri. 2009. *Health Physics* 96:617–628.

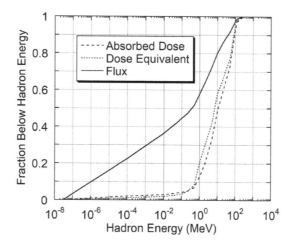

FIGURE 6.21

Fraction of the omnidirectional flux, entrance absorbed dose, and maximum dose equivalent (1973 System) below a given hadron kinetic energy as a function of hadron energy for the region between zero and 450 cm depth and between 300 cm and 750 cm radius calculated for 1000 GeV/c protons incident on the face of a solid concrete cylinder. (Adapted from Van Ginneken, A., and M. Awschalom. 1975. *High energy particle interactions in large targets: Volume I, hadronic cascades, shielding, and energy deposition.* Batavia, IL: Fermi National Accelerator Laboratory.)

7

Induced Radioactivity in Accelerator Components

7.1 Introduction

In this chapter the production of induced radioactivity at accelerators is described. The discussion begins with a review of the basic principles of the production of radioactivity. It proceeds with a discussion of the activation of accelerator components. In the discussion of this chapter, "accelerator components" includes also the associated radiation shielding. Generalizations are included that may be used for practical health physics applications related to external radiation exposures. Chapter 8 continues this discussion for materials of interest from the standpoint of internal radiation protection for both occupational workers and members of the public who are in proximity to an accelerator facility.

7.2 Fundamental Principles of Induced Radioactivity

Induced radioactivity can be produced at nearly all accelerators that produce particles and ions above all but the most minimal energies. When the accelerated beam particles interact with a nucleus, the resultant nuclear reactions can convert it into a different nuclide, one that may or may not be radioactive. The *activity* of a given radionuclide refers to the number of atoms that decay per unit time. The customary unit of activity is the *Curie* (Ci) and its submultiples. One Curie was historically defined to be the activity of 1.0 g of natural radium. It is currently precisely defined to be 3.7×10^{10} decays per second. The SI unit of activity is the *becquerel* (Bq)—1.0 decay per second, with multiples such as GBq (1.0 GBq $= 10^9$ Bq) commonly used.

Also important is the *specific activity*, the activity per unit volume (e.g., Bq m^{-3}, Bq cm^{-3}, etc.) or the activity per unit mass (e.g., Bq kg^{-1}, Bq g^{-1}). Attention needs to be paid to the units used to identify if this quantity is defined in terms of volume or mass.

Radioactive decay is a random process characterized by a *mean life* (units of time) denoted by τ, and its reciprocal the *decay constant* (units of inverse time) λ, with $\lambda = 1/\tau$. One should be aware of context because the symbol λ is also used for attenuation lengths, mean-free paths, etc. If a total of $N_{tot}(t)$ atoms of a radionuclide are present at time t, the total activity $A_{tot}(t)$ is determined from the random nature of radioactive decay to be

$$A_{tot}(t) = -\frac{dN_{tot}(t)}{dt} = \frac{1}{\tau}N_{tot}(t) = \lambda N_{tot}(t) \qquad (7.1)$$

If one starts at time $t=0$ with $N_{tot}(0)$ atoms of the radionuclide present, then this simple differential equation has the solution at some later time $t=T$, with $T>0$:

$$A_{tot}(T) = \lambda N_{tot}(0)\exp(-\lambda T) = A_{tot}(0)\exp(-\lambda T) \tag{7.2}$$

Usually, the time required to decay to half of the original activity, the *half-life* $t_{1/2}$, is tabulated and is related to the mean-life by the following:

$$\tau = \frac{1}{\ln 2} t_{1/2} = \frac{1}{0.693} t_{1/2} = 1.443 t_{1/2} \tag{7.3}$$

In this book values of half-lives are taken from the National Nuclear Data Center (NNDC 2018) as of the time of publication.

The simplest activation situation at accelerators is that of the steady irradiation of some material by a spatially uniform flux density of particles that begins at time $t=0$ and continues at a constant rate for an *irradiation period* that ends at $t=t_i$. This is followed by a decay period called the *cooling time* that is denoted t_c, a period of time that begins at $t=t_i$ and ends at $t=t_i+t_c$. For this simple situation, self-absorption of the incident particles by the target itself is ignored, as also is the likely possibility that a spectrum of particles of different types or energies might be incident. With these simplifications, the process of producing the radioactivity is characterized by a single average cross section σ. In more generalized situations the value of this cross section must be obtained by averaging over the energy spectra of the particles incident.

The number of atoms of the radionuclide of interest per unit volume will thus be governed by the following equation during the period of the irradiation:

$$\frac{dn(t)}{dt} = -\lambda n(t) + N\sigma\phi \tag{7.4}$$

where $n(t)$ is the number density of atoms of the radionuclide of interest at time t (cm^{-3}), N is the number density of "target" atoms (cm^{-3}) as calculated by Equation 1.4, σ is the production cross section (cm^2) for the radionuclide in question, and ϕ is the flux density (cm^{-2} sec^{-1}) of incident particles. On the right-hand side of Equation 7.4, the first term represents the *loss* of radionuclides through decay during the irradiation, while the second term is the *gain* of radionuclides through the production reaction under consideration. If one starts at $t=0$ with no radionuclide present, $n(0)=0$, this equation has the following solution for $0<t<t_i$:

$$n(t) = \frac{N\sigma\phi}{\lambda}\{1-\exp(-\lambda t)\} \tag{7.5}$$

Thus, the specific activity induced in the material as a function of time during the irradiation is given by
$a(t) = \lambda\, n(t)$; hence,

$$a(t) = N\sigma\phi\{1-\exp(-\lambda t)\}(\text{Bq cm}^{-3}) \quad \text{for } 0<t<t_i \tag{7.6}$$

To obtain specific activity in units of Curies cm^{-3}, one must simply divide the result by the conversion factor 3.7×10^{10} Bq Ci^{-1}. At the instant of completion of the irradiation ($t=t_i$), the specific activity is

$$a(t_i) = N\sigma\phi\{1 - \exp(-\lambda t_i)\} \ (\text{Bq cm}^{-3}) \tag{7.7}$$

so that the specific activity as a function of time is characterized by a buildup from zero to the *saturation concentration* value of $N\sigma\phi$ for infinitely long irradiations. After the irradiation has ceased ($t > t_i$), the specific activity as a function of the cooling time t_c will obviously decay exponentially and be given by the *activation equation*:

$$a(t_c) = N\sigma\phi\{1 - \exp(-\lambda t_i)\}\{\exp(-\lambda t_c)\} \ (\text{Bq/cm}^3) \tag{7.8}$$

where t_c is the cooling time, the time period subsequent to cessation of the irradiation, and

$$t_c = t - t_i \tag{7.9}$$

To obtain total activities where uniform flux densities of particles of constant energy are incident on a homogeneous target, one can simply multiply by the volume of the target. In more complex cases, numerical integrations are needed.

For γ-ray emitters typical of those emitted by accelerator-produced radionuclides in the range of from about 100 keV to 10 MeV, many textbooks in health physics demonstrate that the absorbed dose rate dD/dt (rad h^{-1}) at a distance r (meters) from a *point source* is approximately given in terms of the source strength S (Ci) and the photon energies present $E_{\gamma i}$ (MeV) by

$$\frac{dD}{dt} = 0.4 \frac{S}{r^2} \sum_i E_{\gamma i} \tag{7.10}$$

The summation is over all γ-rays present, including appropriate *branching fractions* if multiple photons are emitted in the decay process. The branching fractions are used to account for the fact that in some radioactive decay processes, the decay may proceed through different nuclear excited states or even different intermediate radionuclides in "decay chains" until a stable "progeny" nucleus is reached. A simple process with only one decay has a branching fraction of unity. No branching fraction exceeds unity.

If dD/dt is desired as an approximate absorbed dose rate in Gy h^{-1} at a distance r (meters) from a source strength S in gigabecquerels, the constant 0.4 is replaced with 1.08×10^{-4}. For nonpoint sources, an appropriate spatial integration must be performed. It is noted that GBq (10^9 Bq) or MBq (10^6 Bq) are often better units of activity for practical work than is the relatively small becquerel.

7.3 Activation of Components at Electron Accelerators

7.3.1 General Phenomena

At electron accelerators, as was described in Chapter 3, the direct interactions of electrons in material result in the copious production of photons. Through various nuclear reaction channels, these photons then proceed to produce charged particles and neutrons that then interact further with materials to produce radioactivity. In general, if the facility is properly shielded against prompt radiation, the radioactivity hazard will be confined to accelerator components and the interior of the accelerator enclosure shielding. This is consistent with the experience at most accelerators. Usually most of the radiation dose received by the workers results from maintenance activities on radioactivated components, handling and moving of activated items, radiation surveys, and radioactive waste handling rather than as a result of exposure to the prompt radiation fields. An understanding of the production of radionuclides can help reduce personnel exposures through the selection of more appropriate machine component materials and the optimization of decay ("cool-down") times recommended after the beam has been turned off. Some familiarity with the relevant cross sections is extremely useful. "Global" data (i.e., data spanning the periodic table and a large domain of energies) were compiled in the treatise by Barbier (1969). Sample results from Barbier's treatise are provided here as Figures 7.1 through 7.3.

7.3.2 Results for Electrons at Low Energies

Data such as those presented in Section 7.3.1 form the basis of detailed activation calculations. Swanson (1979a) utilized the methodology of Approximation B of the analytical shower theory of Rossi and Greisen (1941), discussed in Section 3.4, to estimate saturation activities rates in various materials. Since the energy domain below about 35 MeV is characterized by rapidly varying cross sections, Swanson provided energy-dependent results and only considered reactions of the type (γ,n), (γ,p), (γ,np), and $(\gamma,2n)$. Other reactions were ignored due to higher energy thresholds, and smaller cross sections, with the inference that these would be less important. Swanson points out that the dependence of the induced activity as a function of energy above a given nuclear reaction threshold will generally follow that of the neutron yields (see Figure 3.5). In Swanson's calculations, the material in question absorbs *all* of the beam power and has been irradiated for an *infinite* time with no cooldown; $t_i = \infty$, $t_c = 0$ in Equation 7.8. Thus, *saturation activities* are calculated, normalized to the incident electron beam power (kW). These can readily be corrected for other values of t_i and t_c by means of Equation 7.8. Results of these calculations, accounting for the natural isotopic abundances and assuming energies well above the reaction thresholds, are given in Table 7.1.

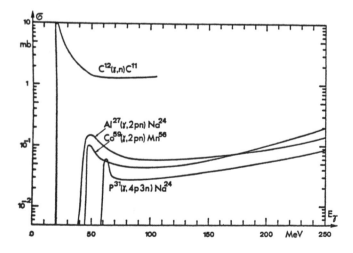

FIGURE 7.1

Plot of cross section σ as a function of photon energy for some photoneutron reactions. Some of these reactions are also useful for monitoring incident photon fluences. (Barbier, M: *Induced radioactivity*. 1969. Copyright Wiley-VCH Verlag GmbH & Co. KGaA. Reproduced with permission.)

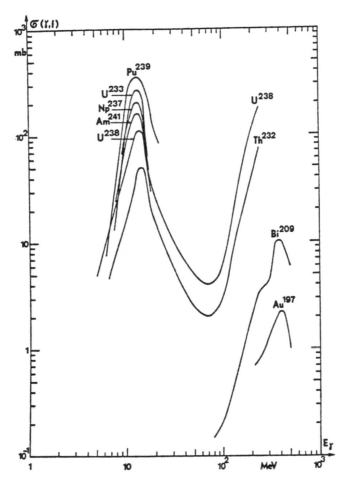

FIGURE 7.2
Plot of cross section σ as a function of photon energy for some photofission reactions for photons incident on heavy elements. (Barbier, M: *Induced radioactivity*. 1969. Copyright Wiley-VCH Verlag GmbH & Co. KGaA. Reproduced with permission.)

The tabulations are likely accurate to within approximately ±30%. At these low energies, the distribution of the radioactivity can often be approximated as a point source for calculating the residual absorbed dose rates plausibly using Equation 7.10, taking the summation over all of the γ-ray emitters presented at a given time. In such tabulations, values are listed for radionuclides that decay from the lowest (i.e., "ground") nuclear states and separately for *isomeric* states, excited states of certain radionuclides that have sufficiently lengthy half-lives to merit special consideration. Such isomeric states are denoted with the superscript letter "m" following mass number. For example, ^{26m}Al (half-life = 7.17×10^5 years) denotes an isomeric state of ^{26}Al (half-life = 6.346 s).

Table 7.1 also provides the *specific γ-ray constant Γ* for each tabulated radionuclide. These constants connect activity with the *absorbed dose rates* at a distance of 1.0 m from a point source of a given activity, accounting for all the photons emitted by the decaying radionuclide and including those emitted secondarily, such as internal bremsstrahlung and annihilation radiation due to β^+ emission manifested in the form of a pair of 0.511 MeV photons. For point sources, absorbed dose rates at other distances can be calculated by

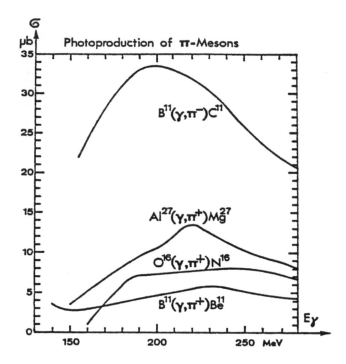

FIGURE 7.3

Plot of cross section σ as a function of photon energy for the photoproduction of pions on light elements. (Barbier, M: *Induced radioactivity*. 1969. Copyright Wiley-VCH Verlag GmbH & Co. KGaA. Reproduced with permission.)

simply using the inverse square law. In this context, absorbed dose rate is loosely connected with a somewhat obsolete unit of measure, the *exposure rate* expressed in roentgens h^{-1} ($R\ h^{-1}$). Exposure rate measures the liberation of ions in air by photons and is only defined for photon radiation fields. The hourly liberation of one *electrostatic unit* of charge cm^{-3} of air at STP is $1.0\ R\ h^{-1}$. Thus, $1.0\ R = 2.58 \times 10^{-4}$ Coulomb kg^{-1}. An exposure rate of $1.0\ R\ h^{-1}$ is approximately equal to 0.95 rad h^{-1} of absorbed dose rate in tissue placed in the radiation field under consideration. In Table 7.1, energy thresholds are the approximate values above which the production nuclear reactions proceed. Even for electron-induced reactions, the reaction thresholds can be calculated as discussed in Section 4.2.2.1 for protons and ions.

7.3.3 Results for Electrons at High Energies

For higher-energy electrons, more possible reactions, sometimes called reaction channels, become available, but the energy dependence is diminished. Swanson (1979a) has also performed calculations of the production of radionuclides in this energy domain, and the results are provided here in Table 7.2.

    The results are valid to within an approximate factor of two for any beam energy E_o that is somewhat above the nuclide production threshold. Saturation activities and absorbed dose rates normalized to the beam power are provided. These quantities are useful because of the dominance, and lack of energy dependence, of the photoneutron production process, as discussed in Chapter 3. The electrons are assumed to be totally absorbed in the material, and no self-shielding effects are considered. The distribution of radioactivity within the material is also not taken into account. Saturation conditions imply $t_i = \infty$, $t_c = 0$ in Equation 7.8. As at for lower energies, results for other values of t_i and t_c can be obtained using

TABLE 7.1

Estimations of Saturation-Specific γ-Ray Constants and Activities at 1.0 Meter for Electrons Having Energies above the Indicated Threshold Energies Incident on Various Target Materials of Naturally Occurring Isotopic Abundances Normalized to the Beam Power

Target Material	Nuclide	Half-Life[a]	Threshold (MeV)	Specific γ-Ray Constant, Γ [(mGy h^{-1}) × (GBq m^{-2})$^{-1}$]	[(rad h^{-1}) × (Ci m^{-2})$^{-1}$]	Saturation Activity per Unit Beam Power (GBq kW^{-1})	(Ci kW^{-1})
Al	^{24}Na	15.00 h	23.7	0.48	1.8	1.1	0.03
	^{26m}Al	6.35 s	13.0	0.16	0.59	330	8.8
Fe	^{54}Mn	312.2 d	20.4	0.32	1.2	22.0	0.59
	^{56}Mn	2.579 h	10.6	0.23	0.86	1.18	0.032
	^{53}Fe	8.51 min	13.6	0.18	0.67	27.0	0.74
Ni	^{56}Ni	6.075 d	22.5	0.43	1.6		
	^{56}Co[b]	72.24 d		0.62	2.3	2.6[c]	0.07[c]
	^{57}Ni	35.60 h	12.2	0.34	1.4		
	^{57}Co[b]	217.7 d		0.35	1.3	155[c]	4.2[c]
Cu	^{61}Cu	3.339 h	19.7	0.19	0.71	32.2	0.87
	^{62}Cu	9.673 min	10.8	0.16	0.60	407	11
	^{64}Cu	12.70 h	9.91	0.10	0.38	185	5.0
W	^{182m}Ta	15.84 min	7.15	0.04	0.15		
	^{182}Ta[b]	114.7 d		0.16	0.61	13.3	0.36
	^{183}Ta	5.1 d	7.71	0.04	0.15	23.3	0.63
	^{181}W	121.2 d	7.99	0.02	0.09	340	9.1
	^{185m}W	1.67 min	7.27	0.05	0.18		
	^{185}W[b]	75.1 d		No γ-ray[d]	No γ-ray[d]	300[c]	8.1[c]
Au	^{195m}Au	30.5 s	14.8	0.04	0.16		
	^{195}Au[b]	186.0 d		0.02	0.07	204[c]	5.5[c]
	^{196m}Au	9.6 h	8.07	0.03	0.12		
	^{196}Au[b]	6.167 d		0.08	0.30	1520[c]	41[c]
Pb	^{203m}Pb	6.21 s	8.38	0.05	0.19	17.4	0.47
	^{204m}Pb	1.116 h	14.8	0.32	1.20	44	1.2

Source: Adapted from Swanson, W. P. 1979a. *Radiological safety aspects of the operation of electron linear accelerators.* International Atomic Energy Agency: IAEA Technical Report No. 188. Vienna, Austria.

Note: In this table the letter "m" following a mass number (e.g., ^{26m}Al) indicates this entry is for an isomeric nuclear state of this radionuclide.

[a] Values of half-lives are those published by NNDC (2018).
[b] This radionuclide is the progeny of the radionuclide above it.
[c] This is the activity of the progeny radionuclide.
[d] The term "no γ-rays" is applied to radionuclides having no, or very rare, emission of photons in their decay.

Equation 7.8. Radionuclides contributing less than 0.1 (mGy h^{-1}) (kW m^{-2})$^{-1}$ or with $t_{1/2} < 1.0$ minute have been excluded as have products of thermal neutron capture reactions.

Cooling curves have been published by Barbier (1969) for high-energy electrons incident on various materials for an infinite irradiation at the rate of one electron per second. A pertinent example is given in Figure 7.4 for an infinite irradiation time $t_i = \infty$. In this figure, results are given for the absorbed dose rates per electron s^{-1} per megaelectron volt (MeV) of electron kinetic energy assuming the applicability of point source conditions. For this purpose, an "infinite" irradiation time is conventionally taken to be one of about eight

TABLE 7.2

Estimations of Saturation Activities and Absorbed Rates at 1.0 Meter in Various Materials, Assuming "Point Source" Conditions for High-Energy Electrons

Produced Radionuclide		Saturation Activity per Unit Beam Power		Saturation Absorbed Dose per Unit Beam Power	
Half-Life[a]	Threshold (MeV)	(GBq kW^{-1})	(Ci kW^{-1})	(mGy h^{-1}) × (kW m^{-2})$^{-1}$	(rad h^{-1}) × (kW m^{-2})$^{-1}$
Material: Natural Aluminum					
^{7}Be 53.22 d	33.0	4.8	0.13	0.04	0.004
^{11}C 20.36 min	33.5	1.9	0.051	0.3	0.03
^{15}O 2.04 min	33.4	2.5	0.07	0.4	0.04
^{18}F 1.830 h	34.4	5.2	0.14	0.8	0.08
^{22}Na 2.602 y	22.5	9.2	0.25	3.0	0.3
^{24}Na 15.00 h	23.7	10.4	0.28	5.0	0.5
^{26m}Al 6.35 s	13.0	321	8.8	26	2.6
Material: Natural Iron					
^{46}Sc 83.79 d	37.4	7.4	0.2	2.0	0.2
^{48}V 15.97 d	25.9	15.0	0.4	8.0	0.8
^{51}Cr 27.70 d	19.7	15.0	0.4	3.0	0.3
^{52}Mn 5.591 h	20.9	1.5	0.04	0.4	0.04
^{52m}Mn 21.1 min	20.9	1.3	0.036	0.23	0.023
^{54}Mn 312.2 d	20.4	22.0	0.59	7.0	0.7
^{56}Mn 2.579 h	10.6	1.1	0.03	0.3	0.03
^{52}Fe 8.725 h	24.1	2.2	0.06	0.4	0.04
^{53}Fe 8.51 min	13.6	27.4	0.74	4.9	0.49
^{55}Fe 2.744 y	11.2	490	13.3	90	9
Material: Natural Copper					
^{58m}Co 9.10 h	41.8	24.4	0.66	4.0	0.4
^{58}Co 70.86 d	41.8	24.4	0.66	2.0	0.2
^{60}Co 5.271 y	18.9	24.0	0.65	8.0	0.8
^{63}Ni 101.2 y	17.1	16.6	0.45	No γ-rays	No γ-rays
^{61}Cu 3.339 h	19.7	32.2	0.87	6.0	0.6
^{62}Cu 9.673 min	10.8	407	11	65	6.5
^{64}Cu 12.70 h	9.9	185	5	19	1.9
Material: Natural Tungsten					
^{182m}Ta 15.84 min	7.15	13.3	0.36	0.3	0.03
^{182}Ta 114.7 d	7.15	13.3	0.36	1.1	0.11
^{183}Ta 5.11 d	7371	22.9	0.62	0.9	0.09
^{184}Ta 8.7 h	14.9	1.78	0.048	0.4	0.04
^{185}Ta 49.4 min	8.39	20.7	0.56	0.6	0.06
^{181}W 121.2 d	8.00	330	8.9	8.0	0.8
^{185m}W 1.67 min	7.27	300	8.1	7.3	0.73
^{185}W 75.1 d	7.27	300	8.1	No γ-rays	No γ-rays
Material: Natural Lead					
^{204}Tl 3.783 y	14.83	0.92	0.025	No γ-rays	No γ-rays
^{206}Tl 4.202 min	7.46	37	1.0	No γ-rays	No γ-rays

(Continued)

TABLE 7.2 (*Continued*)

Estimations of Saturation Activities and Absorbed Rates at 1.0 Meter in Various Materials, Assuming "Point Source" Conditions for High-Energy Electrons

Produced Radionuclide			Saturation Activity per Unit Beam Power		Saturation Absorbed Dose per Unit Beam Power	
Half-Life[a]		Threshold (MeV)	(GBq kW^{-1})	(Ci kW^{-1})	(mGy h^{-1}) × (kW m^{-2})$^{-1}$	(rad h^{-1}) × (kW m^{-2})$^{-1}$
^{207m}Tl	1.33 s	8.04	93	2.5	9.1	0.91
^{207}Tl	4.77 min	8.04	93	2.5	No γ-rays	No γ-rays
^{202m}Pb	3.54 h	15.3	2.2	0.06	0.3	0.03
^{202}Pb	5.25×10^4 y	15.3	2.2	0.06	No γ-rays	No γ-rays
^{203m}Pb	6.21 s	8.38	31	0.83	1.3	0.13
^{203}Pb	2.163 d	8.38	31	0.83	0.7	0.07
^{204m}Pb	1.116 h	14.8	89	2.4	14	1.4
Material: Typical Ordinary Concrete						
^{15}O	2.04 min	15.7	96	2.6	15	1.5
^{22}Na	2.602 y	12.4	3.7	0.1	1.2	0.12
^{27}Si	4.15 s	17.2	74	2.0	12	1.2
^{38}K	7.636 min	13.1	3.7	0.1	1.5	0.15

Source: Adapted from Swanson, W. P. 1979a. *Radiological safety aspects of the operation of electron linear accelerators.* International Atomic Energy Agency: IAEA Technical Report No. 188. Vienna, Austria.

Note: Results have been summed over the naturally occurring isotopic composition of the tarmaterials; In this table the letter "m" following a mass number (e.g., ^{26m}Al) indicates this entry is for an isomeric nuclear state of this radionuclide.

[a] Values of half-lives are those published by NNDC (2018).

half-lives or more. As discussed in Chapter 3, the lack of strong energy dependence and the simplicity of the photoneutron spectra make possible these rather uncomplicated results.

7.4 Activation of Components at Proton and Ion Accelerators

7.4.1 General Phenomena

Protons having energies above about 10 MeV will produce radioactivity upon interacting with matter. In some special cases, radioactivity can be produced at much lower energies due to exothermic nuclear reactions that either produce radionuclides directly or emit neutrons capable of inducing radioactivity through their secondary interactions. For some nuclear reactions the thresholds for individual nuclear reactions can be much less, as discussed in Chapter 4, so that nuclear reaction thresholds, perhaps with the tools provided by the NNDC (2018), should be utilized to verify the possible nuclear reactions that can produce radioactivity.

Radioactivation can also result from interactions by other ions aside from protons above a specific energy of roughly and not exclusively 10 MeV nucleon^{-1} and by other hadrons, particles subject to the "strong" (i.e., "nuclear") interaction. As with electron accelerators, if a given accelerator is properly designed with respect to the shielding against prompt

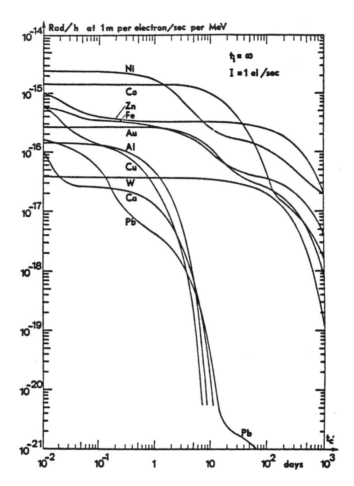

FIGURE 7.4

Examples of total photon absorbed dose rates in units of rads h⁻¹ (1.0 rad h⁻¹ = 10⁻² Gy h⁻¹) per MeV of incident electron energy due to radioactive nuclei produced in thick targets of various materials irradiated by an electron current of 1.0 electron sec⁻¹, as a function of time since the cessation of the irradiation. The irradiation was assumed to have occurred for an infinitely long period of time. The absorbed dose rates are those found at 1.0 m from a point source containing all of the radioactive nuclei produced. (Barbier, M: *Induced radioactivity*. 1969. Copyright Wiley-VCH Verlag GmbH & Co. KGaA. Reproduced with permission.)

radiation and has proper access controls to avoid direct beam-on exposure to people, the induced radioactivity is very likely to be the dominant source of occupational radiation dose.

For the lower incident energies, perhaps below about 30 MeV, one is first concerned with production of radionuclides by such processes as (p,γ) and single- and multinucleon transfer reactions. While the details of the total cross sections for such reactions are complex, the systematics and approximate energy dependencies are globally well understood. In general, one has endothermic nuclear reactions that have a threshold E_{th} below which the process is forbidden by conservation of energy (see Section 4.2.2.1).

The treatise by Barbier (1969) has addressed activation by many types of particles. These results are provided in Figures 7.5 through 7.10.

Some of the results for the light elements are especially important for environmental radiation considerations and support the discussions of Chapter 8, while those for iron

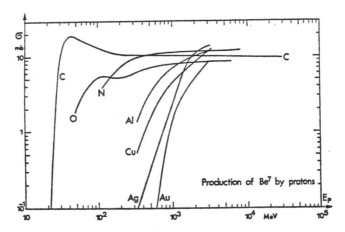

FIGURE 7.5
Plot of cross section σ as a function of proton energy for production of ^{7}Be by protons incident on selected elements. (Barbier, M: *Induced radioactivity.* 1969. Copyright Wiley-VCH Verlag GmbH & Co. KGaA. Reproduced with permission.)

and copper targets, Figures 7.9 and 7.10, respectively, are of great importance due to the ubiquitous presence of those elements in accelerator components.

Thick target yields of radionuclides for materials having a range of atomic numbers have been systematically studied by Cohen (1978) for a number of nuclear processes spanning the periodic table. Figure 7.11 is a representative qualitative plot of the general features of such

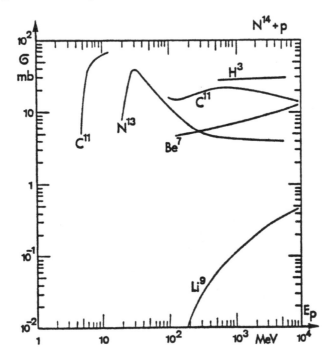

FIGURE 7.6
Plot of cross section σ as a function of proton energy for production of selected radionuclides by protons incident on ^{14}N. (Barbier, M: *Induced radioactivity.* 1969. Copyright Wiley-VCH Verlag GmbH & Co. KGaA. Reproduced with permission.)

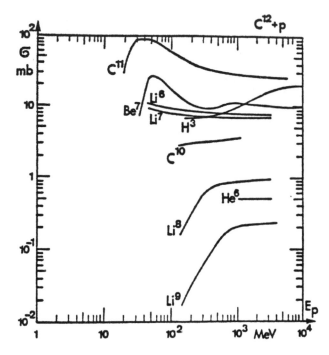

FIGURE 7.7
Plot of cross section σ as a function of proton energy for production of selected radionuclides by protons incident on ^{12}C. (Barbier, M: *Induced radioactivity.* 1969. Copyright Wiley-VCH Verlag GmbH & Co. KGaA. Reproduced with permission.)

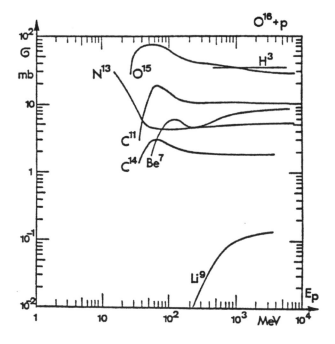

FIGURE 7.8
Plot of cross section σ as a function of proton energy for production of selected radionuclides by protons incident on ^{16}O. (Barbier, M: *Induced radioactivity.* 1969. Copyright Wiley-VCH Verlag GmbH & Co. KGaA. Reproduced with permission.)

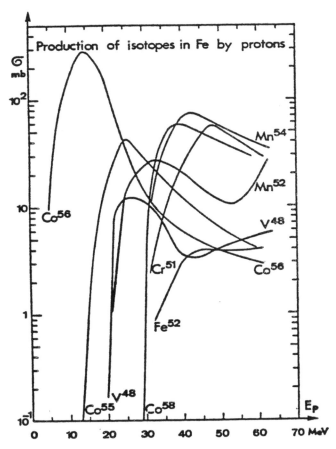

FIGURE 7.9
Plot of cross section σ as a function of proton energy for production of selected radionuclides by protons incident on iron. (Barbier, M: *Induced radioactivity*. 1969. Copyright Wiley-VCH Verlag GmbH & Co. KGaA. Reproduced with permission.)

excitation functions of such nuclear reactions. Specific processes may vary considerably from this behavior since "resonances" at specific nuclear excited states have been ignored.

Table 7.3 lists a variety of such nuclear reactions along with the range of values of energy above threshold at which the radioactivity production rate has risen to 0.1% of the saturation value and also the range of saturation values for the production of radioactivity. It is assumed that the target thickness comfortably exceeds the range of the incident ion and that the irradiation period greatly exceeds the half-life of the radionuclide of interest. For shorter bombarding periods t_i, one needs to multiply by the factor $[1-\exp(-\lambda t_i)]$. Over the energy domain of these curves, the importance of activation by secondary particles is small compared to that encountered at higher energies. In Table 7.3, the ranges of energies are listed at which the production yields are at approximately 0.1% of the tabulated saturation values. The "low/high" values that approximately bound the extent of the saturated activity are also given.

7.4.2 Methods of Systematizing Activation due to High-Energy Hadrons

For proton and ion accelerators of higher energy, the neglect of secondary reactions and the restriction to few- and multinucleon transfer reactions can be a serious deficiency in

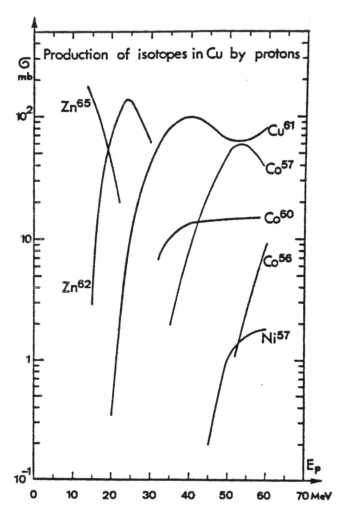

FIGURE 7.10

Plot of cross section σ as a function of proton energy for production of selected radionuclides by protons incident on copper. (Barbier, M: *Induced radioactivity*. 1969 Copyright Wiley-VCH Verlag GmbH & Co. KGaA. Reproduced with permission.)

the accuracy of estimation of induced radioactivity because of the rise in importance of such processes as spallation. Below a kinetic energy of about 40 MeV only few-nucleon transfer reactions are generally prolific. The variety of radionuclides that can be produced increases with increasing bombarding energy because more nuclear reaction thresholds are exceeded. As a general rule, at high energies ($E_o \approx 1.0$ GeV or greater), one must consider that all radionuclides in the periodic table that have mass numbers less than those of the material exposed to the flux of hadrons may be produced. Of course, many of these are of little significance due to short half-lives and small production cross sections.

Table 7.4 lists radionuclides typically encountered at high-energy proton accelerators and their half-lives.

In this table only radionuclides with half-lives between 10 minutes and 5 years are listed. Also, all "pure" β^- (electron) emitters are excluded. Pure β^- emitters are those radionuclides that emit no γ-rays in their decays. They generally present minimal external

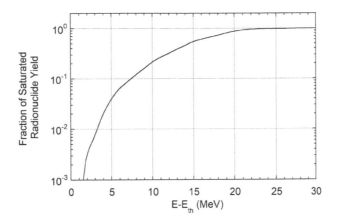

FIGURE 7.11

Typical behavior of radionuclide production by (p,γ) or few-nucleon transfer reactions for energies not far above the reaction threshold E_{th}. This behavior is typical of the nuclear reactions tabulated in Table 7.3. For detailed calculations, data related to specific reactions on particular target materials should be used. (Adapted from Cohen, B. L. 1978. In *Handbook of radiation measurement and protection, Section A: Volume I Physical science and engineering data*, ed. A. Brodsky, 91–212. West Palm Beach, FL: CRC Press/Taylor and Francis Group.)

TABLE 7.3

Tabulation of Generalized Parameters for the Production of Radionuclides by Means of Low-Energy Nuclear Reactions That Span the Periodic Table

Reaction	0.1% Yield-low $(E\text{-}E_{th})$ (MeV)	0.1% Yield-high $(E\text{-}E_{th})$ (MeV)	Sat. Yield-low $(\mu Ci/\mu A)$	Sat. Yield-high $(\mu Ci/\mu A)$	Reaction	0.1% Yield-low $(E\text{-}E_{th})$ (MeV)	0.1% Yield-high $(E\text{-}E_{th})$ (MeV)	Sat. Yield-low $(\mu Ci/\mu A)$	Sat. Yield-high $(\mu Ci/\mu A)$
(p,γ)	4	9	3×10^2	10^3	$(^3He,\gamma)$	4	6	1	2
(p,n)	0	6	3×10^5	8×10^5	$(^3He,n)$	3	12	10^2	3×10^2
$(p,2n)$	1	4	3×10^5	10^6	$(^3He,2n)$	2	7	3×10^2	4×10^3
$(p,3n)$	1	6	3×10^5	10^6	$(^3He,3n)$	2	5	2×10^3	3×10^4
$(p,4n)$	5	8	2×10^5	10^6	$(^3He,2p)$	4	12	2×10^2	10^4
$(p,5n)$	5	10	10^5	2×10^6	$(^3He,\alpha)$	6	14	2×10^2	10^3
(p,pn)	2	5	2×10^5	2×10^6	$(^3He,p3n)$	10	15	10^4	4×10^5
$(p,p2n)$	3	8	3×10^5	2×10^6	(α,γ)	10	13	3	20
(d,γ)	5	7	30	100	(α,n)	1	9	3×10^2	10^4
(d,n)	2	7	4×10^3	3×10^5	$(\alpha,2n)$	1	4	5×10^3	4×10^4
$(d,2n)$	2	5	2×10^5	6×10^6	$(\alpha,3n)$	1	6	3×10^3	7×10^5
$(d,3n)$	1	4	3×10^5	10^6	$(\alpha,4n)$	5	8	3×10^3	4×10^4
$(d,4n)$	4	8	2×10^5	6×10^5	$(\alpha,5n)$	5	8	10^4	3×10^5
$(d,5n)$	6	10	10^5	10^6	(α,p)	5	8	6×10^2	2×10^4
(d,p)	2	7	4×10^4	3×10^5	(α,pn)	3	12	3×10^3	8×10^4
$(d,p2n)$	2	10	10^5	2×10^6	$(\alpha,p2n)$	5	15	3×10^3	7×10^4
$(d,p3n)$	8	15	10^5	2×10^6	$(\alpha,p3n)$	7	15	10^4	3×10^4
$(d,2p)$	5	15	3×10^3	4×10^4	$(\alpha,2p)$	5	10	10^2	3×10^3
(d,α)	4	7	10^4	3×10^4	$(\alpha,\alpha n)$	6	16	3×10^3	3×10^4
$(d,\alpha n)$	5	15	2×10^4	10^5					

Source: Adapted from Cohen, B. L. 1978. In *Handbook of radiation measurement and protection, Section A: Volume I Physical science and engineering data*, ed. A. Brodsky, 91–212. West Palm Beach, FL: CRC Press.

TABLE 7.4

Summary of Radionuclides Commonly Identified in Materials Irradiated around Accelerators

Target Material	Radionuclides	Approximate Threshold (MeV)	Half-Life	Production Cross Section (High-Energy Limit) (mb)
Plastics and Oils	^{3}H	11	12.32 y	10
	^{7}Be	2	53.22 d	10
	^{11}C	20	20.36 min	20
Al, Concrete	As above, plus			
	^{18}F	40	1.830 h	6
	^{22}Na	30	2.602 y	10
	^{24}Na	5	15.00 h	10
Fe	As above, plus			
	^{42}K		12.36 h	
	^{43}K		22.3 h	
	^{44}Sc		3.97 h	
	^{44m}Sc		2.442 d	
	^{46}Sc		83.79 d	
	^{47}Sc		3.349 d	
	^{48}Sc		1.820 d	
	^{48}V	20	15.97 d	6
	^{51}Cr	30	27.70 d	30
	^{52}Mn	20	5.591 d	30
	^{52m}Mn		21.1 min	
	^{54}Mn	30	312.2 d	30
	^{52}Fe	30	8.725 h	4
	^{55}Fe		2.744 y	
	^{59}Fe		44.50 d	
	^{56}Co	5	77.24 d	30
	^{57}Co	30	271.7 d	30
	^{58}Co	30	70.86 d	25
Cu	As above, plus			
	^{57}Ni	40	35.60 h	2
	^{65}Ni		2.518 h	
	^{60}Co	30	5.271 y	15
	^{60}Cu		23.7 min	
	^{61}Cu	20	3.339 h	100
	^{62}Cu		9.673 min	
	^{64}Cu		12.70 h	
	^{62}Zn	15	9.186 h	60
	^{65}Zn	0	243.9 d	100

Source: Adapted from NCRP. 2003. NCRP Report No. 144: *Radiation protection for particle accelerator facilities.* Bethesda, MD; Barbier, M: *Induced radioactivity.* 1969. Copyright Wiley-VCH Verlag GmbH & Co. KGaA. Reproduced with permission with half-lives from NNDC. 2018. National Nuclear Data Center at Brookhaven National Laboratory. http://www.nndc.bnl.gov/ (accessed October 17, 2018).

exposure hazards at accelerators as compared with γ-ray emitters in routine maintenance activities, since the radionuclides are produced throughout the materials comprising accelerator components, with resultant self-shielding of most of the electrons compared with the less effective shielding of the more penetrating γ-rays. In contrast, β^+ (positron) emitters are included in this table due to the generation of the pairs of photons that result from annihilation of each positron with electrons in matter. Each annihilation releases two photons each of energy 0.511 MeV. Approximate thresholds and high-energy cross sections for production of these radionuclides by protons are also provided. Cross sections for production of radionuclides with mass numbers far below that of the target are usually less and sometimes much less than those listed here. In Table 7.4 approximate cross sections for their production at the high-energy limit and approximate thresholds are given for selected radionuclides.

A systematic way of addressing the great multiplicity of radionuclides produced in accelerator components by high-energy particles is highly desirable since it is often not practical to handle them all separately. Global properties of the distribution of radionuclides are found to be useful. Sullivan and Overton (1965) have treated this problem in an elegant manner generally followed here. The initial starting point is a modification of Equation 7.8 describing the dose rate as a function of irradiation and cooling times t_i and t_c:

$$\delta(t_i, t_c) = G\phi[1 - \exp(-\lambda t_i)]\exp(-\lambda t_c) \qquad (7.11)$$

where $\delta(t_i, t_c)$ is the absorbed dose rate; ϕ is the flux density; and G is a collection of many contributing factors including the production cross section, the energy of the beam, the types of secondary particles produced, the isotopic composition of the irradiated component, the geometric configuration, the energy of the γ-rays produced, and the attenuation coefficients for the γ-rays produced.

If the number of radionuclides produced by the irradiation that have decay constants in the interval between λ and $\lambda + d\lambda$ is represented by the differential dm, then the corresponding increment in absorbed dose rate due to them $d\delta(t_i, t_c)$ is given by

$$d\delta(t_i, t_c) = dmG\phi\left[1 - \exp(-\lambda t_i)\right]\exp(-\lambda t_c) \qquad (7.12)$$

If it is assumed that the value of G is independent of λ, or its dependence on λ is small compared to other factors, then making the somewhat surprisingly accurate assumption that, on average, the radionuclide production cross sections under consideration are independent of both the half-lives and the particle energies, one can integrate

$$\delta(t_i, t_c) = G\phi \int_{\lambda_0}^{\infty} d\lambda \frac{dm}{d\lambda}[1 - \exp(-\lambda t_i)]\exp(-\lambda t_c) \qquad (7.13)$$

Here λ_0 is the shortest decay constant, corresponding to the longest mean-life, to be considered. For a given material of atomic mass number A, a study of the chart of the nuclides reveals that there is a finite number of radionuclides within a reasonable domain of half-lives that can be produced in that material (i.e., with mass number $<A$). This was done by Barbier (1969). Figures 7.12 and 7.13 are plots of the number of radionuclides as a function of half-life $t_{1/2}$ that have half-lives less than $t_{1/2}$ for four choices of the upper limit on atomic mass number A of the target material. This corresponds to the distribution

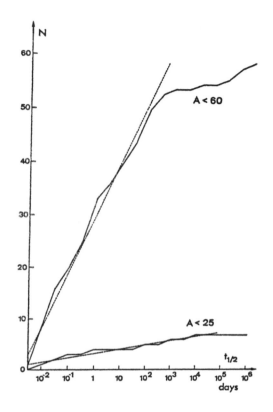

FIGURE 7.12

Total number of radionuclides having half-lives up to a given half-life $t_{1/2}$ as a function of $t_{1/2}$ for target mass numbers A less than those indicated. (Barbier, M: *Induced radioactivity*. 1969. Copyright Wiley-VCH Verlag GmbH & Co. KGaA. Reproduced with permission.)

half-lives of radionuclides that could be produced in a target of mass number A irradiated by high-energy hadrons.

Sullivan and Overton observed that the cumulative distributions are well described over the approximate domain $10^{-3} < t_{1/2} < 10^{3}$ days by the following functional form:

$$N(t_{1/2}) = a + b\ln(t_{1/2}) \tag{7.14}$$

where $N(t_{1/2})$ is the number of radionuclides with half-lives less than the value of $t_{1/2}$, and a and b are fitting parameters. Because of the one-to-one correspondence between values of $t_{1/2}$, τ, and λ in this *Sullivan-Overton approximation* one can just as well write

$$m(\lambda) = a + b\ln\lambda \tag{7.15}$$

where $m(\lambda)$ is the number of radionuclides with decay constants *greater* than λ for the material of concern. Thus,

$$\frac{dm(\lambda)}{d\lambda} = \frac{b}{\lambda} \tag{7.16}$$

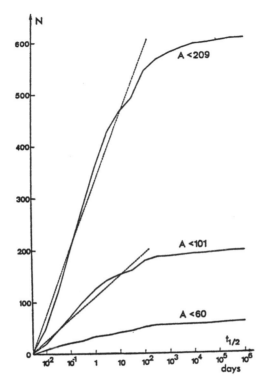

FIGURE 7.13
Total number of radionuclides having half-lives up to a given half-life $t_{1/2}$ as a function of $t_{1/2}$ for target mass numbers A less than those indicated. (Barbier, M: *Induced radioactivity*. 1969. Copyright Wiley-VCH Verlag GmbH & Co. KGaA. Reproduced with permission.)

Substituting into Equation 7.13, one gets

$$
\begin{aligned}
\delta(t_i,t_c) &= Gb\phi \int_{\lambda_0}^{\infty} \frac{d\lambda}{\lambda}[1-\exp(-\lambda t_i)]\exp(-\lambda t_c) \\
&= Gb\phi\left[\int_{\lambda_0}^{\infty} \frac{d\lambda}{\lambda}\exp(-\lambda t_c) - \int_{\lambda_0}^{\infty} \frac{d\lambda}{\lambda}\exp[-\lambda(t_i+t_c)]\right]
\end{aligned}
\tag{7.17}
$$

The changes of variables $\alpha = \lambda t_c$ (first term) and $\alpha' = \lambda(t_i+t_c)$ (second term) are helpful:

$$
\delta(t_i,t_c) = Gb\phi\left[\int_{\lambda_0 t_c}^{\infty} d\alpha \frac{e^{-\alpha}}{\alpha} - \int_{\lambda_0(t_i+t_c)}^{\infty} d\alpha' \frac{e^{-\alpha'}}{\alpha'}\right]
\tag{7.18}
$$

Recognizing that the integrands are identical and simplifying by rearranging the limits of integration,

$$
\delta(t_i,t_c) = Gb\phi \int_{\lambda_0 t_c}^{\lambda_0(t_i+t_c)} d\alpha \frac{e^{-\alpha}}{\alpha}
\tag{7.19}
$$

The integration results in a series expansion found in standard tables of integrals:

$$\int_{x_1}^{x_2} \frac{e^{ax}dx}{x} = \left[\ln x + \frac{ax}{1!} + \frac{a^2 x^2}{2 \times 2!} + \frac{a^3 x^3}{3 \times 3!} + \cdots \right]_{x_1}^{x_2} \tag{7.20}$$

Substituting,

$$\int_{\lambda_0 t_c}^{\lambda_0(t_i+t_c)} \frac{e^{-\alpha}d\alpha}{\alpha} = \left[\ln\alpha - \alpha + \frac{\alpha^2}{4} - \frac{\alpha^3}{18} + \cdots \right]_{\lambda_0 t_c}^{\lambda_0(t_i+t_c)} \tag{7.21}$$

Evaluating, one obtains

$$\delta(t_i,t_c) = Gb\phi\left[\ln\left(\frac{t_i+t_c}{t_c}\right) - \lambda_0 t_i + \cdots \right] \tag{7.22}$$

Since λ_0 approaches zero (corresponding to large mean-lives), the following is obtained:

$$\delta(t_i,t_c) \approx B\phi\ln\left(\frac{t_i+t_c}{t_c}\right) \tag{7.23}$$

where several constants are merged in the new parameter B, and this utilization of only the first term of the series expansions is considered to be sufficiently accurate.

7.4.2.1 Gollon's Rules of Thumb

Gollon (1976) has further elaborated on these principles in stating four very useful "rules of thumb" for high-energy hadron accelerators at which the extranuclear hadron cascade process produces the major fraction of the induced activity.

Rule 1: The dose rate from a point source is given by Equation 7.10.

Rule 2: In many materials, about 50% of the nuclear interactions produce a nuclide with a half-life longer than a few minutes. Further, about 50% of these have a half-life longer than 1 day. Thus, approximately 25% of the nuclear interactions (e.g., the "stars" discussed in Section 4.7.2) produce a radionuclide having a half-life exceeding approximately 1.0 day.

Rule 3: For most common shielding materials, the approximate dose rate dD/dt due to a constant irradiation is given by Equation 7.23. In using this equation, the geometry and material-dependent factor B can often be determined empirically, or estimated by using *Rule 2*, while ϕ is the incident flux density. This expression also works for intermediate-energy heavy ion beams, for example, at 86 MeV nucleon^{-1} (Tuyn et al. 1984).

Rule 4: In a hadronic cascade, each proton or neutron produces about four inelastic interactions for each gigaelectron volt (GeV) of energy.

Some examples can illustrate the use of these rules of thumb. As one illustration, in a short target of 1/10 of an interaction length, approximately 10% of an incident beam of 10^{11} protons s^{-1} will interact. One can assume that this has been occurring for several months (long enough to reach saturation production for many radionuclides) at this constant rate. Using *Rule 2* in conjunction with the previous rate, one determines that the decay rate after 1 day of the shutdown is 2.5×10^9 Bq (68 mCi). If each of these decays produces a 1.0 MeV γ-ray, then *Rule 1* gives an absorbed dose rate of 27 mrad h^{-1} ($\approx$0.27 mGy h^{-1}) at 1.0 m away.

Rule 3 can be used in such a calculation to predict the absorbed dose rate from a point source at some future time after beam shutdown. Furthermore, this rule is not restricted to "point" sources but can be used for larger ones, with suitable adjustments to the geometry factors. Sometimes one can estimate the product $B\phi$ or use a measurement of the exposure or absorbed dose rate early in a shutdown period to determine it empirically to further predict the "cooldown" for later times using Equation 7.23 as a tool for planning radiological work. *Rule 3* also clearly works for extended shields irradiated by secondary particles from a well-developed cascade.

Rule 4 can be used to crudely estimate the activation of a beam absorber by incident high-energy particles when it is coupled with *Rule 2*. For example, a beam of 10^{12} 400 GeV protons s^{-1} ($=$0.16 μA or 64 kW) produces a total of $4 \times 400 \times 10^{12}$ stars s^{-1} in a beam absorber. If 25% of these produce a radionuclide with a half-life >1.0 day (*Rule 2*), then the total amount of the moderately long-lived radioactivity (at saturation) is

$$\frac{(0.25 \text{ atoms/star})(1.6 \times 10^{15} \text{ stars/sec})}{3.7 \times 10^{10} \text{ sec}^{-1} \text{ Ci}^{-1}} = 10.8 \text{ kCi} \qquad (7.24)$$

At a sufficiently large distance (say 10 m), *Rule 1* could be used to calculate an absorbed dose rate from a point source assuming all decays are 1.0 MeV γ-rays:

$$\frac{dD}{dt} 0.4(1.0 \text{ MeV}) \left[\frac{1.08 \times 10^4 \text{ Ci}}{10^2 \text{ m}^2} \right] = 43 \text{ rads h}^{-1} \qquad (7.25)$$

7.4.2.2 Barbier Danger Parameter

A valuable quantity used to quantify the absorbed dose rate dD/dt at the surface of a thick target is the *danger parameter* D_p as developed by Barbier (1969) for a thick object irradiated by a beam having a uniform flux density ϕ. If this source of radioactivity subtends solid angle Ω at the point of concern, then

$$\frac{dD}{dt} = \frac{\Omega}{4\pi} \phi D_p \qquad (7.26)$$

At contact with a semi-infinite slab of uniformly irradiated material, the fractional solid angle factor ($\Omega/4\pi$) subtended by the wall has the intuitively obvious value of 1/2. D_p has the physical interpretation as the absorbed dose rate found inside a cavity of arbitrary form embedded in an infinite volume of a material that has been uniformly irradiated by a unit flux density (1.0 particle s^{-1} cm^{-2}). Figures 7.14 through 7.18 give representative examples of plots of D_p in units of mrad h^{-1} per incident particle s^{-1} cm^{-2} (1.0 mrad h$^{-1} = 10$ μGy h^{-1}) for several substances. These curves thus can be used to predict cooling of various components around accelerators. In these figures the label "D" denotes D_p.

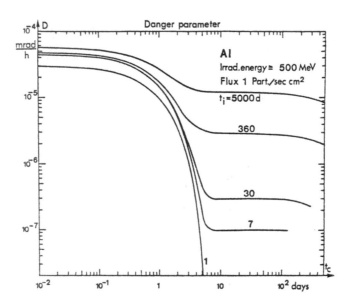

FIGURE 7.14

Values of the Barbier danger parameter D_p for aluminum irradiated by protons having energy greater than 500 MeV. (Barbier, M: *Induced radioactivity.* 1969. Copyright Wiley-VCH Verlag GmbH & Co. KGaA. Reproduced with permission.)

Gollon (1976) has also provided "cooling curves" for iron struck by high-energy protons. These are given in Figure 7.19 and include both calculations by Armstrong and Alsmiller (1969) and empirical measurements at the Brookhaven National Laboratory AGS, the former Fermilab Main Ring Accelerator (a 500 GeV proton synchrotron), and the former Fermilab Neutrino Experimental Area target station.

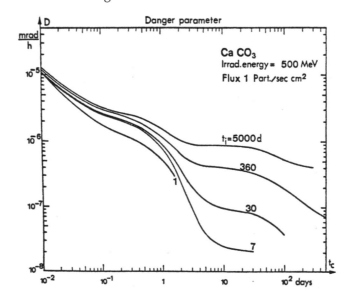

FIGURE 7.15

Values of the Barbier danger parameter D_p for $CaCO_3$ irradiated by protons having energy greater than 500 MeV. (Barbier, M: *Induced radioactivity.* 1969. Copyright Wiley-VCH Verlag GmbH & Co. KGaA. Reproduced with permission.)

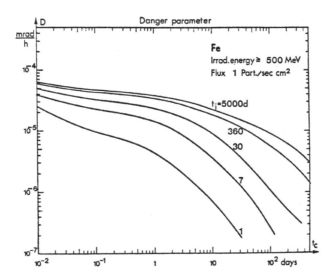

FIGURE 7.16
Values of the Barbier danger parameter D_p for iron irradiated by protons having energy greater than 500 MeV. (Barbier, M: *Induced radioactivity*. 1969. Copyright Wiley-VCH Verlag GmbH & Co. KGaA. Reproduced with permission.)

It is often possible to relate the flux density of high-energy hadrons (i.e., those with energies above the leveling off) to the star density S calculated from such Monte Carlo calculations through the relationship

$$\phi(\vec{r}) = \frac{\lambda}{\rho}\frac{dS(\vec{r})}{dt} \tag{7.27}$$

where $\phi(\vec{r})$, the flux density (cm^{-2} s^{-1}) at position vector $\vec{r}$, is related to the rate of star density production $dS(\vec{r})/dt$ (stars cm^{-3} s^{-1}) at the same location, the rate of delivery of "star"

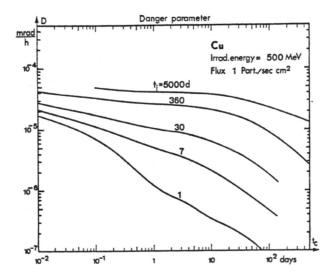

FIGURE 7.17
Values of the Barbier danger parameter D_p for copper irradiated by protons having energy greater than 500 MeV. (Barbier, M: *Induced radioactivity*. 1969. Copyright Wiley-VCH Verlag GmbH & Co. KGaA. Reproduced with permission.)

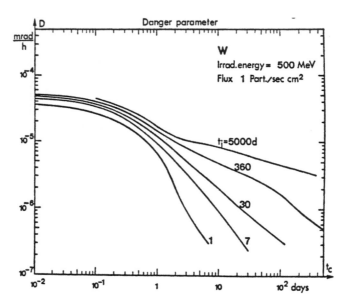

FIGURE 7.18

Values of the Barbier danger parameter D_p for tungsten irradiated by protons having energy greater than 500 MeV. (Barbier, M: *Induced radioactivity*. 1969. Copyright Wiley-VCH Verlag GmbH & Co. KGaA. Reproduced with permission.)

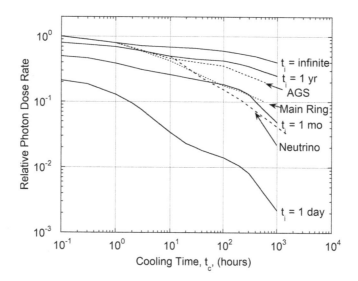

FIGURE 7.19

Cooling curves for various irradiation times for iron struck by high energy protons as calculated by Armstrong and Alsmiller (1969). Also shown are the results of measurements. The one labeled "Main Ring," is the measured average cooling curve for the former Fermilab Main Ring synchrotron after its initial three years of operation at proton energies of 200 or 400 GeV. The curve labeled "Neutrino" is for the now partially "mothballed" Neutrino Experimental Area at Fermilab after eight months of operation with 400 GeV protons. The curve labeled "AGS" is for an extraction splitter in use for many years at the Brookhaven Alternating Gradient Synchrotron at proton energies up to 28 GeV. (Adapted from Gollon, P. J. 1976. *Production of radioactivity by particle accelerators*. Fermi National Accelerator Laboratory: Fermilab Report TM-609. Batavia, IL.)

fluence discussed in Section 4.7.2; ρ (g cm^{-3}) is the density; and λ (g cm^{-2}) is the interaction length. (Here once again one needs to be mindful of the units of measurement being used.) The value of $\phi(\vec{r})$ so determined could, in principle, be substituted into Equation 7.26 for calculating absorbed dose rate due to residual activity using the Barbier danger parameter D_p, if one makes suitable adjustments in the solid angle. These considerations are implicit in the following equation:

$$\frac{dD(\vec{r})}{dt} = \frac{\Omega}{4\pi} \frac{dS(\vec{r})}{dt} \omega(t_i, t_c) \tag{7.28}$$

where the "ω-factor" $\omega(t_i, t_c)$ related to the Barbier danger parameter D_p has been introduced. The choice of the low-energy cutoff used in the Monte Carlo calculation can be important. For a low-energy cutoff of 300 MeV/c (equivalent to a kinetic energy of about 47 MeV for nucleons) using the Monte Carlo code CASIM, Gollon (1976) determined

$$\omega\,(\infty,0) = 9 \times 10^{-6}\,\text{rad h}^{-1}/(\text{star cm}^{-3}\,\text{s}^{-1}) = 9 \times 10^{-8}\,\text{Gy h}^{-1}/(\text{star cm}^{-3}\,\text{s}^{-1}) \tag{7.29}$$

(infinite irradiation time, zero cooling time), and

$$\omega\,(30\,\text{d},1\,\text{d}) = 2.5 \times 10^{-6}\,\text{rad h}^{-1}/(\text{star cm}^{-3}\text{s}^{-1}) = 2.5 \times 10^{-8}\,\text{Gy h}^{-1}/(\text{star cm}^{-3}\text{s}^{-1}) \tag{7.30}$$

(30 days irradiation time, 1.0 day cooling time)

Equation 7.29 is useful for installations that have been in steady operation for periods of years or even decades, while Equation 7.30 can be used for facilities that operate more intermittently or with frequent changes of configuration. The calculations leading to Equations 7.29 and 7.30 were for an "equilibrium" spectrum in a well-developed hadronic cascade. Estimates of ω-factor values for other choices of values of t_i and t_c for irradiated iron can be made by scaling results obtained by Armstrong and Alsmiller (1969) and Gabriel and Santoro (1973) as done, again for irradiated iron, by Cossairt (1998). The choice of the pair $t_i = 30$ d, $t_c = 1.0$ d has become a generally accepted standard condition for further calculations.

Stevenson (2001) has provided similar ω-factors for a variety of materials common at accelerators as provided in Table 7.5.

TABLE 7.5

ω-Factors for Converting Star Density Production Rate to Residual Dose Rate for $t_i = 30$ d, $t_c = 1.0$ d in Various Materials

Material	ω-Factor (Gy h^{-1})/(star cm^{-3}s^{-1})
Iron	1.0×10^{-8}
Copper	1.0×10^{-8}
Stainless steel	1.3×10^{-8}
Aluminum	2.0×10^{-9}
Lead	1.5×10^{-8}
Tungsten	1.1×10^{-8}
Normal concrete	3.0×10^{-9}
Marble	6.0×10^{-10}

Source: Adapted from Stevenson, G. R. 2001. *Radiation Protection Dosimetry* 96:373–380.

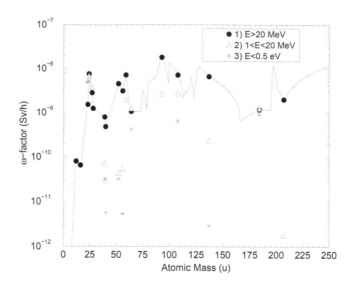

FIGURE 7.20

ω-factor dependence on mass of a target nucleus for three energy groups with $t_i = 30$ d, $t_c = 1.0$ d determined using the MARS code. Normalization is per star cm^{-3}s^{-1} for $E > 20$ MeV, and per neutron cm^{-2}s^{-1} for the lower energy groups. The symbols represent the result using FLUKA of a previous study (Huhtinen 2003) and the curve is an interpolation of the results of that study and those of an earlier one for the high energy group (Huhtinen 1998). (Reproduced from Mokhov, N. V. et al. 2006. *Residual activation of thin accelerator components.* Fermi National Accelerator Laboratory: Fermilab Report FERMILAB FN-0788-AD. Batavia, IL.)

Such ω-factors for residual activation of thin objects, those with linear dimensions less than a fraction of a nuclear interaction length, have been obtained by Mokhov et al. (2006) using the MARS code. As explained by these authors, the lower-energy (or momentum) cutoff is especially important because of the importance of contributions from spallation reactions in the 20–50 MeV domain. Detailed calculations for three different energy groups are provided in Figure 7.20. Huhtinen has performed related calculations (1998, 2003).

Given the obvious variability, these results should be used with some degree of caution. They can readily be used to predict the relative "cooling" rates of various components around accelerators with a fair degree of accuracy. Their use in the precise prediction of absolute absorbed dose rates near activated accelerator components requires additional care. To do this, the geometric configuration should be simple and well defined, the flux density of thermal neutrons should be a small component of the prompt radiation field, and the activation of other materials in proximity such as the enclosure walls should be considered. Cracks through the shielding materials can sometimes result in higher dose rates that are difficult to model. The interactions of thermal neutrons in concrete shielding can make a significant contribution to the dose equivalent rate. This phenomenon has been discussed by Armstrong and Barish (1969) and by Cossairt (1996) and is summarized in Section 7.4.3. A good example of a detailed treatment of a specific situation incorporating the neutron energy spectrum, large gaps between shielding components, and comparisons of calculations with measurements is that of Rakhno et al. (2001).

Gollon derived a simple relationship between dose rates involving cooling times different from "standard" ones for which values of D_p and ω are available. As before, the dose rate after irradiation time t_i and cooldown time t_c is

$$\delta(t_i, t_c) = \sum_{\mu} A_{\mu}[1 - \exp(-\lambda_{\mu}t_i)]\exp(-\lambda_{\mu}t_c) \tag{7.31}$$

where the summation over index μ includes all relevant radionuclides with the product of flux density and geometry factors being absorbed (and allowed to vary with radionuclide) in the quantity A_{μ}. Rearranging, Gollon obtained

$$\delta(t_i, t_c) = \sum_{\mu} A_{\mu}[\exp\{-\lambda_{\mu}t_c\} - \exp\{-\lambda_{\mu}(t_i + t_c)\}] = \delta(\infty, t_c) - \delta(\infty, t_i + t_c) \tag{7.32}$$

Thus, the infinite irradiation curve can be used to determine any other combination of the times t_i and t_c. In fact, this formula may be used also with empirical results, such as, for example, radiation survey data, in order to predict future radiological conditions.

A reliable method for connecting the production of "stars" in material (i.e., as calculated by a Monte Carlo code) to the production of atoms of some radionuclide is by the ratios of cross sections. Thus, at some point in space $\vec{r}$ the rate of production of atoms cm^{-3} $n_i(\vec{r})$ of some radionuclide i is approximately given by

$$\frac{dn_i(\vec{r})}{dt} \approx \frac{\sigma_i}{\sigma_{in}}\frac{dS(\vec{r})}{dt} = \frac{\Sigma_i}{\Sigma_{in}}\frac{dS(\vec{r})}{dt} \tag{7.33}$$

where one essentially scales the star density production rate (e.g., stars cm^{-3} s^{-1}) by the ratio of the production (reaction) cross section for the nuclide of interest σ_i to the total inelastic cross section σ_{in} or, equivalently, by the ratio of the macroscopic cross sections (Σ_i/Σ_{in}). The phenomena will obey the usual activation equation. The reason this is *approximate* is due to the familiar concerns about constancy of cross sections with energy, the lack of perfect "matching" of effective reaction thresholds, etc.

7.4.3 Uniform Irradiation of Walls of an Accelerator Enclosure

Somewhat special considerations may apply to the concrete shielding surrounding accelerators. As was seen before in Table 6.2, ordinary concrete typically contains a partial density of about 0.04 g cm^{-3} of sodium. This "typical" value varies a great deal due to the variety of minerals that might be present in concrete that might be available in a given geographical area. The significance of this seemingly small additive is that the sole naturally occurring isotope is ^{23}Na, thus facilitating the thermal neutron capture reaction ^{23}Na(n,γ)^{24}Na. This reaction has a significant thermal neutron capture cross section, 535 mb (NNDC 2018). It has been reported that that the average thermal neutron flux density ϕ_{th} in a concrete room is approximately given as follows (Patterson and Wallace 1958; NCRP 2003):

$$\phi_{th} = \frac{1.25N_f}{S}(cm^{-2}s^{-1}) \tag{7.34}$$

where N_f is the total number rate of production of fast neutrons in the enclosure, and S is the inside surface area of the enclosure (cm^2). Thus, a substantial flux density of thermal neutrons can be present in an accelerator room, and this flux can produce a significant amount of ^{24}Na with its 15-hour half-life. In view of Equation 7.10, the pair of relatively

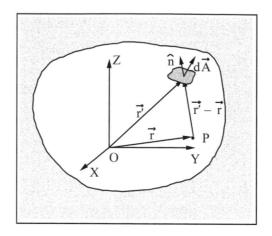

FIGURE 7.21
Geometry for deriving the relationship between a surface of uniform emission and the flux density at any point within it. (Reprinted from Cossairt, J. D. 1996. *Health Physics.* 71:315–319. Rights to this figure are reserved by Fermi Research Alliance, LLC as manager and operator of the Fermi National Accelerator Laboratory.)

energetic photons emitted in each decay of ^{24}Na (1.37 and 2.75 MeV) enhances the residual dose rate. Furthermore, while the dose due to activated components falls off radially with distance, if absorption by the air is not significant, the photon flux density due to activation of the walls of an empty room uniformly irradiated by such thermal neutrons is a constant independent of position within the enclosure. Thus, the absorbed dose rate due to the walls anywhere inside the enclosure will be equal to the absorbed dose rate at the wall. This is true for the interior of all mathematically "well-behaved" closed surfaces (Barbier 1969; Cossairt 1996). This fact can readily be demonstrated by analogy to the Gauss Law in electrostatics as follows by examining the situation in Figure 7.21. The Gauss law of electrostatics has been treated elsewhere, for example, by Konopinski (1981) and Jackson (1998).

Consider a simple, closed surface that emits an omnidirectional flux density of some particle ϕ_o (e.g., particles cm^{-2}s^{-1}) that is constant over the surface. One wants to calculate the flux density at some arbitrary point in space P within the surface. P is located at radius vector $\vec{r}$. Consider further the contributions of the particles emitted by some elemental area $d\vec{A}$ at P, where $d\vec{A}$ is perpendicular to the surface at coordinate vector $\vec{r}$. The solid angle subtended at P by $d\vec{A}$ is

$$d\Omega = \frac{d\vec{A}\cdot\hat{n}}{|\vec{r}'-\vec{r}|^2} \tag{7.35}$$

where the unit vector $\hat{n}$ is given by

$$\hat{n} = \frac{\vec{r}'-\vec{r}}{|\vec{r}'-\vec{r}|} \tag{7.36}$$

But the increment of flux density at point P due to elemental area $d\vec{A}$ is given by

$$d\phi = \frac{\phi_o}{4\pi} \frac{d\vec{A} \cdot \hat{n}}{\left| \vec{r}' - \vec{r} \right|^2}$$

Thus, setting up an integral over solid angle Ω,

$$d\phi = \frac{\phi_o}{4\pi} d\Omega \quad \text{and} \quad \int_{4\pi} \frac{\phi_o}{4\pi} d\Omega = \phi_o \tag{7.37}$$

In some configurations it is worthwhile despite the additional costs to minimize the amount of sodium in the concrete ingredients in order to reduce exposures to individuals conducting maintenance on the accelerator. In fact, the phenomena described has been noticed at accelerators and sometimes leads to "disappointment" in how little γ-ray exposure rates are reduced when activated accelerator components are removed from enclosures with equally activated walls. For example, Armstrong and Barish (1969) have calculated residual dose rates inside a cylindrical accelerator tunnel due to both the magnets and the concrete walls for 3.0 GeV protons incident on iron. The study was performed to assess the benefits of employing concrete in shielding with reduced sodium content. In addition to the production of ^{24}Na by means of the cited thermal neutron capture reaction, these authors also included some other reactions due to higher-energy neutrons, such as spallation, that are capable of also producing ^{24}Na from common ingredients of concrete. The results are shown in Figure 7.22 for the surface at the tunnel wall.

Armstrong and Barish concluded the reduction in the sodium content in the concrete could be beneficial. However, in practice this adjustment in the recipe for shielding concrete

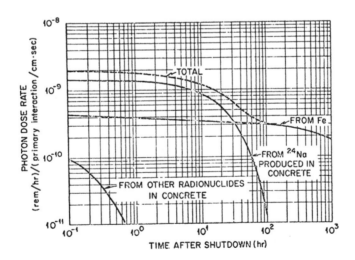

FIGURE 7.22
Photon dose rate at surface of tunnel wall after infinite irradiation time for concrete containing 1.0% sodium by weight by 3 GeV protons incident on an iron cylinder. The *y*-axis represents values for a specific geometrical configuration with resultant actual values of photon dose rate for that condition. However, these results can, reasonably, be used in other situations to obtain *relative* contributions of dose from the ^{24}Na in the walls to that due to the iron directly struck by the beam. (Armstrong T. W., and J. Barish. 1969. *Nuclear Science and Engineering* 38:265–270. Copyright © American Nuclear Society, http://www.ans.org/, reprinted by permission of Taylor & Francis Ltd, http://www.tandfonline.com on behalf of American Nuclear Society.)

is not generally done at large installations because of the probable large increase of cost in purchasing such "special" concrete.

The discussion of the production of radioactivity continues in Chapter 8 with specific emphasis on environmental radiation protection.

PROBLEMS

1. A 1.0 mA beam of 30 MeV electrons is absorbed by an aluminum target. Calculate the saturation activities of all major radionuclides produced in the target assuming sufficient kinetic energy to avoid threshold effects such as those illustrated in Figure 3.5. Assuming no self-absorption and an infinitely long irradiation period, what will be the absorbed dose rate at a distance of 2.0 m away immediately after beam shutdown, and then 1.0 hour later? The target can be assumed to be a point source for this estimate.

2. A copper beam stop has been bombarded with high-energy hadrons for 30 days and exhibits a dose rate of 100 mrem hr^{-1} at 1.0 m away 1.0 day after the beam is turned off. Maintenance work needs to be scheduled in the vicinity within the next 6 months. Using both Gollon's Rule 3 and the Barbier danger parameter curves, predict the cooling curve, and determine when the dose rate is less than a 20 mrem hr^{-1} maintenance work criteria. Make a table of dose rate versus cooling time in days for both methods. How well do the two methods agree? (Hint: Use the initial value of the dose rate to scale values of D_P)

3. A 100 GeV beam (10^{12} protons s^{-1}) strikes the center of a large solid *iron* cylinder 30 cm in radius for 30 days. Use the star density curves from Appendix and the "ω" factors calculated by Gollon to estimate the residual dose rate after 1.0 day cooldown at contact with the side of the cylinder in the "hottest" spot. Using Gollon's Rule 3, how long must the repair crew wait to service this item in a contact radiation field of absorbed dose rate <10 rad hr^{-1}?

4. A copper target is bombarded with high-energy protons such that 10 stars per incident proton are produced. If the incident beam is 10^{11} protons s^{-1}, what is the specific activity (average) of ^{54}Mn that is produced after 2 years of operation? ^{54}Mn has a high-energy spallation production cross section of about 20 mb for protons incident on Cu. The target is a cylinder, 10 cm radius by 15 cm long. The half-life of ^{54}Mn is 312.2 days. Express the answer in both Bq cm^{-3} and Ci cm^{-3}. (Hint: This problem is best handled if the calculation is done at saturation and then corrected for the noninfinite irradiation time. Also, one needs to use the inelastic cross section given, for example, in Chapter 4.)

8

Induced Radioactivity in Environmental Media

8.1 Introduction

Chapter 7 provided the primary tools needed to address the subject of induced radioactivity. In this chapter, the discussion of induced radioactivity at accelerators proceeds to address its production and propagation in environmental media such as air, soil, rock, and water. Aspects pertinent to both occupational radiation safety and environmental protection are covered. Also included is introductory material on meteorology and hydrogeology related to the propagation of this radioactivity in environmental media.

8.2 Airborne Radioactivity

8.2.1 Production

Thomas and Stevenson (1988) and Swanson and Thomas (1990) have comprehensively reviewed the production of radioactivity in air. The principal source of radioactivity in air at accelerators results from the interaction of primary and secondary particles directly with the constituent target nuclei in the air in accelerator enclosures. Activated dust and gaseous emission from activated liquids are usually of much less importance. Table 8.1 gives the abundances and number densities of atoms N_j of the most common *stable* nuclides found in the atmosphere using the elemental volumetric abundances of the U.S. Standard Atmosphere (Lide 2000) with the isotopic abundances of National Nuclear Data Center (NNDC 2018).

Patterson and Thomas (1973) have expanded Equation 7.8, the activation equation, to obtain the total specific activity S (Bq cm^{-3}) of an enclosed, sealed volume of radioactive air:

$$
S = C \sum_i \left\{ \sum_j N_j \bar{\sigma}_{ij\gamma} \phi_\gamma + \sum_j N_j \bar{\sigma}_{ijTH} \phi_{TH} \right.
$$

$$
\left. + \sum_j N_j \bar{\sigma}_{ijHE} \phi_{HE} \right\} \{1 - \exp(-\lambda_i t_{irrad})\} \exp(-\lambda_i t_c)
$$

(8.1)

where ϕ_γ, ϕ_{TH}, and ϕ_{HE} represent the average photon, thermal neutron, and high-energy particle flux densities, respectively. To avoid confusion with the summation over the i

TABLE 8.1

Atomic Number Densities of the Most Prominent Elements and Stable Nuclides in the Atmosphere at Sea Level at Standard Temperature and Pressure (STP, 273.15°K, 760 mm Hg)

Element or Isotope	Percentage (%) by Volume	Percentage (%) Isotopic Abundance	N_j (Atoms cm^{-3})
Nitrogen (N$_2$)	78.084		4.1959×10^{19}
^{14}N		99.636	4.1806×10^{19}
^{15}N		0.364	1.5273×10^{17}
Oxygen (O$_2$)	20.9476		1.1256×10^{19}
^{16}O		99.757	1.1229×10^{19}
^{17}O		0.038	4.2774×10^{15}
^{18}O		0.205	2.3075×10^{16}
Argon (Ar)	0.934		2.5094×10^{17}
^{36}Ar		0.3336	8.3715×10^{14}
^{38}Ar		0.0629	1.5784×10^{14}
^{40}Ar		99.6035	2.4995×10^{17}

Source: Adapted from Lide, D. R. ed. 2000. *CRC handbook of chemistry and physics.* Boca Raton, FL: CRC Press/Taylor and Francis Group, with isotopic abundances from NNDC. 2018. National Nuclear Data Center at Brookhaven National Laboratory. http://www.nndc.bnl.gov/ (accessed October 17, 2018).

produced radionuclides, in this equation t_{irrad} is the irradiation time, while t_c represents the decay time. The $\bar{\sigma}_{ijk}$ values are the corresponding cross sections averaged with the flux density over energy:

$$\bar{\sigma}_{ijk} = \left[\int_{E_{\min}}^{E_{\max}} dE \sigma_{ijk}(E)\phi_k(E) \right] \left[\int_{E_{\min}}^{E_{\max}} dE \phi_k(E) \right]^{-1} \tag{8.2}$$

where the limits of integration correspond to the three broad phenomenological energy domains in the summation. The constant C is the conversion to specific activity, equal to unity for activity in Bq cm^{-3} if the units of length implicit in the quantities in Equation 8.1 are in centimeters. The outer sum over index i includes all possible radionuclides produced, and the sum over the index j is over the parent nuclei found in air. The flux densities are, without further information, the average over some relevant spatial volume.

Table 8.2 lists the radionuclides commonly produced from the principle constituents in air along with the reaction mechanisms associated with their production and an estimate of the average production cross section.

The large cross sections for (n_{thermal}, γ) and (n_{thermal}, p) reactions are for captures of neutrons of thermal energies ($<E_n> \approx 0.025$ eV), while the remaining cross sections are generally the saturation cross sections found in the region above approximately a few 10s of MeV. The photon-induced reactions (γ, n) are present at virtually all accelerators and at most energies. The corresponding cross sections will, of course, be energy dependent, especially at energies just above the reaction thresholds. The values tabulated here can be said to be "representative" values.

TABLE 8.2

Radionuclides with Half-Lives Greater than 1.0 Minute That Can Be Produced in Air at Accelerators

Radionuclide	Half-Life	Emission	Parent Element	Production Mechanism	High-Energy Production Cross Section (mb)
^{3}H	12.32 years	β^-	N	Spallation	30
			O	Spallation	30
^{7}Be	53.22 days	γ, e$^-$ capture	N	Spallation	10
			O	Spallation	5.0
			Ar	Spallation	0.6
^{11}C	20.36 minutes	β^+	N	Spallation	10
			O	Spallation	5
			Ar	Spallation	0.7
^{14}C	5700 years	β^-	N	$(n_{thermal}, p)$	1640
^{13}N	9.965 minutes	β^+	N	Spallation	10
			N	(γ, n)	10
			O	Spallation	9
			Ar	Spallation	0.8
^{14}O	1.177 minutes	β^+, γ	O	Spallation	1.0
			Ar	Spallation	0.06
^{15}O	2.04 minutes	β^+	O	Spallation	40
			O	(γ, n)	10
			Ar	Spallation	
^{18}F	1.830 hours	β^+	Ar	Spallation	6.0
^{24}Ne	3.38 minutes	β^-, γ	Ar	Spallation	0.12
^{22}Na	2.603 years	β^+, γ	Ar	Spallation	10
^{24}Na	15.00 hours	β^-	Ar	Spallation	7.0
^{27}Mg	9.458 minutes	β^-, γ	Ar	Spallation	2.5
^{28}Mg	20.92 hours	β^-, γ	Ar	Spallation	0.4
^{28}Al	2.245 minutes	β^-, γ	Ar	Spallation	13
^{29}Al	6.56 minutes	β^-, γ	Ar	Spallation	4.0
^{31}Si	2.623 hours	β^-, γ	Ar	Spallation	6.0
^{30}P	2.498 minutes	β^+, γ	Ar	Spallation	4.4
^{32}P	14.27 days	β^-	Ar	Spallation	25
^{33}P	25.35 days	β^-	Ar	Spallation	9
^{35}S	87.51 days	β^-	Ar	Spallation	23
^{34m}Cl	32.0 minutes	β^-, γ	Ar	Spallation	0.6
^{38}Cl	37.24 minutes	β^-, γ	Ar	(γ, pn)	4.0
^{39}Cl	56.2 minutes	β^-, γ	Ar	(γ, p)	7.0
^{41}Ar	1.827 hours	β^-, γ	Ar	$(n_{thermal}, \gamma)$	660

Source: Adapted from Swanson and Thomas (1990), with values of half-lives from NNDC. 2018. National Nuclear Data Center at Brookhaven National Laboratory. http://www.nndc.bnl.gov/ (accessed October 17, 2018).

Note: In this table the letter "m" following a mass number (e.g., ^{34m}Cl) indicates that this entry is for an isomeric nuclear state of this radionuclide.

8.2.2 Accounting for Ventilation

Adjustments for the presence of ventilation can be quite conveniently made for a given radionuclide by using an effective decay constant λ' that includes the physical decay constant λ along with a ventilation term r:

$$\lambda' = \lambda + r, \quad \text{with} \quad r = \frac{F}{V} \tag{8.3}$$

where F is the ventilation flow rate in air volume per unit time, and V is the enclosure volume.

Thus, r is the number of air changes per unit time. The applicable differential equation, an extension of Equation 7.4 with ventilation explicitly included, is

$$\frac{dn'}{dt} = -\lambda n'(t) - rn'(t) + N\sigma\phi = -\lambda' n'(t) + N\sigma\phi \tag{8.4}$$

The middle term on the right-hand side adds in the removal of the activity by means of the air leaving the room, where n' is the atom number density of the radionuclide being considered, with ventilation present. After an irradiation time t_i with no initial activation, the solution, analogous to Equation 7.5, is

$$n'(t_i) = \frac{N\sigma\phi}{\lambda + r}\{1 - \exp[-(\lambda + r)t_i]\} \tag{8.5}$$

Thus, the specific activity *including* ventilation mixing $a'(t_i)$ is given by

$$a'(t_i) = \lambda n'(t_i) = \frac{\lambda N\sigma\phi}{\lambda + r}\{1 - \exp[-(\lambda + r)t_i]\} \tag{8.6}$$

But $N\sigma\phi$ is just the saturation concentration a_{sat} after an infinitely long irradiation period as in Equation 7.7. Hence, including ventilation mixing, the saturation concentration a'_{sat} is

$$a'_{sat} = \frac{\lambda\, a_{sat}}{\lambda + r} \tag{8.7}$$

Particle beams at low-energy accelerators must be contained in continuous vacuum systems to avoid losing the beam to ionization loss and multiple Coulomb scattering. This serves to greatly minimize the activation of air at such facilities. In contrast, at high-energy accelerators, it is quite common to have air gaps at certain interface points to accommodate devices associated with beam targetry or beamline diagnostics that render continuous vacuum impractical, and in some situations impossible. Such air gaps are only found in external beam lines and possibly in linear accelerators. The beam in a circular accelerator or storage ring is, of necessity, contained in continuous vacuum. Such air gaps and their necessary vacuum "windows," if traversed numerous times by the orbiting beam particles, would result in an unacceptable rate of beam loss by interactions in the air and the windows. At higher-energy accelerators the large multiplicity of secondary particles produced as a part of both electromagnetic and hadronic cascade processes will produce airborne radioactivity external to the beamline vacuum.

If the accelerator enclosures were completely sealed, there would be no releases to the outside world, and the hazard of these airborne radionuclides would be entirely restricted to those who might have to enter the enclosures. This would, however, allow the longer-lived radionuclides to build up. Also, ventilation is generally needed to provide cooling of components and fresh breathing air for workers. Typically, the average residence time of air in accelerator enclosures is limited to a range of between approximately 30 minutes and not much longer than an hour or two. Thus, in equilibrium the airborne radionuclides in the accelerator environment will have half-lives only up to the order of approximately 1.0 hour. The residence time of the air in conjunction with the cross sections determines the radionuclides of importance.

At some facilities releases of airborne radionuclides to the outdoors are minimized by greatly restricting the release rate of air during accelerator operations. Subsequent to cessation of operations, when personnel enter such enclosures, the ventilation rate must then be increased to levels consistent with good industrial hygiene practice and perhaps also to purge out the airborne radionuclides.

Airborne radioactivity releases can also be reduced by requiring long pathways from the point of production to the ventilation "stacks," allowing time for radioactive decay to occur in transit, and by minimizing air gaps in the beam.

8.2.3 Propagation of Airborne Radionuclides in the Environment

A consideration concerning airborne radioactivity is that associated with the dose delivered to members of the general public when radionuclides are released to the atmosphere external to the accelerator enclosure. The U.S. Environmental Protection Agency (USEPA) has placed an annual limit of 10 mrem on dose equivalent, as defined by the 1973 System, to members of the general public due to the operations of U.S. Department of Energy (DOE) facilities, comparable to limits applied to other facilities by U.S. federal and state regulations. U.S. regulations are used here as examples. Other nations with particle accelerators commonly have similar, but certainly not identical, regulations. The USEPA has also specified the methods for measuring such releases (CFR 1989). An annual dose equivalent of such small value is usually difficult or impossible to measure. Thus, the standard practice is to measure the activity released and then use calculational models to estimate the maximum dose equivalent that actual members of the public could receive. The regulations prescribe the specific computer codes that must be used to perform these calculations by utilizing a *Gaussian plume model* that combines input data on the release of radioactivity with meteorological information. A short synopsis of such a plume model is given here with detailed presentations deferred to the references.

8.2.3.1 Meteorological Considerations

Once radioactive materials of any type are released to the atmosphere, meteorological considerations become of premier importance in governing their dispersal. The treatises edited by Slade (1955 and 1968), notably the 1968 version, remain of high value. Cember and Johnson (2009) present an eloquent discussion of these considerations. The discussion here is based on all three of these references. The basic meteorology is illustrated in Figure 8.1.

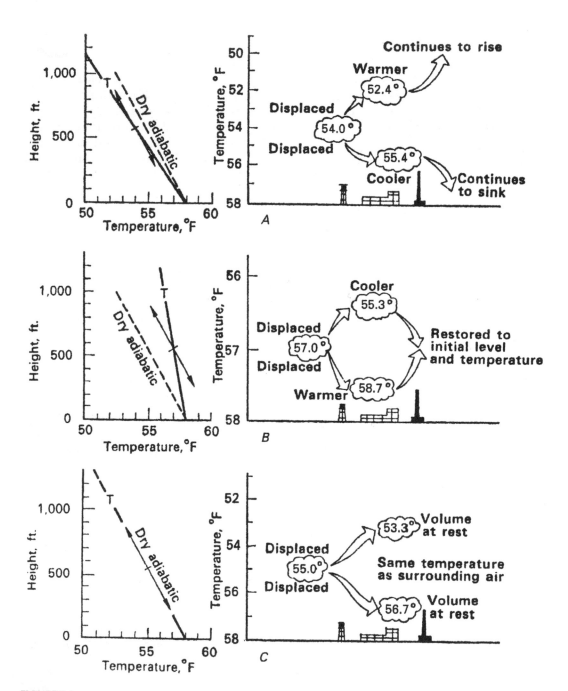

FIGURE 8.1
Effect of atmospheric temperature gradient or lapse rate on a displaced volume of air for various conditions: (a) unstable lapse rate, (b) stable lapse rate, and (c) neutral lapse rate. This figure uses units of measure that remain in common use in the United States. Distances in meters can be determined from those given in feet by multiplying by the factor of 0.3048. Likewise, Celsius (°C) temperatures T_C can be obtained from Fahrenheit temperatures (°F) T_F by means of $T_C = 5(T_F-32)/9$. (Reproduced from Slade, D. A. ed. 1955. *Meteorology and atomic energy*. U.S. Atomic Energy Commission. Washington, DC.)

The concept of *lapse rate* continues to be an important one in the field of meteorology that is directly relevant to environmental radiation protection. Lapse rate is the dependence of air temperature on altitude, a type of gradient. This concept is important because it is connected directly with how the discharged radionuclides propagate from the point of release to their concentrations in the air at distant locations. Table 8.3 gives a summary discussion of atmospheric stability classes related to the lapse rate.

At large particle accelerators, atmospheric gases are generally, but not exclusively, released through fixed ventilation points. Commonly these ventilation points are called *ventilation stacks* or more simply just *stacks* and may or may not be very tall compared with those associated with other types of radiological and even industrial facilities. While diffusion in the atmosphere occurs, the effects of turbulence are usually dominant. Thus, the mathematical models used to describe the phenomena explicitly take the atmospheric stability states into account rather than employ classical diffusion theory.

The most common model in use, indeed that forms the underlying basis of the aforementioned methodologies mandated by regulatory bodies, is that of the Gaussian plume model—one based on straight-line trajectories. Figure 8.2 illustrates the geometrical situation.

In the current discussion it is assumed that the release of radionuclides to the environment is continuous and that the prevailing meteorological conditions remain constant. These conditions are not generally realized for lengthy periods of time corresponding to operational cycles of particle accelerators. Thus, computer models not further discussed here have been developed to perform calculations of the effects of radionuclide releases over long periods.

The straight-line trajectory model in use is commonly referred to as the *Pasquill-Gifford equation* (Slade 1968; Cember and Johnson 2009), adapted here for two different conditions. Some coordinates need to be defined:

- $C(x, y, z)$ (Bq m^{-3} or Ci m^3) is the concentration at point (x, y, z) as defined in Figure 8.2.
- Q (Bq s^{-1} or Ci s^{-1}) is the emission rate of radionuclides.
- σ_y and σ_z (meters) are the horizontal and vertical standard deviations of radionuclide concentration in the plume.

TABLE 8.3

Atmospheric Stability Classes

Superadiabatic	If the rate of decrease of temperature with elevation is greater than that found in adiabatic conditions, an unstable condition results which promotes the vertical dispersion, and hence dilution. A rising parcel does not cool fast enough due to its expansion and therefore remains warmer and continues to rise. Likewise, a falling parcel continues to fall.
Stable	No heat is gained or lost by a parcel of air that rises and expands adiabatically with falling temperature. The adiabatic cooling with rise normally corresponds to a gradient of about 1.0°C/100 m (5.4°F/1000 ft) for dry air and about 0.6°C/100 m (3.5°F/1000 ft) for moist air. If the atmospheric temperature gradient is less than adiabatic, but still negative, stability is achieved because a rising parcel cools faster than its surroundings and then tends to sink. A sinking parcel is warmer than its surroundings and thus is less dense and tends to rise. This restricts the width of the plume and decreases dilution.
Inversion	If the temperature gradient is such that the temperature increases with height, then an inversion occurs. Rising effluent from a stack becomes much denser than its surroundings and thus sinks. The effluent is thus more limited in its ascent, and this limits dilution.

Source: Adapted from Cember H., and T. E. Johnson. 2009. *Introduction to health physics*, fourth edition. New York, NY: McGraw-Hill.

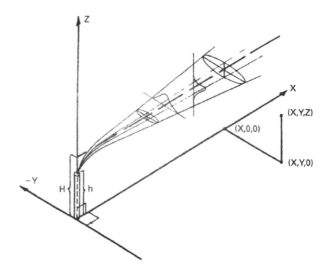

FIGURE 8.2
Gaussian plume dispersion model for a continuous point source. The chosen Cartesian coordinate system (x, y, z) is defined with all units in meters. The emission source, that is, the "stack," is located at the origin of this system; the x-axis is along the prevailing wind direction, the y-axis is the transverse axis, and the z-axis is the vertical axis. h is the physical height of the top of the stack, while H is the effective chimney height. (Reproduced from Cember H., and T. E. Johnson. 2009. *Introduction to Health Physics.* New York, NY: McGraw-Hill. Copyright McGraw-Hill Education.)

- μ (m s^{-1}) is the mean wind speed at the level of the plume center line.
- λ (s^{-1}) is the decay constant of the radionuclide in question (s^{-1}).
- H (m) is the effective chimney height.

In general, the effluent gas emerges from the stack at a significant velocity v (m s^{-1}), commonly an upward one, that results in this effective chimney height exceeding the physical height of the stack h (m). H is determined from

$$H = h + d\left(\frac{v}{\mu}\right)^{1.4}\left(1 + \frac{\Delta T}{T}\right)$$

(8.8)

in which d (m) is the outlet diameter of the stack, and ΔT is the difference between the temperature of the effluent and the ambient outdoor air of absolute (^{0}K) temperature T.

This model uses turbulence types denoted by letters A–F correlated to meteorological conditions as provided in Table 8.4.

In turn, the standard deviations σ_y and σ_z (meters) used in this model are provided in Figures 8.3 and 8.4.

Equation 8.9 gives the concentration as a function of coordinates (x, y, z):

$$C(x,y,z) = \frac{Q}{2\pi\sigma_y\sigma_z\mu}\left\{\exp\left[-\frac{\lambda}{\mu}\sqrt{x^2+y^2+(z-H)^2}\right]\right\}$$
$$\left\{\exp\left(-\frac{y^2}{2\sigma_y^2}\right)\right\}\left\{\exp\left[-\left\{\frac{(z-H)^2}{2\sigma_z^2}\right\}\right] + \exp\left[-\left\{\frac{(z+H)^2}{2\sigma_z^2}\right\}\right]\right\}$$

(8.9)

TABLE 8.4

Relation of Turbulence Types to Weather Conditions

A: Extremely Unstable Conditions	D: Neutral Conditions[a]
B: Moderately Unstable Conditions	E: Slightly stable Conditions
C: Slightly Unstable Conditions	F: Moderately Stable Conditions

Surface Wind Speed, μ (m s^{-1})	Daytime Insolation			Nighttime Conditions	
	Strong	Moderate	Slight	Thin Overcast or $\geq$ 4/8 Cloudiness[b]	$\leq$ 3/8 Cloudiness
<2.0	A	A–B	B		–
2.0	A–B	B	C	E	F
4.0	B	B–C	C	D	E
6.0	C	C–D	D	D	D
>6.0	C	D	D	D	D

Source: Adapted from Slade, D. A. ed. 1968. *Meteorology and atomic energy.* U.S. Atomic Energy Commission, Office of Information Services TID-24190. Washington, DC; Cember H., and T. E. Johnson. 2009. *Introduction to health physics,* fourth edition. New York, NY: McGraw-Hill.

[a] Applicable to heavy overcast, day or night.

[b] The degree of cloudiness is defined as that fraction of the sky above the local apparent horizon which is covered by clouds.

Compared with other types of radiological installations addressed by this model, several of the most common accelerator-produced radionuclides have rather short mean-lives and thus will decay significantly in transit. This decay-in-flight factor is accounted for by the left-most exponential function in Equation 8.9, where a straight-line approximation is made with application of the Pythagorean theorem. This is a conservative approximation that likely overestimates the radionuclide concentrations at the more distant locations. That factor is not included in most presentations of the Pasquill-Gifford equation.

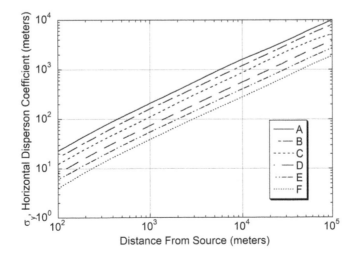

FIGURE 8.3
Horizontal diffusion constants $\sigma_y(x)$ as a function of downwind distance x from the source for turbulence types defined in Table 8.4 for use in Equations 8.9 and 8.10. (Adapted from Slade, D. A. ed. 1968. *Meteorology and atomic energy.* U.S. Atomic Energy Commission. Washington, DC.)

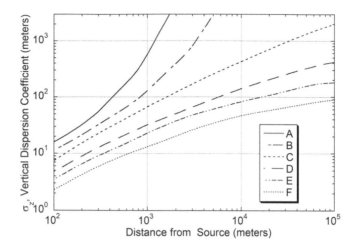

FIGURE 8.4
Horizontal diffusion constants $\sigma_z(x)$ as a function of downwind distance x from the source for turbulence types defined in Table 8.4 for use in Equations 8.9 and 8.10. (Adapted from Slade, D. A. ed. 1968. *Meteorology and atomic energy*. U.S. Atomic Energy Commission. Washington, DC.)

Equation 8.10 gives the results for the more common situation where the receptor of concern is at ground level, $z = 0$:

$$C(x,y,0) = \frac{Q}{\pi \sigma_y \sigma_z \mu} \left\{ \exp\left[-\frac{\lambda}{\mu} \sqrt{x^2 + y^2 + H^2} \right] \right\} \left\{ \exp\left[-\left(\frac{y^2}{2\sigma_y^2} + \frac{H^2}{2\sigma_z^2} \right) \right] \right\} \tag{8.10}$$

In Equation 8.10 the presence of the ground as a "barrier" to the air is taken into account. There are other nuances to these calculations that are beyond the scope of the present discussion but are covered by the references (Slade 1968; Cember and Johnson 2009).

8.2.4 Radiation Protection Standards for Airborne Radioactivity

8.2.4.1 Radiation Protection Standards for Occupational Workers

The airborne radioactivity hazard must be assessed both for workers exposed to airborne radionuclides at accelerators and for members of the public who are exposed to effluents containing radionuclides. Since the principal radionuclides are of relative short half-lives, the hazard is commonly due to the *immersion* in a "cloud" of *external* dose. At times the term *submersion* is used instead of *immersion*. This differs from the *inhalation* pathway that leads to uptakes of longer-lived radionuclides with the receipt of an *internal* dose, a pathway that is important for some radionuclides.

Regulatory authorities, guided by recommendations of the International Commission on Radiological Protection (ICRP) have established *derived air concentrations* (DACs) for radiation workers. These regulatory standards have replaced the *maximum permissible concentrations* (MPCs) formerly employed for this purpose published by the ICRP (1959) and National Council on Radiation Protection and Measurements (NCRP 1959). Care needs to be taken in the usage of DAC values. In the United States a radiological worker who spends all of their 2000 working hours in a year (40 h wk^{-1} for 50 wk y^{-1}) in an atmosphere having a radionuclide concentration of 1.0 DAC will receive 50 mSv (5000 mrem) of dose equivalent. In many other countries the DAC values are based on an annual limit of 20 mSv (2000 mrem), leading to more restrictive DAC values in those jurisdictions.

The U.S. definition is true for all types of facilities as exemplified for those applicable to U.S. DOE facilities, the loci of most of the larger particle accelerators, with equivalent but not numerically identical regulations promulgated by the individual states and by the U.S. Nuclear Regulatory Commission. As examples, for occupational workers at its facilities DOE published DAC values for the 1990 System in terms of 50 mSv of effective dose, updated most recently in 2017 (CFR 2017). The original version of these DAC values issued in 2007 (CFR 2007) and all subsequent versions superseded those based on the 1973 System that also were tied to an annual limit of 50 mSv annual dose equivalent (CFR 1993).

8.2.4.2 Radiation Protection Standards for Members of the Public

Values of derived concentration standards (DCSs) have been issued by the U.S. DOE (2011a,b) that correspond to the receipt of 1.0 mSv (100 mrem) of effective dose (1990 System) by a member of the public breathing air at 1.0 DCS for an entire year full time (168 h wk^{-1}). The DCS values replace the *derived concentration guides* (DCGs) that correlated with the receipt of 100 mrem dose equivalent (DOE 1990) using the 1973 System by a member of the public who spends an entire year breathing air with these concentrations full time.

8.2.4.3 Example Numerical Values of the Derived Air Concentrations and Derived Concentration Standards

In a book of this type it would be futile to try to provide up-to-date values of the DACs and DCSs. The practitioner is best advised to consult the applicable regulatory authorities. Table 8.5 gives representative values of these circumstance-dependent maximum concentrations C_{max} for accelerator-produced radionuclides in air taken from U.S. DOE requirements documents for workers for the superseded 1973 System and the 1990 System, respectively. DOE was chosen as the source of example regulations because it regulates the largest number and diversity of particle accelerators in the world, and the DOE complex includes a large elite cadre of accelerator radiation safety subject matter experts.

The values in Table 8.5 are essentially "worst case" ones taken from the cited references, choosing the most restrictive value from choices of types of exposure where such options are provided. In this table both "customary" and SI units are provided. Since the values ultimately originate from ICRP reports, the SI values should be regarded as the primary ones. Furthermore, in regulatory tables, "round-off" methodologies are not consistent and furthermore are limited to two or even one significant figures. For some radionuclides commonly found at accelerators, as previously discussed, DOE requirements documents give two sets of values under both Radiation Protection Systems, one for air *inhaled* into the lungs and the other for *immersion* in a semi-infinite cloud of γ-emitting radionuclides. As indicated in the footnotes to Table 8.5, for some of the radionuclides commonly found at particle accelerators and subject to the immersion pathway, DAC values are inexplicably not listed in the latest DOE regulations for occupational exposures (CFR 2017), while corresponding DCS values are provided for protection of members of the public (DOE 2011a,b) for these radionuclides. For these, special calculations have been made using the base documents specified in the cited references (DOE 2011a,b; CFR 2017) to determine appropriate working values of immersion concentrations that would deliver an effective dose of 50 mSv (5000 mrem) to workers exposed to them for an entire working year

TABLE 8.5

Airborne Concentration Limits for Radiation Workers and the General Population

	Radiation Worker Derived Air Concentrations (DACs)												
	Inhaled Air Exposure [50 mSv y⁻¹(40 h week⁻¹)]				Immersion Exposure [50 mSv y⁻¹ (40 h week⁻¹)]								
					U.S. DOE Infinite Radius Cloud				Results of Höfert for Immersion Clouds of Selected Radii[c] R (meters)			
	U.S. DOE 1973 System[a]		U.S. DOE 1990 System[b]		1973 System[a]		1990 System[b]		1.0	2.0	4.0	∞
	(μCi m⁻³)	(Bq m⁻³)	(μCi m⁻³)	(Bq m⁻³)	(μCi m⁻³)	(Bq m⁻³)	(μCi m⁻³)	(Bq m⁻³)	(μCi m⁻³)			
³H	20.0	8.0×10^5	20.0	7.0×10^5	NL	NL	NL	NL	NL	NL	NL	NL
⁷Be	9.0	3.0×10^5	10.0	4.0×10^5	NL	NL	NL	NL	NL	NL	NL	NL
¹¹C	200	6.0×10^6	100	6.0×10^6	4.0	1.5×10^5	4.0[d]	1.5×10^5[d]	7.8	6.6	6.2	2.6
¹³N	NL	NL	NL	NL	4.0	1.5×10^5	4.0[d]	1.5×10^5[d]	6.9	5.3	4.7	2.3
¹⁵O	NL	NL	NL	NL	4.0	1.5×10^5	4.0[d]	1.5×10^5[d]	6.4	4.3	3.5	2.0
³⁸Cl	20	6.0×10^5	5.0	2.0×10^5	3.0	1.0×10^5	2.6[d]	9.0×10^4[d]	NL	NL	NL	NL
³⁹Cl	20	8.0×10^5	2.0	1.0×10^5	NL	NL	2.6[d]	1.0×10^5[d]	NL	NL	NL	NL
⁴¹Ar	NL	NL	NL	NL	3.0	1.1×10^5	3.0	1.1×10^5	7.9	6.0	5.4	2.0

(*Continued*)

TABLE 8.5 (*Continued*)

Airborne Concentration Limits for Radiation Workers and the General Population

	U.S. DOE General Population Derived Concentration Guides (DCGs) for Air 1973 System[e] [1.0 mSv year^{-1}(168 h week^{-1})]				U.S. DOE General Population Derived Concentration Standards (DCSs) for Air 1990 System[f] [1.0 mSv year^{-1}(168 h week^{-1})]			
	Inhaled Air Exposure		Immersion Exposure		Inhaled Air Exposure		Immersion Exposure	
	(μCi m^{-3})	(Bq m^{-3})	(μCi m^{-3})	(Bq m^{-3})	(μCi m^{-3})	(Bq m^{-3})	(μCi m^{-3})	(Bq m^{-3})
^{3}H	0.1	3.7×10^3	NL	NL	0.21	7.8×10^3	NL	NL
^{7}Be	0.04	1.5×10^3	NL	NL	0.064	2.4×10^3	NL	NL
^{11}C	1.0	3.7×10^4	0.02	7.4×10^2	0.19	6.9×10^3	0.019	6.9×10^2
^{13}N	NL	NL	0.02	7.4×10^2	NL	NL	0.019	6.9×10^2
^{15}O	NL	NL	0.02	7.4×10^2	NL	NL	0.019	6.9×10^2
^{38}Cl	0.1	3.7×10^3	0.01	3.7×10^2	0.071	2.6×10^3	0.012	4.3×10^2
^{39}Cl	0.1	3.7×10^3	NL	NL	0.070	2.6×10^3	0.012	4.5×10^2
^{41}Ar	NL	NL	0.01	3.7×10^2	NL	NL	0.014	5.2×10^2

Note: NL means that the value was not listed. DOE, Department of Energy.

[a] (CFR 1993).
[b] (CFR 2017).
[c] (Höfert 1969).
[d] Possible working values, not DACs specifically set forth by DOE (CFR 2017). (Quinn et al. 2018).
[e] (DOE 1990).
[f] (DOE 2011a,b).

(Quinn et al. 2018). It is noteworthy that Table 8.5 clearly illustrates the relative stability of the MPC and DAC, and DCG and DCS values over the more than 40 years covered by the implementation of these concepts.

Immersion values are also usually the most limiting (i.e., smallest). A semi-infinite cloud is appropriate for outdoor applications. However, for occupational exposures, the sizes of the "clouds" are not likely to be semi-infinite but will be determined by the dimensions of the accelerator enclosures. The DOE regulations (CFR 2017) explicitly allow for adjustments of the listed DAC values to account for exposures in finite, rather than in the semi-infinite, clouds for which the tabulated values are calculated. Höfert calculated limiting concentrations for immersion clouds of various diameters (1969) using the methodology of ICRP in place at that time (ICRP 1959). While Höfert's calculations are connected with the obsolete MPC values, they remain of importance because they display the *relative* sensitivity to the dimensions of the immersion cloud. Table 8.5 gives Höfert's results for clouds of several radii. This is discussed further by Quinn et al. (2018).

8.2.4.4 Mixtures of Radionuclides

In airborne environments at accelerators, it is far more likely to encounter mixtures of radionuclides than only one radionuclide. This differs from the situations at other types of radiological facilities where perhaps only one radionuclide is present. To account for the presence of multiple radionuclides, the set of individual radionuclide concentrations in the air C_i must satisfy the following inequality:

$$\sum_i \frac{C_i}{C_{\mathrm{max},i}} \qquad\qquad (8.11)$$

where $C_{\mathrm{max},i}$ is the regulatory standard for the ith radionuclide, dependent on the circumstances of the exposure, workers subject to DACs, or members of the public to which the DCS values apply. This equation has been called by some *the weighted sum rule* and, as discussed shortly, is also applied to concentrations of radionuclides in water.

8.2.5 Production of Airborne Radionuclides at Electron Accelerators

At electron accelerators, significant air activation will not occur without bremsstrahlung because the nuclear cross sections of *electrons* are about two orders of magnitude smaller than those of *photons* (Swanson 1979a). The reverse is true for toxic gas production originating from chemical, rather than nuclear, transformations and whose reaction rate is closely proportional to the integral absorbed dose to the air. Such a dose is generally higher if the primary electron beam does not strike a target to produce bremsstrahlung but rather is directly delivered to air. The production of such toxic gases, most notably ozone (O_3), is outside of the scope of this text but has been addressed by Swanson (1979a).

This airborne radioactivity is generally short lived, and the concentrations, as seen in what follows, are usually quickly reduced to levels where the absorbed dose rates (rads h^{-1}) are small compared to those due to the accelerator components. This result is because the radiation length of air is so much longer than that of any solid material (see Table 1.2).

Swanson (1979a) has calculated the saturation activities produced in air normalized to the electron beam power with the results provided in Table 8.6.

TABLE 8.6

Saturation Activities per Unit Path Length and per Unit Beam Power Produced in Air by an Electron Beam Normalized to the Beam Power

Produced Radionuclide		Parent Stable Nuclide			Saturation Activity per Unit Length and Beam Power[a]	
Nuclide	Half-Life	Nuclide	Reaction Type	Threshold (MeV)	(MBq m^{-1} kW^{-1})	(μCi m^{-1} kW^{-1})
^{3}H	12.32 years	^{14}N	$(\gamma, {}^3\text{H})$	22.7		
		^{16}O	$(\gamma, {}^3\text{H})$	25.0	(5.2)	(140)
^{7}Be	53.22 days	^{14}N	$(\gamma, \text{sp})^b$	27.8		
		^{16}O	$(\gamma, \text{sp})^b$	31.9	(1.1)	(30)
^{11}C	20.36 minutes	^{12}C	(γ, n)	18.7		
		^{14}N	$(\gamma, \text{sp})^b$	22.7		
		^{16}O	$(\gamma, \text{sp})^b$	25.9	(11)	(300)
^{13}N	9.965 minutes	^{14}N	(γ, n)	10.6	520	1.4×10^4
^{15}O	2.04 minutes	^{16}O	(γ, n)	15.7	56	1.5×10^3
^{16}N	7.13 seconds	^{18}O	(γ, np)	21.8	(0.02)	(0.5)
^{38}Cl	37.24 minutes	^{40}Ar	(γ, np)	20.6	0.22	6
^{39}Cl	56.2 minutes	^{40}Ar	(γ, p)	12.5	1.5	40
^{41}Ar	1.827 hours	^{40}Ar	$(\text{n}, \gamma)^c$	–	variable	variable

Source: Adapted from Swanson, W. P. 1979a. *Radiological safety aspects of the operation of electron linear accelerators.* International Atomic Energy Agency: IAEA Technical Report No. 188. Vienna, Austria with half-lives from NNDC. 2018. National Nuclear Data Center at Brookhaven National Laboratory. http://www.nndc.bnl.gov/ (accessed October 17, 2018).

[a] Normalized per bremsstrahlung path length in air (m) and electron beam power (kW) incident on a high-Z target, summed over individual contributing reactions. Values in parentheses are rough estimates.

[b] Spallation reaction.

[c] Thermal neutron capture reaction where high neutron fluences are moderated by water or concrete shielding.

The results of these calculations are normalized to unit path length and to beam power. To use them to determine the volume-specific activity (e.g., Bq cm^{-3}, Ci cm^{-3}) in a given accelerator enclosure, one must multiply the tabulated values by the available *bremsstrahlung path length* and divide by the enclosure volume. The bremsstrahlung path length is determined by the physical dimensions of the room or, for a large room, by the attenuation length of the bremsstrahlung radiation in air.

The results found in this table were calculated in a manner completely analogous to those given in Table 7.2 for materials aside from air. For energies close to the threshold of an individual reaction, the rise of activity with beam energy E_o (see Section 7.3.2 and Figure 3.5) must be considered. ^{41}Ar is produced in the thermal neutron capture (n, γ) reaction most copiously where there are high fluences of moderated neutrons present, typically near water-cooled targets and in concrete enclosures.

After calculating the production rates, one can then apply the general methodology presented in this chapter to determine the concentrations within the accelerator enclosure and to estimate the effective dose equivalent rates at offsite locations as well as the status of compliance with applicable regulations.

8.2.6 Production of Airborne Radionuclides at Proton Accelerators

At proton accelerators, the energy dependencies of the cross sections for the possible nuclear reactions listed in Table 8.2 exemplified by those shown in Figures 7.5–7.8 become

TABLE 8.7

Examples of Measured Radionuclide Compositions of Typical Airborne Releases at Proton Accelerators

Situation	Radionuclides (Activity %)					
	^{11}C	^{13}N	^{15}O	^{38}Cl	^{39}Cl	^{41}Ar
CERN 28 GeV protons (Thomas and Stevenson 1988)	31.0	47.0	8.0			14.0
Fermilab 800 GeV protons (Butala et al. 1989)						
(No gap between iron and concrete walls)	46.0	19.0	35.0			
(Gap between iron and concrete walls)	42.0	14.0	0.0	0.0	10.0	34.0
Fermilab 120 GeV protons (Vaziri et al. 1993)	58.5	37.9		1.0	1.1	1.5
Fermilab 120 GeV protons (Vaziri et al. 1996)	64.6	30.5				5.0

important. In general, the positron emitters ^{11}C, ^{13}N, and ^{15}O along with ^{41}Ar (produced by thermal neutron capture), are the nuclides most frequently seen. Work at Fermilab described by Butala et al. (1989) and Vaziri et al. (1993, 1996) has also confirmed these identifications and, additionally, detected ^{38}Cl and ^{39}Cl. The determination of the relative contributions of the various positron emitters present must principally be done by fitting measured decay curves with a sum of exponential functions, each term of which represents one of the possible radionuclides present. This is a result of the fact that their γ-ray spectra are all dominated by 0.511 MeV photons from positron annihilation. The results of analyses of such decay curves have been discussed in additional references (Thomas and Stevenson 1988; Swanson and Thomas 1990).

It was concluded by Butala et al. (1989) that the geometry of target stations significantly can affect the composition. For example, high-intensity targets immediately surrounded with large volumes of iron and concrete (in contact with the iron) produced much less ^{41}Ar than did other targets where the bulk iron shield was located in an open room with a layer of air between the iron and the concrete. Presumably, the open space provided opportunity for the large flux of low-energy neutrons expected external to a pure iron shield (see Section 6.3.5) to "thermalize" and thus enhance the production of ^{41}Ar in the air space. The large cross section for the $^{40}Ar(n, \gamma)^{41}Ar$ reaction at thermal neutron energies ($\sigma_{th} = 660$ mb) also may possibly have provided the photons necessary to enhance the (γ, p) and (γ, pn) reactions required to produce significant quantities of ^{39}Cl and ^{38}Cl, respectively. Some typical percentages of the various radionuclides, by activity concentration, released from high-energy proton accelerators are given in Table 8.7.

As with the situation at electron accelerators, after calculating the production rates, one can then apply the general methodology presented in this chapter to estimate the effective dose equivalent rates as well as the status of compliance with applicable regulations.

8.3 Water and Geological Media Activation

At accelerators appropriate measures need to be taken to assure protection of groundwater resources from contamination with radionuclides. Radioactivity can be produced in soil or rock, in the water it contains, and in water that migrates in these media. Radioactivity produced in water can also be a matter of concern for occupational workers. In practice, it is not always a simple matter to separate these two areas of concern. One can, in principle, initiate calculations of water activation at accelerators by starting from "first principles" using Equation 7.8.

8.3.1 Water Activation at Electron Accelerators

As discussed earlier at several points, questions of radioactivation are generally less complex at electron accelerators. As was done for atmospheric activation, Swanson (1979a) has provided the results of calculations to address the production of radionuclides in water at electron accelerators. Such activation will principally occur in water used to cool magnets and beam absorbers. Depending on applicable regulations, this water can possibly be required to be classified as radioactive waste. The results are, again, in the form of saturation activities normalized to the electron beam power absorbed in the water volume. These are given for infinite irradiation periods with no time allowed for decay. The results are given in Table 8.8 and include specific γ-ray constants Γ useful for calculating absorbed dose rates near point sources.

From these observations it is clear that, aside from short-lived positron emitters, only ^3H and ^7Be are of importance. Activity concentrations can be obtained by assuming rapid mixing of the saturated activity in the available volume of water. Table 8.8 gives the results due to interactions with the ^{16}O found in water. In principle ^3H could be produced from the hydrogen in water by means of two sequential thermal neutron capture reactions, a rare "two-step" process: $^1H(n, \gamma)^2H$ followed by $^2H(n, \gamma)^3H$. However, this two-step process is of limited importance since the cross section for the first reaction is 0.33 barn, while that for the second is 0.52 millibarn.

In practice, due to the compactness of the shielding at electron accelerators compared with that found at proton and ion accelerators, soil activation is generally negligible except perhaps in the vicinity of beam absorbers.

TABLE 8.8

Saturation Activities per Unit Beam Power Produced in ^{16}O by an Electron Beam Normalized to the Beam Power

Produced Radionuclide		Reaction Parameters		Specific γ-Ray Constant Γ		Saturation Activity per Unit Beam Power	
	Half-Life	Reaction	Threshold (MeV)	[(mGy h^{-1}) × (GBq m^{-2})$^{-1}$]	[(rad h^{-1}) × (Ci m^{-2})$^{-1}$]	(GBq kW^{-1})	(Ci kW^{-1})
^3H[a]	12.32 years	$(\gamma, {}^3H)$	25.0	–	–	7.4	0.2
^7Be	53.22 days	$(\gamma, 5n4p)$	31.9	0.008	0.03	1.5	0.04
^{10}C	19.31 seconds	$(\gamma, 4n2p)$	38.1	0.29	1.06	3.7	0.1
^{11}C	20.36 minutes	$(\gamma, 3n2p)$	25.9	0.17	0.62	14.8	0.4
^{13}N	9.965 minutes	$(\gamma, 2np)$	25.0	0.17	0.62	3.7	0.1
^{14}O	1.177 minutes	$(\gamma, 2n)$	28.9	0.45	1.7	3.7	0.1
^{15}O	2.04 minutes	(γ, n)	15.7	0.17	0.62	330	9

Source: Adapted from Swanson, W. P. 1979a. *Radiological safety aspects of the operation of electron linear accelerators.* International Atomic Energy Agency: IAEA Technical Report No. 188. Vienna, Austria, with half-lives from NNDC. 2018. National Nuclear Data Center at Brookhaven National Laboratory. http://www.nndc.bnl.gov/ (accessed October 17, 2018).

[a] Does not present an external radiation hazard.

8.3.2 Water and Geological Media Activation at Proton Accelerators

8.3.2.1 Water Activation at Proton Accelerators

At proton and ion accelerators, as with electron accelerators, radioactivity can be produced directly in water as a result of both proton and neutron interactions. Values for some of the relevant cross sections were given in Chapter 7. Equipped with knowledge of the beam energy and information about the energy spectra of neutrons that are present, one can proceed to calculate the activity produced. In general, the most important radionuclides, as is the situation with electron accelerators, result from the interactions of the hadrons with the oxygen present in the water. As at electron accelerators, the production of ^{3}H from the hydrogen present in the water is possible, but is generally considered to be insignificant. For such calculations, the production of ^{3}H in water from the oxygen in the water is of special importance. Konobeyev and Korovin (1993) have developed a method of globally fitting the existing cross-sectional data on the production of ^{3}H due to neutron interactions with a variety of target elements, including oxygen, with the results shown in Figure 8.5. The results for proton interactions in producing ^{3}H are similar.

8.3.2.2 Geological Media Activation

In addition to ^{3}H, ^{22}Na is usually the most important radionuclide produced in geological media. Figure 8.6 gives cross sections for producing ^{22}Na by interactions of hadrons with the various elements comprising soil due to Van Ginneken (1971). This figure is a companion to Figure 8.5. When possible, specific measurements should be made for specific media.

While calculating the production of radionuclides in soil, and in the water it contains, directly from known cross sections has an appeal due to its simplicity, in practice such calculations have been done more frequently by analyzing data obtained using irradiated samples. The work of Borak et al. (1972) is of singular importance in this regard. These

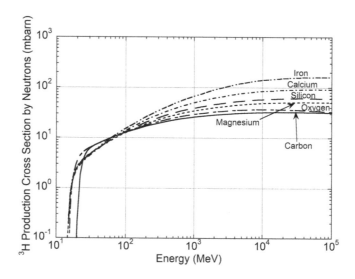

FIGURE 8.5

Energy dependence of the cross sections of the production of ^{3}H due to neutron bombardment of materials commonly found in soil and rock as a function of neutron energy. The calculations have been performed following the method presented by Konobeyev and Korovin (1993). Results for aluminum are quite similar to those found for silicon, and the results for sodium are quite similar to those found for magnesium.

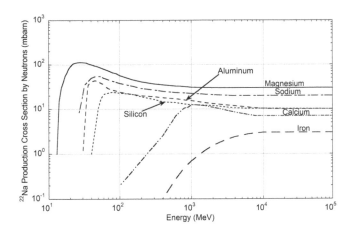

FIGURE 8.6
Energy dependence of the cross sections of the production of ^{22}Na due to neutron bombardment of materials commonly found in soil and rock as a function of neutron energy. Results for potassium are quite similar to those found for calcium. (Adapted from Van Ginneken, A. 1971. ^{22}Na production cross section in soil. Fermi National Accelerator Laboratory: Fermilab Report TM-283. Batavia, IL.)

workers measured the radioactivity produced in soil by high-energy hadrons by radiochemical analysis of soil samples irradiated near high-energy synchrotrons: the 12 GeV Argonne Zero Gradient Synchrotron (ZGS) and the 28 GeV Brookhaven Alternating Gradient Synchrotron (AGS). The radionuclides ^{3}H, ^{7}Be, ^{22}Na, ^{45}Ca, ^{46}Sc, ^{48}V, ^{51}Cr, ^{54}Mn, ^{55}Fe, ^{59}Fe, and ^{60}Co were identified. Experiments were then performed to determine which radionuclides, and what fractions of them, could be leached, loosely speaking "rinsed out" by water. This study determined macroscopic production cross sections and ion velocities relative to groundwater flow in soil. Of these nuclides only ^{3}H, ^{22}Na, ^{45}Ca, and ^{54}Mn were observed in leach waters. The entirety of ^{3}H in its form of HTO molecules was assumed to be all leachable and was measured by driving it out of the sample by baking. Radionuclides with half-lives exceeding 15 days were the only ones considered. The results were based on the elemental composition of soil typical of conditions at the Fermi National Accelerator Laboratory given in Table 8.9.

Borak et al. measured specific activities at saturation A_i (Bq g^{-1}) which are related to the microscopic cross sections by means of the following equation:

$$A_i = \phi \sum_j n_j \sigma_{ij} \tag{8.12}$$

In Equation 8.12 ϕ is the flux density (cm^{-2} s^{-1}), n_j is the number density of target nuclei of the jth nuclide (g^{-1}) of the soil sample, and σ_{ij} (cm^2) is the effective cross section for the transformation from target nucleus j to radionuclide i. The summation in Equation 8.12 is taken over the soil constituents. Borak et al. were able to directly measure these summations—the *total macroscopic cross sections* summed over the soil constituents for each radionuclides of interest. Table 8.10 gives the results of the measurements of these cross sections, denoted Σ (cm^2 g^{-1}), for each of the radionuclides identified in the various types of soils analyzed.

Borak et al. also obtained data related to the *leachabilities* of the various elements from the soils studied. Leachability measures the ability of water to remove a given radionuclide

TABLE 8.9

Composition of Soils Typical of the Fermilab Site

	Elemental Composition of Soil[a]	
Element	Z, Atomic Number	Percentage (%) by Weight
Silicon	14	14.47
Aluminum	13	2.44
Iron	26	1.11
Calcium	20	7
Magnesium	12	3.79
Carbon	6	5.12
Sodium	11	0.34
Potassium	19	0.814
Oxygen	8	≈64

Source: Adapted from Borak, T. B. et al. 1972. *Health Physics* 23:679–687.

[a] The mean moisture percentage was 13.15 ± 4.45%, and the mean pH was 7.6 ± 0.1.

from the soil material. It is not related to nuclear properties but rather is related to chemical properties and processes. A qualitative summary of the results of Borak et al. (1972) is as follows:

- ³H: The leaching process was able to collect all the tritium as measured by a bake-out process. The average value of the macroscopic cross section in soil was found to be 5.1×10^{-3} cm^2 g^{-1} of water. An important conclusion is that the tritium will migrate with the same velocity as any other water in the soil. The slightly larger

TABLE 8.10

Macroscopic Cross Section for Soil Normalized to Unit Flux of Hadrons with Kinetic Energies Greater than 30 MeV

Nuclide	Glacial Till Σ (cm^2 g^{-1})	Gray Sandy Clay Σ (cm^2 g^{-1})	Red Sandy Clay Σ (cm^2 g^{-1})	Gray Clay Σ (cm^2 g^{-1})
^{7}Be	2.9×10^{-4}	3.7×10^{-4}	3.2×10^{-4}	2.7×10^{-4}
^{51}Cr	1.7×10^{-5}	3.7×10^{-5}	2.8×10^{-5}	3.1×10^{-5}
^{22}Na	2.1×10^{-4}	2.3×10^{-4}	2.0×10^{-4}	1.6×10^{-4}
^{54}Mn	5.9×10^{-5}	4.1×10^{-5}	3.5×10^{-5}	3.7×10^{-5}
^{46}Sc	3.0×10^{-5}	1.3×10^{-5}	9.6×10^{-6}	1.1×10^{-5}
^{48}V	4.1×10^{-6}	1.1×10^{-5}	6.7×10^{-6}	7.4×10^{-6}
^{55}Fe	9.3×10^{-5}	1.2×10^{-4}	7.0×10^{-5}	2.1×10^{-4}
^{59}Fe	3.2×10^{-6}	1.7×10^{-6}	1.3×10^{-6}	1.6×10^{-6}
^{60}Co	3.3×10^{-5}	1.4×10^{-5}	1.1×10^{-5}	1.3×10^{-5}
^{45}Ca	1.6×10^{-4}	2.0×10^{-5}	3.0×10^{-5}	1.6×10^{-5}
^{3}H	8.2×10^{-4}	1.1×10^{-3}	3.3×10^{-4}	5.2×10^{-4}
^{3}H[a]	5.9×10^{-3}	5.9×10^{-3}	4.1×10^{-3}	4.4×10^{-3}

Source: Adapted from Borak, T. B. et al. 1972. *Health Physics* 23:679–687.

[a] Cross sections per gram of water in soil.

molecular weight of the HTO molecule compared with that of the H_2O molecule is of no significant effect.

- ^{22}Na: Typically, 10%–20% of this nuclide was found to be leachable. On average, it appeared that the migration velocity of this nuclide is approximately 40% of that of water through the soil due to ion exchange processes.
- ^{45}Ca: At most 5% of this nuclide was leached from the soil. The migration velocity was determined to be extremely small.
- ^{54}Mn: At most 2% of this nuclide was leached from the soil. It was determined that this nuclide will not migrate significant distances.

Thus, based on leachability considerations, ^{3}H and ^{22}Na are the most important leachable radionuclides that can be produced in environmental media such as soil.

One can thus calculate the quantities of radionuclides that might pose a risk to groundwater in the environs of an accelerator. This can be done by using the cross sections directly, or as demonstrated by Gollon (1978) for high-energy protons, by performing, for example, Monte Carlo calculations in which the total stars (i.e., total inelastic nuclear interactions above some threshold) produced in some volume of earth shielding is determined. As in Equation 7.33, the total number of atoms K_i of the ith nuclide that can be produced per star in that same volume is given by

$$K_i = \frac{\Sigma_i}{\Sigma_{in}} \tag{8.13}$$

where Σ_i is the macroscopic cross section (cm^2 g^{-1}) for producing the ith radionuclide, and Σ_{in} is the total macroscopic inelastic cross section (cm^2 g^{-1}) for soil. Gollon inferred a value of $\Sigma_{in} = 1.1 \times 10^{-2}$ cm^2 g^{-1} for typical Fermilab soil, largely classified by geologists as glacial till, from the results of Borak et al. (1972). One can calculate values for tritium and ^{22}Na, respectively, as K_3 and K_{22}:

$$K_3 = \frac{8.2 \times 10^{-4}}{1.1 \times 10^{-2}} = 0.075 \tag{8.14}$$

and

$$K_{22} = \frac{2.1 \times 10^{-4}}{1.1 \times 10^{-2}} = 0.020 \tag{8.15}$$

One can then calculate the total number of atoms of radionuclides produced during some time interval in some volume by simply multiplying these factors by the number of stars (or inelastic interactions) in the same volume. The number of atoms then can be converted to activity using the decay constant. The previous values of K_i are applicable to soils such as those found in soils at Fermilab. For other soil compositions one may need to use cross sections for producing the radionuclides of interest in various target elements and integrate over the energy spectrum of incident hadrons, or resort to direct measurements similar to those of Borak et al. (1972).

Some Monte Carlo codes of more recent development can now calculate these quantities directly from the energy-dependent production cross sections. However, given the limited energy dependence at high energies, working with the total stars remains worthwhile as a means to achieve results rapidly, or as a "quality check" on the more complex computations.

8.3.3 Regulatory Standards

The quantity of ultimate concern, of course, is the resultant concentration in water. The water could be an actual or potential drinking water resource that might well be subject to specific regulatory requirements. The regulations may differ between different governing jurisdictions. The requirements, generally not developed for application to the operations of particle accelerators, need to be understood by facility management personnel. The standards generally differ for drinking water supplies and surface water discharges. The allowable concentrations for surface waters may be larger due to the likelihood that such discharges will most certainly be diluted significantly prior to the consumption by individuals. However, in some jurisdictions this may not be the case, and the appropriate authorities must be consulted.

In the United States, for public drinking water supplies, the USEPA (CFR 1976, 2000) limits such concentrations to those that would result in an annual dose equivalent of 4.0 mrem (40 μSv) and gives a specific numerical limit of 20 pCi cm^{-3} for tritium based on the methodologies, now considered to be obsolete, previously established by ICRP (1959) and NCRP (1959). The dose is calculated to be that received by a person who uses this water as their household water supply on a full-time basis. An explicit limit for ^{22}Na is not specified by the USEPA. For surface water discharges, the DOE (2011a,b) has, as for airborne radioactivity, established DCSs for water. These are the concentrations that would result in members of the public each receiving 100 mrem in a year should they use such water for their household needs. Using the DOE DCSs one obtains a value of 76 pCi cm^{-3} for ^{3}H and 0.4 pCi cm^{-3} for ^{22}Na in drinking water to correspond to an annual effective dose of 4.0 mrem. However, the USEPA's standard numerical value for ^{3}H in drinking water is considered as legally preeminent for applications in the United States. Table 8.11 lists the concentration limits $C_{max,i}$ for radionuclides of most concern in water.

For purposes of this discussion, surface water discharges include those to streams, ponds, etc., while *drinking water standards* apply to water that could potentially end up in a source of drinking water such as public, or even private, wells. In the United States, local jurisdictions in some cases have applied drinking water standards to all discharges.

TABLE 8.11

Concentration Limits for ^{3}H and ^{22}Na in Surface Water Discharges and in Drinking Water

Radionuclide	Half-Life (Years)[a]	Concentration Limit $C_{max,i}$ (pCi cm^{-3})	
		Surface Water	Drinking Water
^{3}H	12.32	1900[b]	20[c]
^{22}Na	2.602	10[b]	0.4[b]

[a] Half-lives are from NNDC (2018).
[b] Value taken from (DOE 2011b).
[c] Value taken from (CFR 1976, 2000). A value of 76 pCi cm^{-3} is implied by DOE (2011b), based on the annual dose limit of 4 mrem (40 μSv) specified by the USEPA.

In parallel with the situation found with airborne radioactivity (Equation 8.11), to account for the presence of multiple radionuclides, the set of radionuclide concentrations in the water C_i must satisfy the following inequality, where $C_{max,i}$ is the regulatory standard for the ith radionuclide for the particular circumstances of exposure:

$$\sum_i \frac{C_i}{C_{max,i}} \leq 1 \tag{8.16}$$

8.3.4 Propagation of Radionuclides through Geological Media

8.3.4.1 General Considerations

The methods for calculating these concentrations in actual environmental media vary with the regulatory authority and the "conservatism" of the institution. The most conservative assumption is to assume that saturation concentration values of production are reached. This is equivalent to assuming that the accelerator will operate "forever" in a static configuration and that the water in its vicinity never moves. This assumption is an extremely unrealistic one in terms of operations. Also, it is highly questionable that the "motionless" water in such a medium actually comprises a potential source of usable drinking water. For an irradiation over a finite period of time, the activity concentration C_i of radionuclide i in leaching water under such conditions can be calculated by means of the following formula:

$$C_i = \frac{N_p K_i L_i S_{ave}}{1.17 \times 10^6 \, \rho w_i} \{1 - \exp(-t_{irrad}/\tau_i)\} \exp(-t_c/\tau_i) \, (\mathrm{pCi\,cm}^{-3}) \tag{8.17}$$

where

- N_p is the number of incident particles delivered per year.
- K_i is as previously determined.
- L_i is the fraction of the radionuclide of interest that is leachable.
- S_{ave} is the average star density (stars cm^{-3}) in the volume of interest per incident particle.
- ρ is the density of the medium (g cm^{-3}).
- w_i is the mass (grams) of water per unit mass (grams) of medium required to leach some specified fraction of the leachable radioactivity and is thus linked to the value of L_i.
- t_{irrad} is the irradiation time expressed in units consistent with those used for the mean-lives τ_i.
- t_c is the "cooling" time once the irradiation is suspended expressed in units consistent with those used for the mean-lives τ_i.
- τ_i is the mean-life of the ith radionuclide.

The constant in the denominator contains the unit conversions for results in pCi cm^{-3}. For a given medium, the ratio L_i/w_i should be determined by measurements specific to the local media.

An important quantity is the *effective porosity p* that represents the volume fraction of the material that is available to water movement. It is given by

$$p = \rho w_i \qquad (8.18)$$

as verified by the following unit analysis:

$$p = \rho w_i \rightarrow \frac{\text{gm (rock)}}{\text{cm}^3\text{(rock)}} \frac{\text{gm (water)}}{\text{gm (rock)}} = \frac{\text{gm (water)}}{\text{cm}^3\text{(rock)}} = \frac{\text{cm}^3\text{(water)}}{\text{cm}^3\text{(rock)}} \qquad (8.19)$$

The effective porosity is essentially equal to the pore volume of the material for soils. For consolidated materials (i.e., rock), it does not include sealed pores through which movement does not occur. This provides a means by which "worst-case" estimates may be made. For realistic estimates water movement must be taken into account.

8.3.4.2 Simple Single Resident Model

At Fermilab, a simple, and overly simplistic, *single resident model* model allowing for some movement and further dilution of water was employed for many years (Gollon 1978). In this model the vertical migration of water was, as determined from local measurements conducted in the prevailing glacial till regime conservatively, assumed to be 2.2 m yr^{-1}. In the standard clays present at Fermilab, this velocity is likely conservative (i.e., large) by at least an order of magnitude. In a rather imprecise manner the use of this model allowed for the presence of cracks and fissures through which more rapid propagation of water might be possible. The tritium vertical velocity was taken to have this value, while the results of Borak et al. (1972) were used to obtain a value of about 1.0 m yr^{-1} for ^{22}Na. Only the leachable fraction of the ^{22}Na was included, making the implied assumption that the nonleachable fraction is precluded from entering the water. The procedure then allowed for decay during the downward migration of the total inventory of radionuclides produced in 1.0 year, integrated over the entire volume of the irradiated material, to the highest aquifer below the location of the irradiation. At that point, it was assumed that the radionuclides were rapidly, in effect instantaneously, transported horizontally to a shallow well where it was presumed that the flow of water collecting the radionuclides is entirely used by a lone user (memorialized in the name of the model) who consumes a volume of 150 L day^{-1}. This value, a minimal one, was taken from results reportedly achieved by municipalities that have needed to ration public water consumption during conditions of severe drought. Thus the annual production transported vertically with radioactive decay included, was diluted into the 5.5×10^7 cm^3 yr^{-1} that this represents. This simple model is generally conservative, as it did not take advantage of the fact that the radionuclides are initially distributed over a considerable volume as they are produced and thus subject to an initial dilution at the point of production. The model is not completely conservative as it did neglect that fact that the water movement may *not* be uniform from year to year. It also does not properly account for the presence of "sand lenses," sand and gravel deposits that exist in fissures in clay soils that can provide pathways for quite rapid transport of contaminants, including radionuclides.

8.3.4.3 Concentration Model

It is clear that better methods are warranted, and a better model has been developed for use at Fermilab (Malensek et al. 1993). The *concentration model* now in use at Fermilab calculates the production of the radionuclides of concern in accordance with Equation 8.17. Variations of this approach are in use elsewhere. The result provides an initial concentration that is available for further migration, decay, and dilution. The concentration subsequent to migration is then calculated by using up-to-date modeling techniques to calculate the reduction in the concentration due to dilution, diffusion, and radioactive decay. At the point of concern, usually the location of an aquifer producing water suitable for consumption as a supply of drinking water, the concentrations calculated are then substituted into Equation 8.16 in order to determine if a proposed shielding design is adequate.

To do these calculations properly requires a detailed knowledge of the media involved. Some principles are given here, but many details are left to the references (Fetter 1988; Batu 1998). Anderson (2007) has provided an excellent introductory article on this subject. In situations where a definite potential gradient, often called the *hydraulic gradient dh/dx*, is applied to water in a medium, the rate of flow is said to be *advective*. Under such conditions and in situations where only one dimensional coordinate is important, the average linear velocity (or seepage velocity) *v* is given by the application of *Darcy's law*:

$$v = \frac{K}{p}\frac{dh}{dx} \tag{8.20}$$

where the effective porosity *p* was defined previously. More complicated situations involving two and three dimensions are addressable using vector calculus. The derivative is the gradient of the *hydraulic head* in the material. *K* in this equation represents the *hydraulic conductivity*. This quantity is a function of the material and its moisture content. All of the factors in this equation can, and generally should, be determined empirically for the medium and location under consideration. Typical values of *K* are given in Table 8.12 (Batu 1998).

TABLE 8.12

Examples of Typical Values of Hydraulic Conductivity

Group	Porous Materials	Range of K Values (cm s^{-1})
Igneous rocks	Weathered granite	$(3.3–52) \times 10^{-4}$
	Weathered gabbro	$(0.5–3.8) \times 10^{-4}$
	Basalt	$(0.2–4250) \times 10^{-6}$
Sedimentary materials	Sandstone (fine)	$(0.5–2250) \times 10^{-6}$
	Siltstone	$(0.1–142) \times 10^{-8}$
	Sand (fine)	$(0.2–189) \times 10^{-4}$
	Sand (medium)	$(0.9–567) \times 10^{-4}$
	Sand (coarse)	$(0.9–6610) \times 10^{-4}$
	Limestone and dolomite	$(0.4–2000) \times 10^{-7}$
	Karst limestone	$(1–20,000) \times 10^{-4}$
	Gravel	$(0.3–31.2) \times 10^{-1}$
	Silt	$(0.09–7090) \times 10^{-7}$
	Clay	$(0.1–47) \times 10^{-8}$
Metamorphic rocks	Schist	$(0.002–1130) \times 10^{-6}$

Source: Batu, V: *Aquifer hydraulics*. 1998. New York, NY: John Wiley and Sons. Copyright Wiley-VCH Verlag GmbH & Co. KGaA. Reproduced with permission.

    Darcy's law can then be used to determine the rate of migration of a contaminant, in this case, radioactivity, from one point to another. During the time of migration, the concentration would be *decreased* by radioactive decay and dilution of the plume while possibly being *increased* by any ongoing radioactivation process, active while the accelerator is operational. One often encounters the problem of calculating the concentration of radionuclides at some location as a function of time during, or after, a period of irradiation comparable to the mean-lives of the radionuclides of concerns. At a given location in such a medium, denoted by the coordinate x, one needs to solve the following continuity equation, an extension of Equation 7.4, for situations where the velocity of water movement v can be thought of as slowly varying or a constant over time and some volume of space:

$$\frac{\partial C_i}{\partial t} = \frac{L_i}{w_i'} Q_i(x,t) - v \frac{\partial C_i}{\partial x} - \lambda_i C_i(x,t) \tag{8.21}$$

where all variables are as in Equation 8.17, with the refinements that λ_i is the decay constant of the ith radionuclide, x is the spatial coordinate, t is the time, and w_i' is the water content of the media per unit volume of media. The quantity $Q_i(x,t)$ represents the production of the ith radionuclide and is equivalent to the factor $N_p S_{ave}/(1.17 \times 10^6 \rho)$ in Equation 8.17. It includes any time dependence in the delivery of beam. The middle term in the right-hand side of the equation takes care of movement from a point of one concentration to another at the seepage velocity v. As seen elsewhere in this book, one can commonly describe the spatial dependence of the production factor in a thick shield as an exponential function where ξ is the reciprocal of the absorption length for some process:

$$Q_i(x,t) = Q_{oi}(t) \exp(-\xi x) \tag{8.22}$$

    Mokhov (1997, private communication to J. D. Cossairt) has solved this equation for the typical initial conditions of $C_i(x, 0) = 0$ and $x \geq 0$, $t \geq 0$. In general,

$$C_i(x,t) = \frac{L_i}{w_i} \int_0^t dt' Q_i(x - vt', t') \exp(-\lambda t') \tag{8.23}$$

and for an exponential spatial dependence like that in Equation 8.22, a ubiquitous description of radiation shielding phenomena, this becomes

$$C_i(x,t) = Q_{oi}(t) \frac{L_i}{w_i} \frac{1}{\eta_i} \exp(-\xi x)[\exp(\eta_i \tau) - 1] \tag{8.24}$$

    Here $\eta_i = \xi v - \lambda_i$, $\tau = t$ for $t < x/v$, and $\tau = x/v$ for $t \geq x/v$. $C_i(x,t)$ has a maximum at $x_{i,max}$ given by

$$x_{i,max} = -\frac{v}{\lambda_i} \frac{\ln\left(\frac{\xi v}{\lambda_i}\right)}{1 - \frac{\xi v}{\lambda_i}} \tag{8.25}$$

In using these results, one must take care that the algebraic signs of the coordinates x relative to that of v are properly taken into account. In situations where the seepage velocity is extremely slow, *diffusion* becomes the dominant mechanism for water flow and dilution. Mathematically, a second partial derivative with respect to the spatial coordinate is added to Equation 8.21. Examples are provided by Fetter (1988). Computer software has been written to address this topic, such as that of Sudicky et al. (1988).

8.3.4.4 Example of Application: Jackson Model

As a further example of how these methodologies can be employed in solving such problems, Jackson et al. (1987) estimated the dilution for a shallow uncased well in an aquifer a distance r from a beam loss point also in the aquifer. The loss point was assumed to be within the drawdown zone of the well. This was performed for a simple geology that involved a single uniform stratum of earth above some level of impervious stratum. Figure 8.7 shows the situation described by this model.

Here, a given well is modeled by using the profile of the depth of water $h(r)$ as a function of r. The value of $h(r)$ is determined by the depth of a test well at radius r from the well under consideration and represents the hydraulic potential. The well is assumed to supply a volume Q of water per some chosen unit of time. The flux of water is determined by the gradient relation, equivalent to Darcy's law:

$$S_r = k\frac{dh(r)}{dr} \tag{8.26}$$

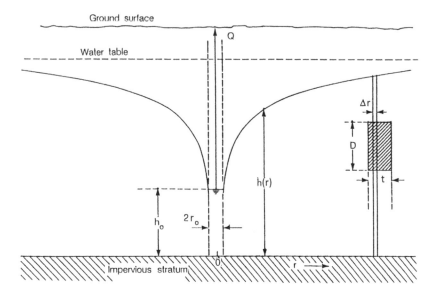

FIGURE 8.7

Hydrogeological model of a shallow well in proximity to an accelerator tunnel where a beam loss occurs. The radioactivated region is represented in cross section by the cross-hatched rectangle to the right. h represents the elevation of the water table, as perturbed by the well, above the impervious stratum as a function of radial distance from the well r, while the water table is a distance H above the impervious stratum where the water table is not perturbed by wells as indicated by the dashed line. (Reprinted from Jackson, J. D. et al. 1987. *SSC environmental radiation shielding.* Superconducting Super Collider Central Design Group: Report SSC-SR-1026. Lawrence Berkeley National Laboratory, Berkeley, CA.)

where S_r is the inward flux at radius r, and k is a constant with dimensions of volume per unit time per unit area and is characteristic of the soil. Conservation of water taking into account its incompressibility yields the steady-state equation:

$$Q = 2\pi rh(r)S_r = 2\pi rkh\frac{dh}{dr} = \pi k\frac{d(h^2)}{d(\ln r)} \tag{8.27}$$

The quantity $2\pi rh(dh/dr)$ corresponds to the rate of change of volume of the cylindrical shell of height h (i.e., the hydraulic head) with respect to r. This equation has the following solution:

$$Q\ln\left(\frac{r}{r_0}\right) = \pi \; k\{ h^2(r) - h_0^2\} \tag{8.28}$$

where r_0 is the radius of the well, and h_0 is the height of water above the impervious stratum at the well. If H is the depth of the impervious layer below the water table in an asymptotic region unperturbed by any wells, the radius of influence R of the well can be defined by the following relation:

$$\ln\frac{R}{r_0} = \frac{\pi k\left\{H^2 - h_0^2\right\}}{Q} \tag{8.29}$$

R is the distance beyond which the well has no effect on the water table. However, the detailed solution is not necessary. Suppose that this well is a distance r away from the region of deposition of radioactivity near an accelerator. One also assumes that the activation zone lies below the water table and that the deposition region lies within the radius of influence of the well. This assumption leads to higher concentrations than would be obtained if the activation zone were totally, or partially, above the water table. The amount of activity drawn into the well is determined by the volume rate of pumping Q and the necessary total flow through a cylinder of radius r and height $h(r)$, as we have seen. Let ΔV be the volume of soil yielding Q gallons of water. The cylindrical shell providing this amount of water will be of radial thickness Δr, where $\Delta V = 2\pi rh(r)\Delta r$.

The fraction F of the volume of activity included in this shell can be said to be given by

$$F = \frac{\Delta r}{t} = \frac{2\pi \; rh\Delta r}{2\pi \; rht} = \frac{\Delta V}{2\pi \; rht} \tag{8.30}$$

provided that $\Delta r < t$.

If the activated region contains leachable activity A (either total activity or that of a particular radionuclide of interest), the corresponding specific activity a in water drawn from the well is thus given by

$$a = F\frac{A}{Q} = F\frac{A}{p\Delta V} = \left[\frac{\Delta V}{2\pi \; r \; h \; t}\right]\frac{1}{p}\left[\frac{1}{\Delta V}\right]A = \frac{1}{2\pi \; r \; t \; D}\frac{f}{p}A \tag{8.31}$$

where $f = D/h$ is the fraction of the total height of the cylindrical shell occupied by the activated region, and p is the effective porosity of the soil. The pumping volume Q is implicit in f. Porosity values vary considerably but in general are in the range of

$$0.2 < p < 0.35 \tag{8.32}$$

Thus, this formula may be used to obtain an estimate of the specific activity as a function of distance from the well, although it is perhaps not too useful for applications to beam losses far from the well. By definition, $f \leq 1$ and one can use the lower value of porosity to obtain upper limit estimates of the concentration. It must be emphasized that this model depends on uniformity of water conduction by the strata. The presence of cracks, voids, localized deposits of readily permeable materials such as sand, or more complex geological strata can provide much more rapid movement that is not well addressed by this simple model. In glacial tills such as those found at Fermilab, the sand deposits are colloquially called "sand lenses" because they can provide pathways for movements of water that are quite rapid compared with that possible in adjacent soils such as clays (see Table 8.12).

PROBLEMS

1. A 20 m long air gap has a beam of 10^{12} s^{-1} of high-energy protons passing through it. First, calculate the production rate of ^{11}C in the gap at equilibrium if one approximates air in the gap by nitrogen and assumes $\sigma(^{11}$C$) = 10$ mb. Take the density to be that of air, not nitrogen, at NTP. Assume that there are no significant losses of beam by interaction after checking to see that this assumption is, in fact, true. Table 1.2 contains helpful information.

   a. If the air gap is in a $10 \times 10 \times 20$ m^3 enclosure with *no* ventilation, calculate the equilibrium concentration of ^{11}C in the room (in units of μCi m^{-3}) assuming extremely rapid mixing (i.e., no time allowed for decay while mixing occurs) of the enclosed air. Compare the concentration with the most restrictive derived air concentration (DAC) value for workers under the 1990 System in Table 8.5 and calculate, using simple scaling, the dose equivalent to a worker who spends full time in this room. Comment on the effect of the finite room dimensions. (This is a purely hypothetical scenario due to the much larger hazards of the intense direct beam.)

   b. Calculate the concentration if 2.0 air changes hr^{-1} are provided.

   c. Assume the exhaust of the ventilation described in part (b) is through a 10 cm radius stack 3.0 m tall. Calculate the air speed in the stack and the emission rate in Ci s^{-1}. Then estimate the concentration directly downwind at ground level, and hence the effective dose (1990 System) 1.0 km away with stable meteorological conditions and an average wind speed of 10 km hr^{-1}.

2. In soil conditions similar to those at Fermilab, a volume of soil around a beam absorber approximately 10 m wide by 10 m high by 20 m long is the scene of a star production rate (averaged over the year) of 0.02 stars proton^{-1} at a beam intensity of 10^{12} protons s^{-1}.

   a. Calculate the annual production of ^{3}H ($t_{1/2} = 12.3$ years), the saturated activity (in Bq and Ci), and the average saturated specific activity in the previous volume's water (assume 10% water content by volume).

   b.   Use the older Fermilab single residence model to calculate the concentration at the nearest well. Assume the activation region (beam loss point) is 50 m above the aquifer and the usual migration velocities.

   c.   "Conservatively" apply the "Jackson model" to estimate the concentration at a well 100 m distant from the center of the activation region.

3. The method of accounting for ventilation presented in Section 8.2.2 can readily be generalized to include other mechanisms that "remove" airborne radionuclides such as absorption, filtration, etc. Assume that an arbitrary total number "j" of such mechanisms is present and that the irradiation has gone on sufficiently long to have come to equilibrium between the production of radionuclides and all modes of removal. Following termination of the irradiation, determine the fraction of the total activity that is removed from the air volume by each of the "j" mechanisms. It is safe to assume that all of the atoms of the radionuclide produced are removed by one of the processes. The solution of this problem has some importance for the more long-lived radionuclides, for it leads to a method of estimating the total activity expected to be found on, say, filter media.

9

Radiation Protection Instrumentation at Accelerators

9.1 Introduction

In this chapter instruments and dosimeters currently used in the environment of particle accelerators to measure and characterize the radiation fields are discussed. The emphasis is on instrumentation that addresses those aspects of accelerator radiation fields that pose special problems perhaps somewhat unique to this branch of radiation protection. Thomas and Stevenson (1988) and Swanson and Thomas (1990) also discuss these matters. Cember and Johnson (2009) cover the basics of radiation measurement instrumentation, while Knoll (2010) has written a most comprehensive treatise on the subject. Virtually all particle detection techniques that have been devised by physicists have to some degree been employed in radiation measurements at accelerators. Often, the specialized instruments used to characterize the accelerator radiation fields are found to be of value to the researcher in the understanding of experiment "backgrounds." In this chapter the focus is on the special attributes of accelerator radiation fields that need to be addressed.

9.2 Counting Statistics

Many of the detection techniques employed to measure radiation fields are dependent on the counting of individual events such as the passage of charged particles through some medium or the decay of some particle or radionuclide. Radioactive decays are randomly occurring events having a sampling distribution that is correctly described by the *binomial distribution* given by the following expansion:

$$(p+q)^n = p^n + np^{n-1}q + \frac{n(n-1)}{2!}p^{n-2}q^2 + \frac{n(n-1)(n-2)}{3!}p^{n-3}q^3 + \cdots + \tag{9.1}$$

where p is the mean probability for occurrence of an event, q is the mean probability of nonoccurrence of the event ($p + q = 1$), and n is the number of chances of occurrence. The probability of exactly n events occurring is given by the first term, the probability of $(n - 1)$ events is given by the second term, etc. For example, in the throwing of a dice, the probability of throwing a "1" is 1/6 while that of throwing a "1" three times in a row ($n = 3$) is

$$p^n = \left(\frac{1}{6}\right)^3 = \frac{1}{216} \tag{9.2}$$

In three throws, the probabilities of throwing two "ones," one "one," and zero "ones" are given by the second, third, and fourth terms; 15/216, 75/216, and 125/216, respectively.

This distribution becomes essentially equivalent to the *normal* or *Gaussian distribution* when n has a value of about 30 or larger. The Gaussian distribution is as follows:

$$p(n) = \frac{1}{\sigma\sqrt{2\pi}} \exp[-(n-\bar{n})^2/(2\sigma^2)] \tag{9.3}$$

where $p(n)$ is the probability of finding exactly n, $\bar{n}$ is the mean value, and σ in this context is the standard deviation, not a reaction cross section.

Radioactive decays or particle reactions usually can be characterized as highly improbable, i.e. "rare," events. For such events, the binomial distribution approaches the *Poisson distribution* where the probability of obtaining n events is given by

$$p(n) = \frac{(\bar{n})^n e^{-\bar{n}}}{n!} \tag{9.4}$$

where $\bar{n}$ is the mean value. For example, consider $10^{-3}\ \mu\text{Ci}$ (10^{-9}Ci, 37 Bq) of activity so $\bar{n} = 37$ decays s^{-1}. The probability of exactly observing this number of events in any one second is

$$p(37) = \frac{(37)^{37} e^{-37}}{37!} = 0.0654 \tag{9.5}$$

As in the case of the normal distribution, 68% of the events lie within one standard deviation of the mean, 96% of the events would lie within two standard deviations of the mean, etc. For Poisson statistics the *standard deviation* is given by

$$\sigma = \sqrt{n} \tag{9.6}$$

with the *relative error* σ/n thus given by $\sqrt{n}/n$.

Often, when dealing with instrumentation, the *counting rate* is important. For this quantity the following holds:

$$r \pm \sigma_r = \frac{n}{t} \pm \frac{\sqrt{n}}{t} \tag{9.7}$$

where r is the counting rate per unit time, σ_r is its standard deviation, and t is the counting time during which the rate is measured. The quantity t could even approximate the integration time constant of an instrument. It follows that

$$\sigma_r = \frac{\sqrt{n}}{t} = \sqrt{\frac{n}{t} \cdot \frac{1}{t}} = \sqrt{\frac{r}{t}} \tag{9.8}$$

Usually, counting events due to various background radiations are present and must be dealt with. The *standard deviation of the net counting rate* σ_n is

$$\sigma_n = \sqrt{\sigma_g^2 + \sigma_{bg}^2} = \sqrt{\frac{r_g}{t_g} + \frac{r_{bg}}{t_{bg}}} \tag{9.9}$$

where the subscripts g refer to the measurement of the *gross counting rate*, while the subscripts bg refer to the measurement of the *background counting rate*. The time durations (i.e., approximate time constants) of the measurements of the rates r_g and r_{bg} are t_g and t_{bg}, respectively.

Another quantity that often is important is the *resolving time* or *dead time*, of an instrument. This is the period of time during which the detector, following an event, is incapable of correctly measuring a second event while it is processing the first. It is a function of both electronic characteristics and the physical process inherent in the detection mechanism. It can be measured by exposure to two different sources of radiation where the instrument has a measured background rate of R_{bg} and responds to the first source alone with a rate R_1 and to the second source alone with a rate R_2, where both R_1 and R_2 include the background. When exposed to both sources simultaneously, the measured rate is R_{12}. According to Cember and Johnson (2009), the resolving time τ is given by

$$\tau = \frac{R_1 + R_2 - R_{12} - R_{bg}}{R_{12}^2 - R_1^2 - R_2^2} \tag{9.10}$$

However, it is often more straightforward to determine τ from the physical properties of the detection mechanism or from the electronic time constants of the measurement circuitry. With a finite resolving time τ and a measured counting rate of R_m, the "true" counting rate R that would be observed with a *perfect* instrument having $\tau = 0$ is given by

$$R = \frac{R_m}{1 - R_m \tau} \tag{9.11}$$

Knoll (2010) gives a more detailed discussion of this topic.

9.3 Special Considerations for Accelerator Environments

There are several important features of accelerator radiation fields that merit attention in choosing instrumentation or measurement techniques that should be discussed here.

9.3.1 Large Range of Flux Densities, Absorbed Dose Rates, etc.

The *dynamic range* of quantities to be measured encountered at accelerators can extend from fractional μSv yr^{-1} encountered in environmental monitoring to large values of absorbed dose of up to megagray (MGy) that can be of concern for radiation damage. One should be reminded that it is customary in the field of health physics to quantify intense radiation fields, those well above those encountered in routine personnel protection, in terms of absorbed dose, rather than dose equivalent or effective dose in recognition of the increased importance of deterministic effects relative to stochastic effects in short-term exposures at higher dose rates. Thus, absorbed dose is used here for a more reliable basis of comparison that is consistent with the universal goal of managing effective doses as low as reasonably achievable (ALARA) at accelerators.

9.3.2 Possible Large Instantaneous Values of Flux Densities, Absorbed Dose Rates, etc.

Certain accelerators such as linear accelerators (linacs), rapid cycle synchrotrons, and "single-turn" extracted beams from synchrotrons can have very low average intensities but extremely high instantaneous rates. Such circumstances arise when the *duty factor*, the fraction or percentage of the time the beam is actually present due to the operational characteristics of the accelerator, is small. Thus, the effects of dead time must be taken into account or the apparent measured values of radiological quantities such as flux densities or dose rates can be misleadingly low. Some instruments can even be completely paralyzed by high instantaneous rates and read "zero" in a high radiation field, a potentially dangerous situation.

9.3.3 Large Energy Domain of Neutron Radiation Fields

At any given accelerator capable of producing neutrons, the properties of nuclear interactions make it highly probable that neutrons will be present at all energies from thermal ($\langle E_n \rangle \approx 0.025$ eV, see Section 9.5.1) up to nearly the energy of the beam. As will be seen, the useful methods of detection of neutrons vary considerably over this energy domain. This renders the choice of instrumentation crucial to the success of the measurement. For no other particle type is the energy range of the particles encountered in the accelerator environment so large nor are the types of effective detection techniques so diverse.

9.3.4 Presence of Mixed Radiation Fields

At accelerators, one should consider that any given radiation field external to shielding is likely to be composed of a mixture of photons, neutrons, and at high energies and especially at forward angles, muons and even a multitude of other particles and perhaps ions. In proximity to the beam, the multiplicity of particle types present can be quite large. Furthermore, virtually all neutron fields contain at least some photon component, often due to the capture of thermal neutrons by means of (n, γ) reactions. Moreover, muon radiation fields near proton and ion accelerators commonly contain some neutron component. Thus, the choice of instrumentation is somewhat dependent on what component of the radiation field needs to be characterized.

9.3.5 Directional Sensitivity

Certain instruments intrinsically exhibit directional sensitivity. This feature can be either beneficial or detrimental, depending on the situation. In all instances, it must be understood. It can lead to underestimates in radiation fields where all particles are not monodirectional. Directional sensitivity can be beneficial in certain circumstances to identify the sources of unwanted radiation.

9.3.6 Sensitivity to Features of Accelerator Environment Other than Ionizing Radiation

While the focus of this discussion is on ionizing radiation, other features must sometimes be understood. The most prominent of these is the presence of *radiofrequency* (RF) *radiation*. RF can perturb instruments that can act, sometimes rather effectively, as "antennas." Environmental effects such as temperature and humidity can also be important. In addition, one must use caution when attempting radiation measurements in the presence of magnetic fields. Induced eddy currents might be misinterpreted as "radiation." Instruments may become magnetized, and meter movements may be damaged or "paralyzed." Also, devices based on photomultiplier tubes commonly read "zero" in static magnetic fields of even moderate strength because of severe deflections of the low-energy electrons within the tubes.

9.4 Standard Instruments and Dosimeters

This section reviews instruments and dosimeters. Some of these are commonly available from commercial sources. Such commercial instruments should be used with care at accelerator facilities to be sure that their properties are adequate for usage in the particular radiological and physical environment at hand. One can see that no single commercial instrument "solves all problems" simultaneously, especially for neutron fields. The practitioner is encouraged to utilize a variety of instruments, including some of the special techniques discussed in the following text, to fully understand the radiation fields.

9.4.1 Ionization Chambers

A basic type of instrument commonly used at accelerators to measure absorbed dose rates is the *ionization chamber*. Such ion chambers rely on the collection of charge liberated by particles passing through a gas. Some detectors used in physics research now employ liquids, both room temperature and cryogenic, for the ionization medium. For ion chambers based on gaseous media, a beneficial result that comes directly from atomic physics is that the *energy loss per ion pair W* is nearly a constant over a number of materials and rather independent of type of charged particle as exhibited in Table 9.1. Also listed in this table is the first ionization energy—the energy necessary to remove the most loosely bound electron from outermost atomic structure orbitals.

TABLE 9.1

Values of the First Ionization Energy and the Energy Deposition per Ion Pair W for Different Materials for Fast Electrons and α-Particles in Gaseous Media

Gas	First Ionization Energy (eV)	W (eV/Ion Pair)	
		Fast Electrons	**α-Particles**
Ar	15.7	26.4	26.3
He	24.5	41.3	42.7
H_2	15.6	36.5	36.4
N_2	15.5	34.8	36.4
Air		33.8	35.1
O_2	12.5	30.8	32.2
CH_4	14.5	27.3	29.1

Source: Knoll, G. F: *Radiation detection and measurement.* 2010. Copyright Wiley-VCH Verlag GmbH & Co. KGaA. Reproduced with permission. based on the work of the ICRU. 1979. International Commission on Radiation Units and Measurements. *Average energy required to produce and ion pair.* ICRU Report 31. Bethesda, MD.

Table 9.1 lists values of W for "fast" electrons, those with kinetic energies well above that needed to produce an ion pair, and for alpha particles. With some naivete, one might consider taking the first ionization energy to be the energy dissipation per ion pair. However, the average amount of energy required to liberate an ion pair has to exceed the first ionization energy for the material since there are other excitation processes that serve to elevate electrons into higher, but still bound, excited states that do not release an ion pair. For gases, experimental data support the general conclusion that while the value of W has a clear dependence on the gaseous material and the type of radiation and its energy, it is not a strong function on any of these values. Thus, the listed values of 25–35 eV ion pair^{-1} result. For nongaseous media such as scintillators and semiconductors, the situation is quite different. In those cases strong dependences on the material and the type of radiation and its energy are found. These matters have been discussed in much more detail by Knoll (2010) and by the International Commission on Radiation Units and Measurements (ICRU 1979).

Thus, in a gas with a certain value of W (eV/ion pair), a charged particle depositing an amount of energy ε (MeV) will liberate an average electrical charge Q_{elect} (Coulombs), according to

$$Q_{elect} = \frac{1.602 \times 10^{-13} \varepsilon}{W}$$

(9.12)

The charge Q_{elect} is collected by electrodes biased at some voltage V. The collected charge generates a small change in the voltage ΔV (volts), in accord with the relation,

$$\Delta V = \frac{\Delta Q_{elect}}{C}$$

(9.13)

where C is the capacitance of the total circuit (including that of the chamber) in units of Farads. For typical chambers C is of the order of 10^{-10} Farads. The measured signal originates with ΔV. Knoll (2010) gives many details about the size and form of the electrical signals that can be measured. Such chambers can be operated either in a *current mode*, also called *ratemeter mode*, or alternatively in an *integration mode* in which the charge is collected (integrated) over some measured time period, then digitized into pulses that represent some increment of absorbed dose or dose equivalent.

In the ion chamber mode of operation, the applied voltage is sufficiently small so that *gas multiplication* (charge amplification) does not occur. In the most simple-minded approach, one might believe that for measurements in photon fields one could fill such a chamber with gases that "mimic" tissue and, with suitable calibration, convert the charge collected into absorbed dose. Such *tissue equivalent* materials range from complex mixtures to simple hydrocarbons, depending on the accuracy of the representation of biological tissue that is desired. However, since ion chamber gases are in general much less dense than tissue, one must also capture the energy of the secondary electrons, which in the region of a few megaelectron volts (MeV) have ranges of several meters in such gaseous material. It is thus necessary to use compensation techniques in which the solid material of the walls is chosen because of properties that match those of the gas. This condition can be readily achieved by utilizing any material having an atomic number close to that of the gas, an approximation sufficiently accurate for most practical purposes. Thus, aluminum and especially plastics are reasonably equivalent to tissue and air, at least for use in photon radiation fields. Such walls should be of sufficient thickness to establish *electronic equilibrium*, where the flux of secondary electrons leaving the inner surface of the wall is independent of the thickness.

TABLE 9.2

Thicknesses of Ionization Chamber Walls Required for
Establishment of Electronic Equilibrium

Photon Energy (MeV)	Thickness[a] (g cm^{-2})
0.02	0.0008
0.05	0.0042
0.1	0.014
0.2	0.044
0.5	0.17
1	0.43
2	0.96
5	2.5
10	4.9

Source: Knoll, G. F: *Radiation detection and measurement.* 2010.
Copyright Wiley-VCH Verlag GmbH & Co. KGaA.
Reproduced with permission; based on the results published by the ICRU. 1971. *Radiation protection instrumentation and its application.* ICRU Publication 20. Bethesda, MD.

[a] The thicknesses quoted are based on the range of electrons in
water. The values will be substantially correct for tissue equivalent ionization chamber walls and also for air. Half of the above
thickness will give an ionization current within a few percent of
its equilibrium value.

Table 9.2 gives the wall thicknesses needed to establish electronic equilibrium for photons of various energies as provide by Knoll (2010) based on results by the ICRU (1971).

The measurement of absorbed dose is accomplished by application of the *Bragg-Gray principle* that states that the absorbed dose D_m in a given material can be deduced (with suitable unit conversions) from the ionization produced in a small gas-filled cavity within that material as follows:

$$D_m = WS_mP \tag{9.14}$$

where W is the average energy loss per ion pair in the gas, and P is the number of ion pairs per unit of mass formed. S_m is the ratio of the mass stopping power (i.e., the energy loss per unit density in units of, say, MeV g^{-1} cm^2) of the material of interest to that of the chamber gas. For D_m to be in grays (J kg^{-1}), W must be expressed in Joules per ion pair and P in ion pairs per kilogram.

For accelerator radiation fields that contain neutrons, or mixtures of neutrons with muons and photons, one is commonly able to use an ion chamber to measure the absorbed dose D and determine the dose equivalent H_{equiv} or effective dose H_{eff} by using the average quality factor Q or radiation weighting factor w_R, respectively, as follows (see Chapter 1):

$$H_{equiv} = QD \quad \text{or} \quad H_{eff} = w_RD \tag{9.15}$$

Ion chambers with tissue equivalent walls have been used in this manner at many accelerators. The value of Q or w_R has to be determined by some other means such as those described in this chapter, usually as a separate measurement or even a theoretical calculation.

TABLE 9.3

Descriptions of Ionization Chambers Used at Fermilab

"Old" Chipmunk	A high-pressure gas-filled ionization chamber designed by Fermilab and built by LND™, Inc., with 4 mm thick walls of tissue equivalent plastic. The fill gas is 10 atmospheres of ethane. The chamber is enclosed in a protective box that contains a sensitive electrometer and its associated electronics to measure the current output and convert it to the dose equivalent rate. Switch-selectable quality factors of 1, 2.5, or 5 are available. The instrument is equipped with a visible dose equivalent ratemeter and audible alarms. It provides a remote readout and capability for interface with radiation safety interlock systems.
"New" Chipmunk	These instruments are similar to the Old Chipmunk except for the use of phenolic-lined ionization chamber, filled with propane gas at atmosphere pressure and an electrometer encased in a sealed container. The reduced gas pressure was chosen for safety, and the sealed container was provided to improve reliability over a larger range of temperature and humidity. The ion chambers were supplied by HPI, Inc.™ The latest versions of this instrument also allow for the selection of a quality factor of 10.
"Old" Scarecrow	A high-pressure ionization chamber with bare stainless steel walls filled with 10 atmospheres of ethane gas. The instrument is otherwise similar to the Old Chipmunk but with a fixed quality factor of 4 and capability to measure dose equivalent rates 100 times higher (up to 10 rem h^{-1}). A visible ratemeter, audible alarm, and remote readout capability are present as is the provision for interface to radiation safety interlocks.
"New" Scarecrow	The electronics and functionality are similar to that of the Old Scarecrow, but the ion chamber of the New Chipmunk is used.

Source: Adapted from Freeman W. S., and F. P. Krueger. 1984. *Neutron calibration tests of Fermilab radiation detectors.* Fermi National Accelerator Laboratory: Fermilab Radiation Physics Note No. 48. Batavia, IL.

Awschalom (1972) described the initial use of excellent examples of such instruments, called *chipmunks*, at Fermilab. These instruments have been further discussed in their more recent evolution (Krueger and Larson 2002). The ion chambers within the chipmunks, now updated several times, are commercially produced and read out with Fermilab-designed electronics. The current chamber has a net volume of 3.41 L. The 1.3 mm thick outer wall of aluminum is lined with a 3.2 mm thick layer of phenolic, an insulating resin with approximate tissue equivalency. They are filled with propane gas at about 1.0 atm (absolute), and contain an electrometer encased in a sealed container. Several versions of the instrument, including the higher dose rate version called the *scarecrow*, have been studied by Freeman and Krueger (1984) with properties given in Table 9.3. In Table 9.3, the instruments designated "new" were produced after 1980, while those designated "old" were produced earlier.

Typically chambers of this general type are calibrated using photons and may have a typical "quality factor" built into the electronics. Such chambers are available either as line-powered fixed monitors or as handheld survey instruments. The use of such instruments at accelerators must be done with the assurance that the instrument will respond correctly to the radiation field present. Neutron radiation fields are generally considered to be the most difficult in which to do this successfully. Höfert and Raffnsøe (1980) at CERN reported measurements of the response of various instruments, including tissue equivalent ion chambers. They were able to test such chambers, along with instruments to be discussed later in this chapter, in neutron radiation fields having measured neutron energies ranging from thermal to 280 MeV. Table 9.4 provides the results.

All but one of the neutron radiation fields by Höfert and Raffnsøe originated from reactor and radioactive sources. The lone exception was that the 280 MeV neutrons were obtained

TABLE 9.4

Absorbed Dose Response and Measurement Errors for Tissue
Equivalent Ion Chambers as a Function of Neutron Energy

Neutron Energy (MeV)	Absorbed Dose Response (10^5 Coulombs Gy^{-1})	Error (%)
Thermal	0.446	9.8
0.0245	0.404	12.1
0.1	0.622	6.1
0.25	0.806	7.1
0.57	0.885	5.4
1.0	0.885	5.4
2.5	0.993	6.1
5.0	1.179	5.2
15.5	1.370	5.2
19.0	1.664	12.1
280.0	0.389	10.1

Source: Adapted from Höfert M., and Ch. Raffnsøe. 1980. *Nuclear Instruments and Methods* 176:443–448.

by a time-of-flight technique from 600 MeV protons incident on a target at the CERN Synchrocyclotron. The same neutron fields were used to test other instruments with the results for those instruments displayed in Tables 9.5 and 9.8.

In examining Table 9.4, the "error" in the third column represents the error of the measurement technique, not the deviation of the measured absorbed dose from the true absorbed dose. An "ideal" ion chamber used to measure absorbed dose would achieve the same value of "absorbed dose response" at all energies. Clearly that was not achieved in these measurements. However, over a large energy domain, from about 0.25 to 20 MeV, the absorbed dose response was measured "correctly" within a factor of two, an energy domain that largely dominates the delivery of effective dose in many typical accelerator-produced neutron fields as discussed in Chapter 6.

Measurements have been conducted at Fermilab which indicate that absorbed dose measured in muon radiation fields with ionization chambers of this type is adequately understood using the γ-ray calibration of such instruments (Cossairt and Elwyn 1987). These tests involved comparison with direct measurements of the muon fluence using counter-telescope techniques (see Section 9.5.8) and typically agree within about 10% for the Fermilab-built instruments described previously. This is not surprising since muons at high energies behave essentially as minimum ionizing particles with ionization energy losses quite similar to those of electrons.

Practical problems encountered with such ion chambers are mostly those due to radiofrequency interference, rapidly pulsed radiation fields, and environmental factors such as temperature and humidity extremes.

Cossairt and Elwyn (1987) also determined that air-filled, self-reading pocket ion chambers of the type commonly issued to personnel to allow real-time monitoring of exposure to γ-rays performed very well in *muon* radiation fields, measuring absorbed doses to within about ±15%. This is a result of the fact that the ratio of muon stopping power in tissue to that in air for energies between 1.0 and 800 GeV is within 5% of a value of 1.0 (Stevenson 1983).

9.4.2 Geiger-Müller Detectors

These instruments, among the oldest developed for the detection of radiation, are in prominent use at particle accelerators primarily with respect to detection and measurement of induced activation and removable induced activity (contamination). In rare instances, such instruments can be used to identify, but most certainly not quantify, prompt radiation fields. They are very rugged and remarkably insensitive to environmental effects such as temperature and humidity. However, the typical dead time of 100 μsec or so renders them to be generally useless in prompt radiation fields at accelerators. In a Geiger-Müller instrument, the detector response is in the form of an *avalanche* that is essentially independent of the energy of the incoming particles, including photons.

9.4.3 Thermoluminescent Dosimeters

Swanson and Thomas (1990) and Knoll (2010) have provided discussions of the properties of thermoluminescent dosimeters (TLDs). These dosimeters are now preferred to photographic film particularly to monitor personnel exposures in β and γ radiation fields. Furthermore, pairs of ^{6}LiF and ^{7}LiF TLD crystals have been found to be useful in measuring neutron radiation fields. This exploits the fact that the ^{6}Li(n,α)^{3}H reaction has a large thermal neutron capture cross section of 940 barns in contrast to the much smaller thermal neutron capture cross section of 0.037 barns for ^{7}Li(n,γ)^{8}Li (Knoll 2010). Since a TLD containing either lithium isotope has a nearly identical efficiency for detecting β, photon, or muon radiation, measurement of the response of the two detectors can be used to determine the dose due to thermal neutrons in the presence of photons or muons. These reactions provide tools to use in the detection of fast neutrons if moderation is supplied, as is discussed later.

TLDs operate on the principal that some of the radiation liberated by the ionizing particle is "trapped" in band gaps in the crystal lattice. The process is well described by Knoll (2010). Ionizing radiation elevates electrons from the valence to the conduction band where they are then captured by a "trapping center." At room temperatures, there is only a small probability per unit time that such "trapped" electrons will escape back to the conduction band from the valence band. Thus, exposure to radiation continuously populates the traps. "Holes" are similarly trapped in the valence band. When readout of the dose is desired, the crystal is heated, and this thermally excites the electrons and holes out of the traps. This process is accompanied by the emission of light that can then be measured as a *glow curve*. A number of other materials can function as TLDs, notably, $CaSO_4$:Mn, CaF_2, and CaF_2:Mn. These materials have properties that can be optimized for specific applications. CaF_2:Mn is particularly useful for environmental monitoring purposes, where extraordinarily high sensitivity is required. The large numbers of trapped electrons and holes per unit of dose permit sensitivity to absorbed doses as small as 2×10^{-5} rads. LiF "fades" over time to a lesser degree than most of the other materials at room temperature, and its average atomic number is reasonably close to that of tissue, so it is particularly useful for personnel dosimetry.

TLDs can give valid results for fields as high as 100 rads. Higher doses can be measured under certain conditions if one takes care to use crystals especially calibrated for intense radiation fields, since linearity of the response breaks down in the high-dose region. In fact, these devices exhibit superlinearity. TLDs are not generally susceptible to dose rate problems. However, the readout process is intrinsically "destructive" and usually cannot be repeated.

9.4.4 Optically Stimulated Luminescence Dosimeters

As described in much more detail by Knoll (2010), this newer dosimeter is based on using light rather than heat to release energy deposited in trapping centers in materials which was originally excited by the incoming ionizing radiation. This method is now commonly employed in personnel dosimetry in photon radiation fields. For that application it has the unique advantage that, in contrast to TLDs, the dosimeter can be re-read, as the optical stimulation technique is not completely destructive. Also, since light can be well transported using fiber optics, remote readouts of optically stimulated luminescence (OSL) dosimeters are possible, while remote readouts of TLDs would be considerably more cumbersome. However, unlike for TLDs, no clear equivalent to the use of ^{6}Li, ^{7}Li pairs as used for neutron dosimetry has been found for OSL dosimeters. A typical material used in OSL dosimeter technology is Al_2O_3:C. There are other candidate materials, such as BeO, SiO_2, alkali earth halides, and alkali halides. None of these materials can be regarded as tissue equivalent, requiring correction filters and algorithms for use in dosimetry of low-energy photon fields (Knoll 2010).

9.4.5 Nuclear Track Emulsions

This discussion is summarized from that of Swanson and Thomas (1990). For many years thin ($\approx$25 µm) emulsions have been used for personal dosimetry in fast neutron fields. The technique is based on detection of tracks left by proton recoils in the film. The energy range for which these dosimeters are effective is from roughly 0.5 to 25 MeV, because below that range the tracks are too short to be read out, while above it there are too few tracks because the (n,p) cross section, dominated by elastic scattering, decreases with energy. However, neutrons in this energy range represent an important, if not dominant, source of dose equivalent or effective dose at accelerators. The singular important problem with nuclear track emulsions is that the latent image fades and leads to underestimates of the dose equivalent. The fading time can be a short as 2 weeks. Extreme efforts to keep out the moisture and experience in dry climates give some indication that this problem can be overcome.

Höfert (1983) and Greenhouse et al. (1987) have summarized experience with this type of dosimeter at accelerators. The dose equivalent range from about 10 mrem to a few hundred mrem is that for which this dosimeter can be expected to perform acceptably.

Any technique based on track formation is likely independent of dose rate effects.

9.4.6 Track Etch Dosimeter

Swanson and Thomas (1990) have discussed the use of such dosimeters. In these detectors, the passage of a charged particle through a dielectric material will result in a trail of damaged molecules in the material. These tracks can be made visible upon etching in a strong acid or base solution. The tracks will be etched at a faster rate than the undamaged portions of the material. As with nuclear track emulsions, there is a minimum detectable track length that sets a threshold of about 0.5 MeV for neutron detection. Such detectors have been reviewed extensively by Griffith and Tommasino (1990). Mica, polycarbonates such as Lexan™, and other materials are suitable for this purpose, and electronic methods of readout are available. Repeated readouts of the processed tracks are feasible.

9.4.7 CR-39 Dosimeters

Swanson and Thomas (1990) have provided a discussion of applications of dosimeters that utilize CR-39 (allyl diglycol carbonate) at accelerators. This material, also a "track detector," has largely replaced nuclear track emulsion as a neutron dosimeter. It is a casting resin also commonly used in eyeglass lenses and therefore transparent. It is the most sensitive of the track detectors and registers recoil protons from about 0.1 MeV up to approximately 15 MeV. It is processed either chemically or electrochemically. Repeated readouts of the processed tracks are feasible. The lower limit of detection appears to be superior to that of nuclear emulsions. In this energy domain the sensitivity is adequate, about 7×10^3 tracks cm^{-2} rem^{-1}, but may be as much as a factor of two lower in high-energy spectra. Fading appears to be insignificant. However, natural radon gas can contribute to background readings, and the angle of particle incidence is important.

9.4.8 Bubble Detectors

The use of these detectors at accelerators has also been discussed by Swanson and Thomas (1990) and by Knoll (2010). The bubble damage polymer detector is an innovative dosimeter that is akin in some ways to a classic bubble chamber in that a liquid whose normal boiling point is below room temperature is kept under pressure. When the pressure is released bubbles form along the ionization path of a charged particle that has traversed it. To enhance the effect, superheated droplets of a volatile liquid are dispersed in a gelatinous medium. There are two types of these detectors that have been developed: one type by Apfel (1979) and the other type by Ing and Birnboim (1984). The polymer or gel is supplied in a clear vial. When a neutron interacts in the sensitive material, a bubble is created that expands to optically visible dimensions and can thus be counted. There is no angular dependence, but temperature effects must be considered. The Ing and Birnboim detector was reported to exhibit a constant response over the range $15 < T < 35°C$. The material can be tailored to match a chosen neutron energy threshold that can be as low as 10 keV or less. Indeed, sets have been prepared with arbitrary threshold such as of 0.010, 0.100, 0.500, 1.0, 3.0, and 10 MeV, etc. The range of sensitivity can be adjusted to be between 10 and 300 bubbles per microsievert (μSv), or larger, in a volume of about 4.0 cm^3, and the physical mechanism is not readily sensitive to dose rate effects. Disadvantages include a high unit cost, and the fact that once the vial is opened it is only usable for limited periods of time of dose integration. The materials have been successfully used at accelerator facilities. These detectors could not be expected to give accurate results in high dose rates.

9.5 Specialized Detectors

9.5.1 Thermal Neutron Detectors

Although thermal neutrons are not commonly the major source of neutron dose equivalent at particle accelerators, they are of considerable importance in accelerator radiation protection because of the ability to moderate the fast neutrons into thermal ones, as we see later. Furthermore, because some of the most prominent thermal neutron detectors rely on radioactivation (by neutron capture) as the detection mechanism, they have the advantage that the response is entirely independent of dose rate effects and hence free of dead time effects. An excellent discussion, summarized here, on thermal neutron detectors is given by Knoll (2010).

At the outset, there are some general features concerning thermal neutrons that need to be recalled. The kinetic energy E values of thermal neutrons have the familiar relationship as a function of temperature given by the Maxwell-Boltzmann distribution:

$$f(E) = \frac{2}{\sqrt{\pi}(kT)^{3/2}} E^{1/2} \exp\left\{-\frac{E}{kT}\right\} \tag{9.16}$$

where $f(E)$ is the fraction of particles, in this case neutrons, of energy E per unit energy interval; k is the Boltzmann constant (see Table 1.1); and T is the absolute temperature of the gas. The *most probable energy* E_{mp} is given by

$$E_{mp} = kT \tag{9.17}$$

while the *average energy* $\langle E \rangle$ at any given temperature is

$$\langle E \rangle = \frac{3}{2}kT \tag{9.18}$$

At "room temperature" $T = 293°K$, so that the most probable energy is 0.025 eV. Normally, thermal neutron cross sections are tabulated for this very special value of kinetic energy. Since thermal neutrons are most decidedly nonrelativistic, the most probable velocity v_{mp} at $T = 293°K$ is determined from

$$E_{mp} = \frac{1}{2}mv_{mp}^2 = kT, \quad \text{so that} \quad v_{mp} = 2200\,\text{m s}^{-1} \tag{9.19}$$

As the neutron energy increases above the thermal value (up to about 1.0 keV), unless there are "resonances" (i.e., nuclear excited states) present that perturb the absorption cross section, the absorption cross section σ has been found to be approximately described by the relation

$$\sigma \propto \frac{1}{\sqrt{E}} \propto \frac{1}{v} \tag{9.20}$$

that is known as the $1/v$ *law*. Thus, within the limits of validity of the $1/v$ law, one can scale from the tabulated thermal neutron cross section σ_{th} at the most probable energy E_{mp} to other values of neutron kinetic energy E as follows:

$$\sigma(E) = \sigma_{th}\sqrt{\frac{E_{mp}}{E}} \tag{9.21}$$

Several different nuclear reactions that are initiated by the capture of thermal neutrons are used as the basis of detectors. They all involve particular target nuclei and thus the detector materials sometimes depend on isotopically separated materials to enhance the effectiveness. Several of these are discussed as prominent examples.

9.5.1.1 Boron-10

The ^{10}B(n,α)^{7}Li reaction is exothermic, having a Q-value $Q_v = 2.790$ MeV, and leads either to the ground state of ^{7}Li or its first excited state at 0.482 MeV. The latter occurs for about 94%

of the time when thermal neutrons are incident. Thus, for the dominant transition to the excited state, the reaction imparts about 2.308 MeV to the reaction products. This energy is much larger than the kinetic energy of the incoming *thermal* neutron. Since energy and momenta must be conserved, for the dominant excited state branch the kinetic energy of the α-particle $E(\alpha) = 1.47$ MeV and, accordingly for the recoil nucleus, $E(^7\text{Li}) = 0.84$ MeV. This is because the following must hold:

$$E(^7\text{Li}) + E(\alpha) = 2.308 \text{ MeV} \qquad (9.22)$$

due to energy conservation for the excited state branch, if one neglects the very small kinetic energy of the incident thermal neutron. Also,

$$\sqrt{2m(^7\text{Li})E(^7\text{Li})} = \sqrt{2m(\alpha)E(\alpha)} \qquad (9.23)$$

holds due to conservation of momentum, since the two reaction products emerge in opposite directions. The very small momentum of the thermal neutron is ignored, and one recalls that, nonrelativistically, $p^2 = 2mE$, where m denotes the rest mass of the particle.

The excited state subsequently decays by emission of a photon. For this reaction, at 0.025 eV $\sigma_{th} = 3840$ barns (Knoll 2010). The relatively large natural abundance of ^{10}B is 19.9% compared with 80.2% for the other stable isotope, ^{11}B (National Nuclear Data Center [NNDC] 2018). The large natural abundance of ^{10}B makes this reaction very favorable for thermal neutron detection. In addition, material enriched in ^{10}B is readily available. Also, the reaction products, and thus their deposited energies, being of short ionization range are contained in detectors of reasonable size. Figure 9.1 gives the cross sections as a function of

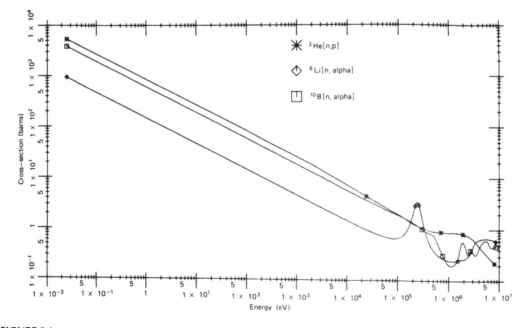

FIGURE 9.1

Cross sections versus neutron energy for some reactions of interest in neutron detection. (Knoll, G. F: *Radiation detection and measurement*. 2010. Copyright Wiley-VCH Verlag GmbH & Co. KGaA. Reproduced with permission. Copyright © 2010, 2000, 1990, 1980.)

neutron energy for this and some other thermal capture reactions discussed in this chapter. It is useful that the boron-10 reaction has a rather featureless cross section and obeys the $1/v$ law quite well even up to an energy of approximately 0.5 MeV.

This reaction has been used principally in BF_3 gas in *proportional counters*. Proportional counters are somewhat similar in concept to ionization chambers except that the applied electric fields are of sufficient strength to accelerate the initial electrons liberated by the ionization to energies above the thresholds for liberating additional secondary electrons. In typical gases at 1 atm, this threshold is of the order 10^6 volts m^{-1}. Under proper conditions, the number of electrons generated in this process can be kept proportional to the energy loss, but the number of electrons released (and hence the size of the signal) can be amplified by an effective signal gain of a factor of many thousands. In proportional chambers, the region in which these secondary electrons are released is kept small compared to the chamber volume. If the voltage is raised to higher levels, then this proportionality is lost, and the counter enters the Geiger-Müller mode. Knoll (2010) has given a detailed exposition on proportional chambers and the gas multiplication process. BF_3 is the best of the boron-containing gases as a proportional counter gas because of its "good" properties as a counter gas along with the high concentration of boron in the gas molecule. Typical BF_3 tubes are biased at 2000–3000 volts with gas ionization multiplications ranging from about 100 to 500. An enriched (96%) BF_3 tube can have an absolute detection efficiency of 91% for 0.025 eV incident neutrons, dropping to 3.8% for 100 eV neutrons. Alternatives with somewhat better gas properties (and cleaner signals) have been achieved by using boron-lined chambers with other gases that have more optimum properties when used in proportional chambers.

9.5.1.2 Lithium-6

The reaction of interest is $^6Li(n,\alpha)^3H$. For this reaction $Q_v = 4.783$ MeV. The process leads only to the ground state of ^3H. As discussed in connection with the $^{10}B(n,\alpha)^7Li$ reaction, conservation of energy and momentum determines that $E(^3H) = 2.73$ MeV and $E(\alpha) = 2.05$ MeV. For incident thermal neutrons at 0.025 eV, $\sigma_{th} = 940$ barns and the natural isotopic abundance of ^6Li is 7.59% (NNDC 2018). Figure 9.1 includes the cross section of this reaction as a function of neutron kinetic energy. The cross section exhibits a significant resonance at about 3×10^5 eV. For use in gas counters, no fill gas containing lithium having suitable properties analogous to those of BF_3 has been found. Instead, ^6Li has been successfully added to scintillators. With the addition, commonly called a "doping," of a small amount (<0.1% of the total atoms) of europium to LiI as LiI(Eu), the light output is as much as 35% of that of a comparable size NaI(Tl) crystal, the latter being a commonly used reference detector known to have very good detection efficiency. Such scintillators have a decay time of approximately 0.3 μs, and hence an example of an intrinsic "dead time" as discussed previously. Of course, ^6LiF is in prominent use as a TLD and employs the same nuclear reaction. The TLD has an advantage over this "live" detector in very intense radiation fields as no instantaneous readout is involved and thus no detector dead-time effects are encountered.

9.5.1.3 Helium-3

This nuclide, gaseous at room temperature, is used as a detector employing the $^3He(n,p)^3H$ reaction. The $Q_v = 0.764$ MeV, so that, as for the other reactions, $E(p) = 0.574$ MeV and $E(^3H) = 0.191$ MeV for incident thermal neutrons. For this reaction $\sigma_{th} = 5330$ barns (Knoll 2010). Although this isotope of helium can be used directly as a detector gas, it has the disadvantage that the natural abundance is only 0.000134% (NNDC 2018), rendering enriched ^{3}He to be costly. There is also currently a global shortage of ^{3}He amplified by its great promise for use in neutron-detection instrumentation. Also, as a technical limitation some of the energy can escape the sensitive volume of a detector of reasonable size because of the relatively long range of the emitted proton, as compared with the α-particles emitted in the ^{10}B(n,α)^{7}Li and ^{6}Li(n,α)^{3}H reactions. Again, the cross section as a function of energy is given in Figure 9.1. As seen, the cross section is quite "well-behaved" as a function of energy. ^{3}He is a reasonable gas for proportional chambers; however, no compounds are available since helium, after all, is a noble gas. In sufficient purity it will work as an acceptable proportional counter gas. Because a proton is the reaction product instead of the short-range α-particle, "wall effects" (i.e., effects in which some energy escapes the counting gas volume) may be somewhat more severe than for BF$_3$. However, proportional chambers filled with ^{3}He can be operated at much higher pressures than are possible with BF$_3$ and can thus have enhanced detection efficiency.

9.5.1.4 Cadmium

This discussion would be incomplete without discussing cadmium. Averaged over the naturally occurring isotopes of cadmium, the thermal neutron capture reaction of form ^{A}Cd(n,γ)$^{A+1}$Cd has a cross section $\sigma_{th} = 2450$ barns. More spectacularly, the reaction ^{113}Cd(n,γ)^{114}Cd has a value of $\sigma_{th} = 19910$ barns. ^{113}Cd has a natural abundance of 12.22% (NNDC 2018). Thus, even without using isotopically enriched material, the thermal neutron cross section is large. This element is not used directly in the detector medium. Rather, it is used to shield other detectors from thermal neutrons, essentially as a "anti-detector," because the large cross section results in the absorption of essentially all neutrons with energies less than about 0.4 eV. Hence, one can do measurements with and without the cadmium inside of some moderator (see Section 9.5.2) and thus identify the thermal component.

9.5.1.5 Silver

Awschalom (1972) was able to use thermal neutron capture on silver as a basis of a moderated detector. As it occurs in nature, silver has two stable isotopes that both capture thermal neutrons via the (n,γ) process: ^{107}Ag (51.8% natural abundance, $\sigma_{th} = 40$ barns) and ^{109}Ag (48.2% natural abundance, $\sigma_{th} = 93.5$ barns) (NNDC 2018). The average value of σ_{th} is 63.6 barns for naturally occurring silver. While the cross sections are not as large as those of some of the other reactions discussed, the material is readily available and enrichment is not needed. Though considered a "precious metal," the cost is sufficiently manageable to make the small amounts needed for detectors feasible. The detector that utilized these capture reactions was a moderated one (see Section 9.5.2) in which the output of a Geiger-Müller tube wrapped with silver that sensed the capture γ-rays was compared with an identical tube wrapped with the same thickness of tin, an element with a mass number reasonable close to that of the silver. Tin has an average value of $\sigma_{th} = 0.63$ barns (NNDC 2018) and is thus comparatively insensitive to thermal neutrons. The response of tin-wrapped tube simultaneously measured in the same radiation field was used to subtract background due to muons, photons, etc.

9.5.2 Moderated Neutron Detectors

As seen previously, many neutron reactions tend to have much smaller cross sections in the MeV region than they have in the "thermal" region. Shortly after the discovery of the neutron, it was observed that surrounding a thermal neutron detector with hydrogenous material enhances detection rates exhibited by a "bare" thermal neutron detector placed in the same radiation field. The reason this occurs with hydrogenous materials is that discussed in connection with Equation 6.1, repeated here for convenience. The fraction ΔE of the incident energy E_o that can be transferred to the target nucleus after a collision where the target nucleus recoils at angle θ is given by

$$\frac{\Delta E}{E_o} = 4\frac{M}{m_n}(\cos^2\theta)\left(1+\frac{M}{m_n}\right)^{-2} \tag{9.24}$$

where M is the mass of the target nucleus, and m_n is the neutron mass. The energy that can be transferred in the reaction is maximized in the head-on collision ($\theta = 0$) and has its maximum value of unity when $M \approx m_n$ (i.e., for hydrogen).

One might naively expect the detection efficiency to improve with the thickness of the moderator. However, as the moderator thickness increases, the probability that a given *moderated* neutron will actually ever reach the detector *decreases*. Figure 9.2 illustrates these trade-offs.

In general, the optimum thickness will for moderators such as polyethylene range from a few centimeters for kiloelectron volt (keV) neutrons to several tens of centimeters for MeV neutrons. Furthermore, for any given thickness, the overall counting efficiency as a function of energy will peak at some energy related to the thickness.

9.5.2.1 Spherical Moderators, Bonner Spheres, and Related Detectors

Bramblett et al. (1960) employed spherical moderators to obtain low-resolution neutron spectra using a method that has become known as the *Bonner sphere technique*. In this technique moderating spheres of different diameters surrounding a thermal neutron detector of some type are placed in a given radiation field. The sets of responses of the spheres normalized to the incident neutron fluence are indicative of the neutron energy spectra. As one might expect, the determination of the efficiency of each sphere as a function of energy is a rather complicated matter. Such *response functions* have been calculated, using techniques such as the Monte Carlo method, by a number of authors over the years since this method was invented. Hertel and Davidson (1985) have calculated the response functions for spheres that possess what has become generally accepted as the "standard" set of diameters. Other response functions, perhaps more accurate in neutron fields of higher energies, have been reported by Awschalom and Sanna (1985). The results of Awschalom and Sanna are given in Figure 9.3 for cylindrical LiI(Eu) detectors of lengths equal to their diameters, which are each 1.27 cm (0.5 inch) in polyethylene of density 0.95 g cm^{-3}.

Awschalom and Sanna provide a variety of response functions for different detector sizes, principally detector volume), as well as moderator thickness and density.

Most of the available efficiency calculations have been made for ^{6}LiI(Eu) scintillators, but also can be used for ^{6}LiF TLD dosimeters. They cannot, in general, be used for other thermal neutron capture reactions used to detect thermal neutrons, as the neutron cross sections needed for the calculation of the responses will differ, as exemplified in Figure 9.1. Experimental verifications of the details of these response functions are rare because of the difficulty of the measurements. Kosako et al. (1985) have successfully verified some of the important response functions using a neutron time-of-flight technique in the especially difficult keV energy region of neutron energy.

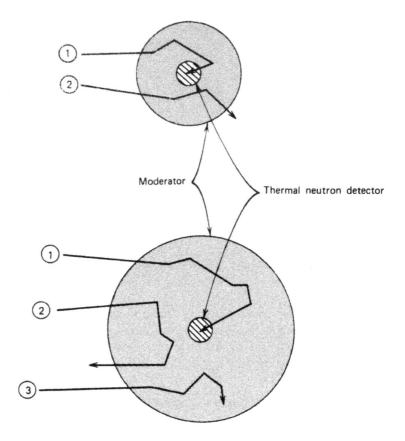

FIGURE 9.2
Neutron tracks in spherically moderated detectors of two different diameters. The thermal neutron detector is the cross-hatched sphere centered in each larger sphere of moderator material. Tracks labeled **1** represent incident fast neutrons that are successfully moderated and detected. Tracks labeled **2** represent those neutrons that are partially or fully moderated, but which escape without reaching the thermal neutron detector and are thus undetected. The track labeled **3** represents those neutrons that are captured by the moderator but which do not reach the detector and thus are also not detected. Larger moderators will tend to enhance process **3** while reducing process **2**. (Knoll, G. F: *Radiation detection and measurement.* 2010. Copyright Wiley-VCH Verlag GmbH & Co. KGaA. Reproduced with permission. Copyright © 2010, 2000, 1990, 1980.)

A Bonner sphere determination of the neutron spectrum is composed of a set of measurements of the responses r for the different spheres of radius C_r, where r has the discrete values based on the available set. In conducting such measurements, the detector is placed in the center of the sphere through an access hole that, to the extent possible due to the need to accommodate signal cables, during the measurement is filled with a cylindrical plug made of the same material as that of the moderating sphere. The measured responses ideally are given by

$$C_r = \int\limits_0^\infty dE \frac{dN}{dE} R_r(E) \tag{9.25}$$

where dN/dE is the differential neutron flux density (the neutron spectrum), and $R_r(E)$ is the energy-dependent response function for the sphere of radius r. One measures C_r for

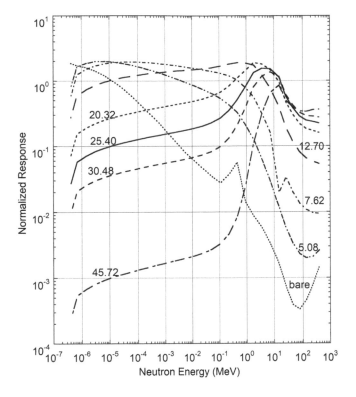

FIGURE 9.3
The calculated responses for the bare 1.27 cm diameter by 1.27 cm long cylindrical LiI detector and for the same detector centered inside 5.08, 7.62, 12.7, 20.32, 25.4, 30.48, 38.1, and 45.72 cm diameter polyethylene spheres of 0.95 g cm⁻³ as a function of neutron energy. The detector is a cylinder having a length equal to its diameter. The "normalization" is that needed to provide the correct responses according to Equation 9.25. (From tabulations in Awschalom, M., and R. S. Sanna. 1985. *Radiation Protection and Dosimetry* 10:89–101.)

the set of spheres and, with *a priori* knowledge of $R_r(E)$, determines dN/dE by *unfolding* the spectrum. In practice, one works with a discrete approximation to the integral:

$$C_r = \sum_{i=1}^{n} \Delta E_i \frac{dN}{dE_i} R_r(E_i) \qquad (9.26)$$

where the index i labels each member of the set of *energy groups* used, n in number. The unfolding procedure is a difficult mathematical problem that, unfortunately, suffers from being underdetermined and mathematically ill conditioned. One has as many "unknowns" as one has energy groups, with typically only eight or nine measurements to determine the response. It is common for 31 energy groups to be used with the objective of achieving "reasonable" energy resolution in the results.

Prominent computer codes in use at accelerators include BUNKI (Lowry and Johnson 1984), LOUHI (Routti and Sandbert 1980), and SWIFT (O'Brien and Sanna 1981). BUNKI uses an iterative recursion method, and LOUHI uses a method of least squares fitting procedure with user-controlled constraints. One essentially starts with an "educated guess" at the spectrum and iterates to fit the responses. As we have seen, a $1/E$ spectrum is a good starting point for an accelerator-produced neutron energy spectrum. SWIFT is based on a somewhat different principle; it is a Monte Carlo program that makes no *a priori* assumptions on the spectrum and can thus provide a "reality check" on results using the other two. It has the disadvantage in that it is known to sometimes produce nonphysical peaks in the unfolded spectrum. In general, the codes agree best with each other for those properties that are determined by integrating over the spectrum such as the total fluence, the average quality or radiation weighting factor, the absorbed dose, and the dose equivalent or effective dose. Typical spectra obtained from such unfolding procedures have been reported by a wide variety of practitioners. Results at Fermilab, as representative examples, have been summarized by Cossairt et al. (1988) and revisited by Cossairt and Vaziri (2009). Similar results are reported by other laboratories. Further discussion of examples of neutron spectrum measurements is given in Chapter 6.

It is sometimes important to verify the reasonableness of the unfolded spectrum. Comparisons can be made with known spectra from radioactive sources such PuBe or AmBe, and such comparisons have been made. The normalized responses C_r can be directly used to check the qualitative "reasonableness" of the unfolded spectrum. For example, this was done for measurement in the labyrinth studied at Fermilab (Figure 5.11) and for the iron leakage measurements also studied at Fermilab (Figure 6.15). The results are shown in Figures 9.4 and 9.5, respectively.

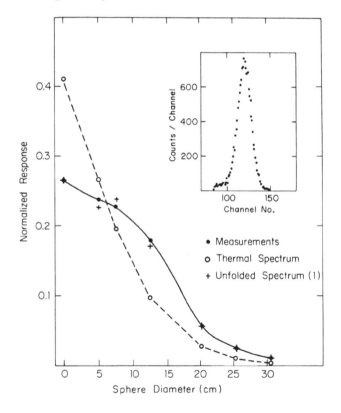

FIGURE 9.4

Normalized response from the Bonner sphere detectors as a function of spherical moderator diameter within the second leg of the labyrinth shown in Figure 5.11. The solid circles are the measurements, while the open circles represent calculated results assuming a purely thermal neutron spectrum. The +'s are the results calculated from the neutron energy spectrum unfolded using the program SWIFT. The solid and dashed curves are drawn to guide the eye. The inset shows a typical gated spectrum of the pulse heights in the ^{6}LiI(Eu) phoswich detector described shortly. (Reprinted with permission from Cossairt, J. D. et al. 1985b. *Health Physics* 49:907–917. https://journals.lww.com/health-physics/pages/default.aspx.)

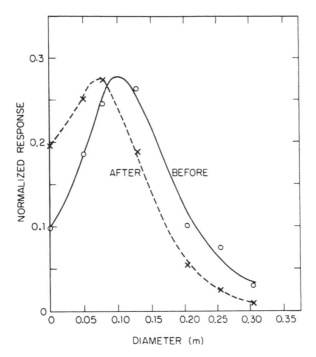

FIGURE 9.5
Normalized detector response as a function of spherical moderator diameter for the situation described in Figure 6.15. The open circles are the measurements *before*, and the X's are the measurements *after* the placement of the additional concrete shielding. (Reproduced with permission from Elwyn, A. J. and Cossairt, J. D. 1986. *Health Physics* 51:723–735. https://journals.lww.com/health-physics/pages/default.aspx.)

In Figure 9.4 the responses measured in the labyrinth are compared with the sphere responses expected for a pure thermal neutron spectrum. The enhanced responses for the intermediate-sized spheres correlate with the somewhat more energetic unfolded neutron spectrum. It is also clear that the unfolded neutron spectrum "played back" well to generate the measured detector responses. For the iron leakage spectrum in Figure 9.5 one can see evidence for the "softening" of the spectrum after the concrete was added.

In the use of ^{6}LiI(Eu) scintillators for such detectors in mixed fields, there are situations in which the signals from photons and/or muons can overwhelm the neutron signal. Awschalom and Coulson (1973) developed a technique in which the ^{6}LiI(Eu) is surrounded by plastic scintillator. The physical configuration of such a *phoswich* detector is shown in Figure 9.6.

The electronic readout circuity for this detector can separate relatively *slow* pulses indicative of α-particles resultant from thermal neutron capture interactions in the ^{6}LiF(Eu) core from the *fast* pulses indicated of muons interacting in the surrounding plastic scintillator that might also traverse the ^{6}LiF(Eu) and produce a signal in both parts of the detector. Interactions of thermal neutrons in the plastic scintillator have a negligible probability occurrence and also much smaller signal sizes than those due to the α-particles emitted in the ^{6}Li(n,α)^{3}H thermal neutron capture reaction.

Typical pulse height spectra obtained by use of this detector in a long exposure to environmental neutrons are given in Figure 9.7.

In this technique, a "fast" discriminator is set to respond to the 2–3 nanosecond decay time of the plastic scintillation signal, while a "slow" discriminator is set to respond to the

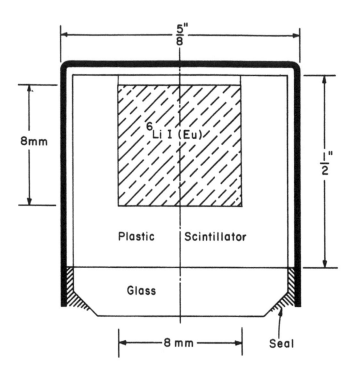

FIGURE 9.6

Cross-sectional view of 8 mm × 8 mm cylindrical phoswich. 1.0 inch (") is equal to 25.4 mm. (Reproduced from Awschalom, M., and L. Coulson. 1973. A new technique in environmental neutron spectroscopy. In *Proceedings of the Third International Conference of the International Radiation Protection Association*. CONF 730907-P2, 1464–1469. Oak Ridge, TN: U.S. Department of Energy Technical Information Center.)

1.4 μsec decay time of the crystal. Selecting the slow counts *not* accompanied by fast counts clearly gives superior discrimination against nonneutron events from environmental radiation (e.g., cosmic rays muons along with photons due to terrestrial sources) or from accelerator-produced muons, which produces coincident pulses in *both* the crystal and the plastic scintillator. In Figure 9.7 the well-defined peak centered on approximately on channel number 115 is interpreted to be due to neutrons captured in the ^{6}LiF(Eu), since neutrons would not produce a "fast" pulse. The success of the technique is illustrated by the much improved peak-to-valley ratio obtained with the utilization of this pulse-shape discrimination. The same detector was used to produce the pulse-height spectrum shown in the inset in Figure 9.4.

In performing Bonner sphere measurements in neutron fields that are suspected of being spatially nonuniform in space, it may be necessary to measure C_r over the set of spheres individually for two reasons: First, arranging them in an array in proximity to each other may result in undesired "cross-talk" between the moderators in which the detectors with smaller moderators begin to respond as if they had larger ones. Second, it is common for the radiation field that is the subject of the measurement to be significantly nonuniform in space. It is obvious that making individual measurements with the set of spheres will require a good method of normalizing the measurements to one another to achieve acceptable input data suitable for application of the unfolding program.

Since accelerator neutron fields are often quite similar to each other, it was noticed that the choice of a single moderator size might well offer the opportunity to construct a *rem-meter*, a device with the ability to, at least approximately, deliver an output signal proportional

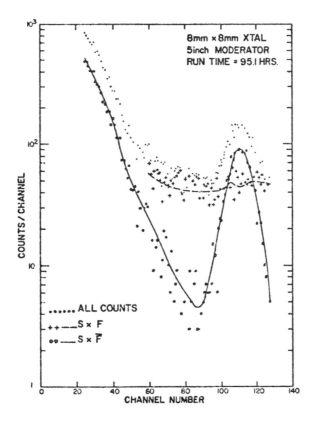

FIGURE 9.7

Pulse height spectra obtained using the phoswich shown in Figure 9.6 in a natural background radiation field and exposed for a period of 95.1 hours. The phoswich was placed in the center of a 12.7 cm (5 inch) diameter polyethylene moderating sphere. The solid curve guides the eye through the spectrum of slow pulses *not* accompanied by fast pulses (S × $\bar{F}$ in the figure, circles), events likely indicative of neutrons. The dashed curve guides the eye through the spectrum of slow pulses accompanied by fast pulses (S × F in the figure, + signs), events likely indicative of "noise" or charged particles in the environment. The small dots show all slow pulses (S × F + S × $\bar{F}$). (Reprinted from Awschalom, M., and L. Coulson. 1973. A new technique in environmental neutron spectroscopy. In *Proceedings of the Third International Conference of the International Radiation Protection Association*. CONF 730907-P2, 1464–1469. Oak Ridge, TN: U.S. Department of Energy Technical Information Center.)

to the dose equivalent or effective dose in a real-time manner. Such instruments have been designed by choosing a given sphere, or moderator of another shape with response function particularly well matched to energy dependence of the dose per fluence factor *P*. The standard example of this is the *Andersson-Braun* detector equipped with a BF_3 detector (Andersson and Braun 1964). This technique was reviewed by Thomas and Stevenson (1988). Generally, the 25.4 cm (10 inch) diameter polyethylene sphere, or other moderators with an equivalent mass of polyethylene surrounding the core thermal neutron detector, is used because its response curve provides the best match to the dose equivalent per fluence function. Höfert and Raffnsøe (1980) have measured the dose equivalent response of such an instrument as a function of neutron energy with the results given in Table 9.5.

In examining Table 9.5, the "error" in the third column represents the error of the measurement technique and not the deviation of the dose equivalent measurement from the true dose equivalent. An "ideal" instrument used to measure the dose equivalent would achieve the same value of "dose equivalent response" at all energies. Commercial versions

TABLE 9.5

Dose Equivalent Response and Measurement Errors for a 25.4 cm
Diameter Polyethylene Moderating Sphere as a Function of
Neutron Energy

Neutron Energy (MeV)	Dose Equivalent Response (10^5 Coulombs Sv^{-1})	Error (%)
Thermal	0.349	10.0
0.0245	3.209	12.1
0.1	1.335	6.8
0.25	1.082	6.1
0.57	0.923	5.2
1.0	0.745	5.2
2.5	0.784	6.1
5.0	0.653	5.2
15.5	0.348	5.2
19.0	0.445	12.2
280.0	0.157	10.1

Source: Adapted from Höfert M., and Ch. Raffnsøe. 1980. *Nuclear Instruments and Methods* 176:443–448.

of this instrument usually operate in the proportional counter mode, a choice that renders them somewhat suspect in accelerator fields with high instantaneous dose rates due to small duty factors. A similar detector using ^{6}LiI(Eu) as the detector has been developed by Hankins (1962). Its response is shown in Figure 9.8 compared with the "Inverse of RPG Curve" that embodies the relative dose equivalent delivered per neutron as a function of neutron energy.

Leake (1968) developed an alternative detector of this type. In this detector a ^{3}He proportional counter is used in a 20.8 diameter sphere to reduce background due to photons along with a cadmium filter against thermal neutrons. It is stated that this detector is effective in photon fields as intense as 20 rads h^{-1}. This type of instrument may seriously underestimate doses due to neutrons above 10 MeV. The impact of the transition to the 1990 System is considered to be a minor one, a simple adjustment of calibration factors.

It is not necessary for radiation protection purposes that a "spherical" moderator be an exact sphere. Awschalom et al. (1971) measured the responses of three polyethylene moderators: a sphere, an octagon of revolution (a "pseudosphere"), and a cylinder. The sphere had a diameter of 25.4 cm and the dimensions of the other moderators were chosen to have the same volume as that of the sphere. Using polyethylene material of the same density, the detector would thus be surrounded by the same mass of moderator. The results of the measurements of Awschalom et al. made using a Pu-Be α-particle emitting source are shown in Figure 9.9.

In examining Figure 9.9, one should note that the ordinate is displayed with a truncated scale. It was found that the alternative moderators have a response only mildly distinguishable from that of the sphere as a function of polar angle. The ideal shape is that of the sphere because of the symmetry; however, that shape is the most difficult to produce by machining. A cylinder is rather trivial to machine but, as seen in Figure 9.9, has the largest deviation from the ideal response of complete independence of the polar angle. The pseudosphere studied as an octagon of revolution is a shape that can also be readily fabricated. Both pseudospheres and cylinders also have the advantage that they can be set on a flat surface without rolling about.

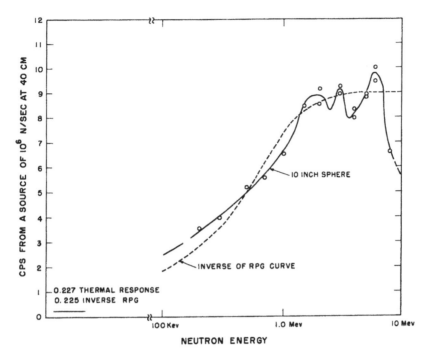

FIGURE 9.8

Sensitivity of detector composed of a 25.4 cm (10 inch) diameter moderating sphere surrounding a 4 × 4 mm^2 cylindrical LiI scintillator in counts s^{-1} at 40 cm distance from a source of 10^6 neutrons s^{-1} plotted as a function of neutron energy. The ordinate is the response in counts per second (CPS) at the specified distance from this neutron source. Also shown is the relative dose equivalent per neutron labeled as "Inverse of RPG Curve." At thermal energies, the response was measured to be 0.227 compared with a value of 0.225 for the "Inverse RPG" curve. (Reprinted from Hankins, D. E. 1962. *A neutron monitoring instrument having a response approximately proportional to the dose rate from thermal to 7.0 MeV.* Los Alamos Scientific Laboratory: Report LA-2717. Los Alamos, NM.)

9.5.2.2 Long Counters

Another type of moderated neutron detector that has been used extensively is the *long counter*. The idea is to adjust the configuration of moderators around some thermal neutron detector in such a manner as to assure that the detection efficiency is approximately independent of energy over as "long" of an energy domain as practical. It has been found empirically that the best detector is a cylinder of moderating materials surrounding a thermal neutron detector (also cylindrical) on the axis. Since a cylindrical detector placed axially in such a cylinder is the optimum configuration, the BF$_3$ proportional counter is a popular choice. Hanson and McKibben (1947) were the pioneers of the technique. An improved version, which has rather widespread use, was developed by DePangher and Nichols (1966). Figure 9.10 shows a cross-sectional view of this detector. The length and diameter are both approximately 41 cm, and the mass of the detector is about 45 kg. It is designed and calibrated for use with the neutrons incident on the "front" face, optimally oriented facing the neutron source if the source of neutrons is a well-defined one.

The version shown contains an optional built-in PuBe neutron source to provide a readily available calibration check source. The presence of the neutron source would not be desirable in an instrument to be used in radiation fields near natural background and not including it also avoids the potential necessity to address nuclear materials accountability requirements.

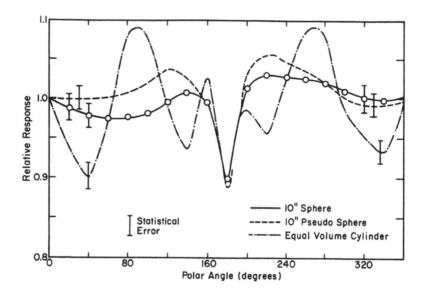

FIGURE 9.9

Relative neutron detection efficiency of three different moderators with a 4×4 mm^2 cylindrical ^{6}LiI(Eu) detector at the center. The efficiencies are plotted as a function of the polar angle. The polar angle is measured from the axis of the light pipe. (Reproduced from Awschalom, M. et al. 1971. *A study of spherical, pseudospherical, and cylindrical moderators for a neutron dose equivalent rate meter.* Fermi National Accelerator Laboratory: Fermilab Report TM-291. Batavia, IL.)

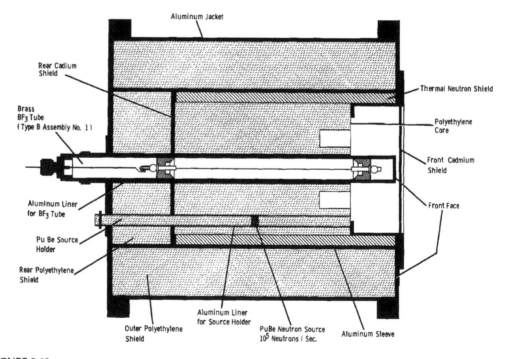

FIGURE 9.10

A DePangher Long Counter in longitudinal cross section. The approximate dimensions and mass of this instrument are given in the text. (Reprinted from DePangher, J., and L. L. Nichols. 1966. *A precision long counter for measuring fast neutron flux density.* Battelle Memorial Institute Pacific Northwest Laboratory: Report BNWL-260. Richland, WA.)

Perhaps the best calibration measurements for this type of long counter are those of Slaughter and Rueppel (1977). They used filtered beams from a reactor ($E_n \approx$ keV) as well as monoenergetic neutron beams from (p,n) and (d,n) reactions at accelerators to cover the energy range from 10 keV to 19 MeV. An average of about 3.5 counts cm^2 neutron^{-1} sensitivity was reported over this energy domain, with deviations of from 5% to 30% from absolute independence of neutron energy. A similar detector has been used to conduct studies of neutron skyshine at Fermilab (Cossairt and Coulson 1985; Elwyn and Cossairt 1986). The large peak in the pulse-height spectrum of the BF$_3$ tube from thermal neutron capture ($Q_v = 2.79$ MeV) renders the detector essentially insensitive, with the application of simple pulse-height discriminator, to all other radiations. Knoll (2010) summarizes results with long counters of alternative designs that achieve better uniformity and higher levels of sensitivity.

9.5.3 Activation Detectors

As we have seen, certain nuclear reactions have relatively sharp thresholds that can be used to determine portions of a hadron spectrum that exceed it, since the "leveling off" of the cross sections is generally well behaved. In addition to information on reaction thresholds provided in Chapter 7, where referral was made to activation threshold techniques, Table 9.6 summarizes some of the useful reactions along with some pertinent information about threshold detectors that have been found to be useful in practical work. Some of these reactions are discussed in the following text. Both maximal (peak) and high-energy limit cross section limits are given. Thomas and Stevenson (1985, 1988) provide a list of other reactions that possibly have useful thresholds.

The reactions that produce ^{11}C from ^{12}C are of special interest because of the fact that plastic scintillators can themselves become activated by hadrons (especially neutrons and protons) exceeding 20 MeV. This technique was first developed by McCaslin (1960). The ^{11}C

TABLE 9.6

Important Characteristics of Various Nuclear Reactions Used as the Basis of Activation Detectors

Detector	Reaction	Energy Range (MeV)	Half-Life	Typical Detector Size	Cross Section Peak (mb)	Cross Section High Energy (mb)	Radiation Detected
Sulfur	^{32}S(n,p)^{32}P	>3	14.26 d	4 g disk	500[a]	10[a]	β^-
Aluminum	^{27}Al(n,α)^{24}Na	>6	15.00 h	16–6600 g	11[b]	9[b]	γ
Aluminum	^{27}Al(n,x)^{22}Na	>25	2.603 y	17 g	30[b]	10[b]	γ
Plastic scintillator	^{12}C → ^{11}C	>20	20.33 min	13–2700 g	90[b]	30[b]	β^+, γ
Plastic scintillator	^{12}C→^{7}Be	>30	53.24 d	17 g	18[b]	10[b]	γ
Mercury	^{198}Hg→^{149}Tb	>600	4.12 h	up to 500 g	2[b]	1[b]	α, γ
Gold	^{197}Au→^{149}Tb	>600	4.12 h	0.5 g	1.6[b]	0.7[b]	α, γ
Copper	Cu→^{24}Na	>600	15.00 h	580 g	4[c]	3.9[c]	γ
Copper	Cu→^{52}Mn	>70	5.59 d	580 g	5[c]	4.6[c]	γ
Copper	Cu→^{54}Mn	>80	312.1 d	580 g	11[c]	11[c]	γ

Source: Values of half-lives are from NNDC. 2018. National Nuclear Data Center at Brookhaven National Laboratory. http://www.nndc.bnl.gov/ (accessed October 17, 2018).
[a] Swanson and Thomas (1990).
[b] Barbier (1969).
[c] Baker et al. (1984, 1991).

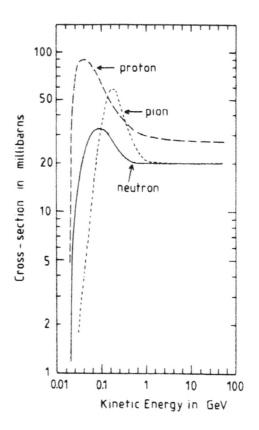

FIGURE 9.11

Plots of cross section as a function of incident kinetic energy for the reactions $^{12}C \rightarrow {}^{11}C$ induced by neutrons, pions, and protons. The arithmetic mean of the positive and negative pion cross sections is shown as the pion curve. (Reprinted with permission from Stevenson, G. R. 1984. *Health Physics* 47:837–847. https://journals.lww.com/health-physics/pages/default.aspx.)

production cross sections, as initiated by several different types of incident particles, are shown in Figure 9.11.

Stevenson (1984) has determined that a value of 28 fSv m^2 (0.28 nSv cm^2) is an appropriate factor to apply to the conversion of the measured fluence of neutrons with $E_n > 20$ MeV to the associated dose equivalent, assuming a typical accelerator spectrum found within thick shields of earth or concrete. Such measurements can be useful to determine the contribution of the high-energy ($E > 20$ MeV) neutrons to the total neutron dose equivalent or effective dose, with suitable use of the dose fluence^{-1} conversion factors P discussed in Chapter 1.

Moritz (1989) found that the use of scintillators of type NE102A, a common general-purpose scintillator, activated by the reaction $^{12}C(n,2n)^{11}C$ can be included as an additional high-energy detector in a Bonner sphere measurement as a means of extending the energy range to higher values. Following Stevenson (1984), Moritz used an average cross section of 22 mb for the $^{12}C(n,2n)^{11}C$ reaction. NE102A has a carbon content of 4.92×10^{22} atoms g^{-1} and a density of 1.032 g cm^{-3} (Knoll 2010). Moritz used a cylindrical detector 5.0 cm in diameter by 5.0 cm long achieving an efficiency of 93% in detecting the 0.511 MeV annihilation γ-rays produced as a result of the ^{11}C decay. In effect, the addition of this reaction reduced the degeneracy of the spectrum unfolding process using the code LOUHI.

Figure 9.12 provides the plots of cross sections as a function of incident neutron energy for some other useful reactions with very high thresholds.

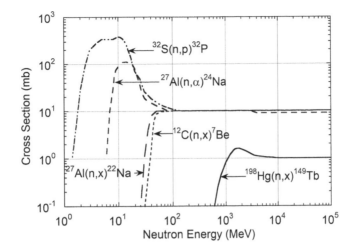

FIGURE 9.12
Plots of cross sections as a function of incident neutron energy for several threshold reactions. In this figure the nuclear reactions identified as "(n,x)" are "inclusive reactions" for which the plotted cross section applies to the totality of all processes that result in the production of the indicated residual nucleus. (Adapted from Gilbert, W. S. et al. 1968. *1966 CERN-LRL-RHEL shielding experiment at the CERN proton synchrotron.* University of California Radiation Laboratory: Report UCRL 17941. Berkeley, CA.)

9.5.4 Special Activation Detectors for Very High–Energy Neutrons

There are two reactions with rather high thresholds that are of special note for employment as activation detectors for high-energy hadrons, protons and neutrons. The cross sections for these are shown in Figure 9.13. For both ^{197}Au and ^{209}Bi, it is fortuitous that these isotopes represent 100% abundance of the naturally occurring chemical elements, eliminating the need for procurement of isotopically enriched materials.

The ^{197}Au $\rightarrow$ ^{149}Tb reaction is a suitable monitor for very high–energy particles and is commonly used as a beam calibrator. However, three additional reactions involving copper targets useful for this purpose have been found by Baker et al. (1984, 1991). These reactions are more convenient because the half-lives of the radionuclide produced are longer than that of 4.1 hours for ^{149}Tb. The cross sections have been measured for energies from 30 to 800 GeV and are included in Table 9.6.

Fission reactions have been exploited as neutron (or hadron) detectors at accelerators. The fission of ^{209}Bi is especially interesting since the reaction has a threshold for obtaining useful results of about 50 MeV and also exhibits strong evidence that the neutron and proton-induced fission cross sections are approximately equal. Bismuth has been used in ionization chambers where the large energy deposited by the fission fragments gives a clear "signature" of this process, indeed recording the presence of hadrons above this relatively high-energy threshold. Thus, akin to the use of ^{11}C, it can provide further information about high-energy neutrons and resolve ambiguities in the unfolding of spectra from Bonner sphere measurements. McCaslin et al. (1968) have presented results obtained using this process.

9.5.5 Proton Recoil Counters

Knoll (2010) describes a variety of techniques for detecting neutrons based on measuring the energy of recoil particles. The ^{3}He(n,p)^{3}H reaction has a reasonable cross section

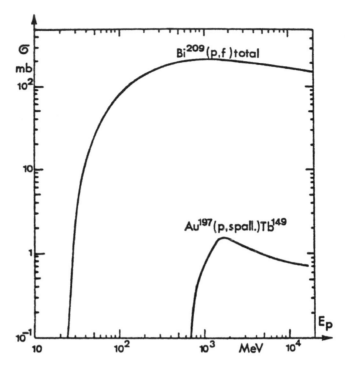

FIGURE 9.13

Cross sections for two different reactions of interest for detecting high-energy nucleons. While these are given specifically for incident protons, the cross sections for neutrons are anticipated to be nearly identical. In this figure, (p,spall.) denotes spallation of the ^{197}Au (gold) nucleus initiated by an incident proton (or neutron). (p,f) denotes fission of the ^{209}Bi (bismuth) nucleus initiated by an incident proton (or neutron). (Barbier, M: *Induced radioactivity*. 1969. Copyright Wiley-VCH Verlag GmbH & Co. KGaA. Reproduced with permission.)

even into the MeV region but suffers from competition with (n,d) processes and elastic scattering. Elastic scattering of neutrons in which the energy of the recoil particle is measured and correlated with the neutron energy has received a great deal of attention. The most obvious recoil particle is the proton, because hydrogenous detector materials (e.g., plastic scintillator) are readily available and the proton can receive the most energy in the recoil process as described in Equation 9.24. Detector designers have been able to exploit the fact that scattering from hydrogen in the region $E_n < 10$ MeV is *isotropic* in the center of mass frame of reference. Knoll has shown that the probability $P(E_r)$ of creating a recoil proton having energy E_r is also independent of angle in the laboratory frame within this energy domain. Thus, the recoil energy is only a function of the incident neutron energy. However, complexities enter the picture because in scintillators carbon is present along with the hydrogen and can also contribute recoil protons. Furthermore, the magnitude of the cross section is a function of neutron energy as is the efficiency of neutron detection in the scintillator. These effects, along with that of finite pulse height resolution, can lead to the need to use unfolding techniques in which the pulse height, indicative of the energy of the recoil proton, is correlated with the average neutron energy that could produce such a pulse. The technique has exhibited some promise in measuring the energy spectra of neutron radiation fields. A good summary is that of Griffith and Thorngate (1985) who were able to determine neutron energy spectra in the region between 2.0 and 20 MeV.

9.5.6 Tissue Equivalent Proportional Chambers and Linear Energy Transfer Spectrometry

In mixed radiation field dosimetry, a useful technique implemented commercially is the *tissue equivalent proportional chamber* (TEPC), sometimes referred to as the "Rossi counter" after its inventor, H. Rossi (Rossi and Rosenzweig 1955). The technique has been further described by Brackenbush et al. (1979). In this instrument tissue equivalent walls are employed to apply the Bragg-Gray principle. In such chambers, the pressure is maintained at low values, only a few torr (1.0 torr = 1.0 mm of Hg = 133.3 pascals); thus the energy deposited is kept small. The energy so deposited will be equal to the linear energy transfer (LET) of the particle multiplied by the path length. At these low pressures, the gas-filled cavity has the same energy loss as does a sphere of tissue of diameter about 1.0 μm, hence representing an "equivalent diameter of 1.0 μm." In principle, determining the absorbed dose from events in such chambers is a straightforward unit conversion from a measured pulse height spectrum (calibrated in energy) to absorbed dose (in tissue):

$$D(\text{Gy}) = 1.602 \times 10^{-10} \frac{C}{\rho V} \sum_{i=i_{\min}}^{i=i_{\max}} iN(i) \qquad (9.27)$$

where the summation is over channels i ($i_{\min} \leq i \leq i_{\max}$, see later) in the pulse height spectrum corresponding to the radiation type of interest, V is the sensitive volume (cm^3), ρ is the density (g cm^{-3}), C converts the channel number to energy in MeV, and $N(i)$ is the number of counts in channel number i. In such instruments C is a constant independent of channel number.

In such chambers, the transition between photon- and neutron-induced events occurs at a pulse height of about 15 keV μm^{-1}. It is possible to determine the quality factor Q or radiation weighting factor w_R from a single TEPC measurement. Under the conditions stated and using a formula derived by Rossi (1968), one can unfold from the pulse height spectrum the distribution of absorbed dose as a function of LET, $D(L)$. The formula is complicated by the fact that one must average over mean chord lengths in the chamber. Such a distribution is used to calculate quality factor or radiation weighting factor, and hence the dose equivalent (1973 System) or effective dose (1990 System). The advent of microprocessors has made portable versions of such instruments possible. Figure 9.14 shows a typical pulse height spectrum obtained with such an instrument.

In higher-energy fields, dose distributions due to other particles with the same characteristic shapes but larger pulse sizes appear as the ^{2}H, ^{3}H, ^{3}He, ^{4}He, and even ^{7}Li "drop points." This obviously adds complexity to the unfolding procedures in the determination of LET spectra. A good discussion of the application of this technique is given by Vasilik et al. (1985).

9.5.7 Recombination Chamber Technique

An adaptation of the tissue equivalent ion chamber that has shown considerable potential for usefulness as a dose equivalent meter in a mixed field of radiation is based on the exploitation of recombination phenomena in such chambers. As charged particles interact in such a chamber, the gas is ionized. The electrodes will collect only those ions that do not recombine before they reach the cathode. The extent of such *columnar recombination* is dependent on the average distance between the ions as well as on the applied voltage. The biasing voltage sets the speed at which the ions migrate to the cathode. For a given voltage,

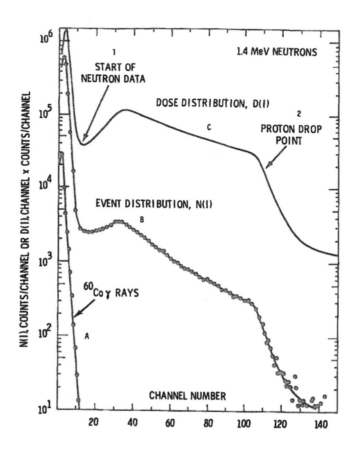

FIGURE 9.14

Pulse-height spectra from a tissue equivalent proportional counter exposed to 1.4 MeV neutrons and ^{60}Co γ-rays. (Reproduced from International Atomic Energy Agency Publication: Brackenbush, L. W. et al. 1979. In *Advances in radiation monitoring. Proceedings of a symposium held by the International Atomic Energy Agency in Stockholm, Sweden*: IAEA, Vienna. IAEA-SM-229/52:231–240.)

a chamber should exhibit more severe recombination for the radiations having high LET (e.g., neutrons, heavy ions, etc.) than for those having low LET (electrons, photons, and muons). In the high LET situation, the slowly moving positive ions are surrounded by a higher density of electrons than they would be in under conditions of low LET. Zielczynski (1963) did the initial work on this topic. The methodology has been further described by Sullivan (1984).

Baarli and Sullivan (1965) reported similar results over a somewhat larger range of values of quality factor Q ($2 < Q < 20$ as defined by the 1973 System) and further refined the technique. It turns out that the current i (or charge, if the measurement of current is integrated over some defined period of time) measured in a given radiation field as a function of the applied voltage V is well approximated by the following expression:

$$i = kV^n \tag{9.28}$$

The parameter n is correlated with the quality factor Q, and k is a constant proportional to the intensity of the radiation field.

Using a special ion chamber engineered for this purpose, Cossairt and Gerardi (2009) documented studies of the correlation of n with Q (1973 System) or of n with w_R (1990

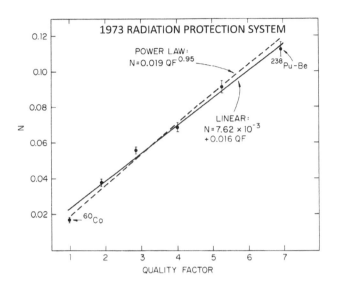

FIGURE 9.15
Response of a recombination chamber as a function of quality factor Q (1973 System) obtained in mixed fields using radioactive sources. (Reprinted from Cossairt, J. D., and M. A. Gerardi. 2009. *Measurements of radiation quality factors using a recombination chamber.* Fermi National Accelerator Laboratory: Fermilab Report TM-1248-REV. Batavia, IL.)

System). The response of the chamber as described by Equation 9.28 was measured in each member of a set of mixed fields staged with combinations of γ-rays from ^{60}Co and neutrons and γ-rays from a ^{238}PuBe source to expose the chamber to different average values of Q or w_R. The results are shown in Figures 9.15 and 9.16.

Adequate fits are provided by linear and power law formulations. The results for the 1973 System were

$$n = 0.00762 + 0.016Q \tag{9.29}$$

and

$$n = 0.019Q^{0.95} \tag{9.30}$$

The results for the 1990 System were

$$n = 1.38 \times 10^{-2} + (8.84 \times 10^{-3})w_R \tag{9.31}$$

and

$$n = 0.017w_R^{0.762} \tag{9.32}$$

Patterson and Thomas (1973) presented similar results over a somewhat larger range of Q ($2 < Q < 20$) under the 1973 System. Typically, the response of such a chamber is measured as a function of applied voltage for the special chamber provided for that purpose over a voltage range domain that might be dependent on the chamber used. A voltage domain of $20 \leq V \leq 1200$ volts is a typical one used. In radiation fields where the intensities are not steady with time, the response must be normalized against some other measured quantity

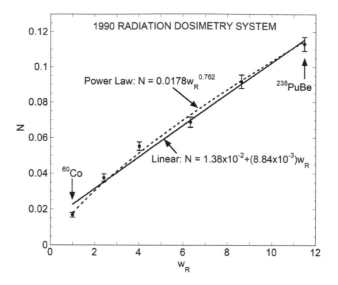

FIGURE 9.16

Response of a recombination chamber as a function of radiation weighting factor w_R (1990 System) obtained in mixed fields using radioactive sources. (Reprinted from Cossairt, J. D., and M. A. Gerardi. 2009. *Measurements of radiation quality factors using a recombination chamber.* Fermi National Accelerator Laboratory: Fermilab Report TM-1248-REV. Batavia, IL.)

that accurately is proportional to the intensity of the radiation field. The method of least squares is then applied to determine n by taking advantage of the fact that Equation 9.28 can be rewritten as

$$\ln i = \ln k + n \ln V \tag{9.33}$$

In typical measurements, such a log-log fit to the data is of moderately good quality. The quality of this fit is dependent on the accuracy of the normalization used and the intensity of the radiation field. Higher radiation intensities result in a better signal-to-noise ratio in the chamber and lead to better measurement results. From the value of n determined by such a fit, the quality or radiation weighting factor can then be determined by using equations like those previously determined for the particular ion chamber being used. Figure 9.17 shows the response measured in a field known to be dominated by high-energy muons and thus known to have $Q = w_R = 1.0$ (Cossairt and Elwyn 1987).

As one can see, the results were completely consistent with the known result of unity for Q and w_R in such radiation fields.

Data taken in the iron leakage spectrum associated with the configuration described in Figure 6.15 are shown in Figure 9.18.

Measurements of this type have been used to check the quality factors (1973 System) obtained in the unfolding of Bonner sphere data. Table 9.7 illustrates the typical agreement between these entirely different techniques for diverse radiation fields.

As one can see, these two entirely different approaches to determining this important parameter of the radiation field for these four spectra agree well within the assigned measurement errors, thus giving the practitioner alternate tools.

Zel'chinskij and Zharnovetskij (1967) proposed using two such recombination chambers placed in the radiation field of interest; one operated at a low voltage and other at a

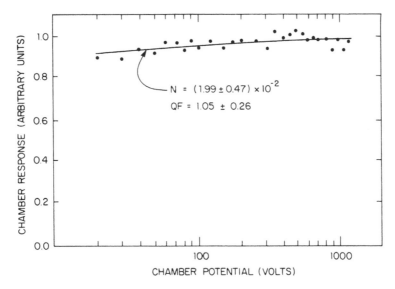

FIGURE 9.17
Recombination chamber response as a function of chamber potential in a radiation field nearly completely consisting of high-energy muons. "QF" is the quality factor Q as defined by the 1973 Radiation Protection System. (Reprinted with permission from Cossairt, J. D., and Elwyn, A. J. 1987. *Health Physics* 52:813–818. https://journals. lww.com/health-physics/pages/default.aspx.)

high voltage. The differences in responses read out by the two chambers would then be proportional to the dose equivalent rate. This approach is appealing because it would in principle provide the dose equivalent or effective dose rate in a single measurement; ostensibly much faster than the technique previously described of collecting data from a set of responses at individual applied voltages V. In practice it turns out that measuring differences in ion chamber currents found in practical chambers is difficult due to the small currents and electrical leakage problems associated with electrical feed-throughs and cable connectors.

Höfert and Raffnsøe (1980) have measured the dose equivalent response of such an instrument as a function of neutron energy and obtained the results in Table 9.8.

In examining Table 9.8, the "error" in the third column represents the error of the measurement technique and not the deviation of the dose equivalent measurement from the true dose equivalent. An "ideal" instrument used to measure dose equivalent would achieve the same value of dose equivalent response at all energies. The recombination chamber technique may be applicable to other ion chambers aside from those manufactured specifically for this purpose if successfully tested in known mixed radiation fields.

9.5.8 Counter Telescopes

Since the dose equivalent per fluence for muons varies so little over a wide range (see Figure 1.4), *scintillation telescopes* provide an attractive method for assessing pure muon radiation fields. At suitable distances and at forward angles, muons will dominate the radiation fields, and the result is that little or no discrimination against other particles is necessary.

At Fermilab a pair of 20.32 cm by 20.32 cm by 0.635 cm thick plastic scintillators has been used routinely (Cossairt 1983). The separation distance between these detectors

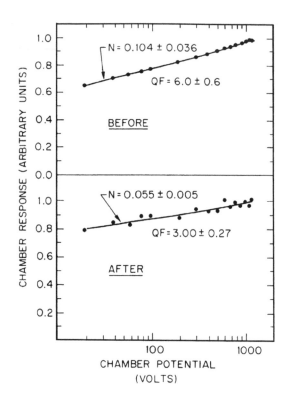

FIGURE 9.18

Recombination chamber response functions measured both *before* (*top frame*) and *after* (*bottom frame*) the placement of additional shielding associated with the configuration described in Figure 6.15. The value of "QF" is the quality factor Q as defined by the 1973 System. (Reprinted with permission from Elwyn, A. J., and Cossairt, J. D. 1986. *Health Physics* 51:723–735. https://journals.lww.com/health-physics/pages/default.aspx.)

provides moderate directional sensitivity when a coincidence is required between the two scintillator paddles in a relatively parallel beam of muons. An aluminum plate, 2.54 cm thick, is employed in the gap between the two scintillators to reduce false coincidences due to recoil electrons ("δ-rays") produced in collisions occurring in the first scintillator being detected by the second, if the aluminum were absent. These plates are mounted in an

TABLE 9.7

Average Quality Factors Q (1973 System) Obtained for Various Neutron Energy Spectra Measurements at Fermilab

	Technique	
Description of Radiation Field	**Unfolding**	**Recombination**
Mixed field of neutrons and muons (Cossairt and Elwyn 1987)	1.4 ± 0.2	1.1 ± 0.3
Iron leakage spectrum before shielding was added (Figure 6.16) (Elwyn and Cossairt 1986)	5.4 ± 0.2	6.0 ± 0.6
Iron leakage spectrum after shielding was added (Figure 6.17) (Elwyn and Cosairt 1986)	2.5 ± 0.3	3.0 ± 0.3
Spectrum in a labyrinth (Figure 6.14) (Cossairt et al. 1985b)	3.1 ± 0.7	3.4 ± 0.1

Source: Adapted from Cossairt, J. D. et al. 1988. *Measurement of neutrons in enclosures and outside of shielding at the Tevatron*. Fermi National Accelerator Laboratory: Fermilab Report FERMILAB-CONF-88/106. Batavia, IL.

TABLE 9.8

Dose Equivalent Response and Measurement Errors for Recombination Chamber as a Function of Neutron Energy

Neutron Energy (MeV)	Dose Equivalent Response (10^5 Coulombs Sv^{-1})	Error (%)
Thermal	0.830	10.0
0.0245	2.579	12.1
0.1	1.451	6.2
0.25	1.585	6.1
0.57	1.215	5.2
1.0	1.215	5.2
2.5	1.112	6.1
5.0	0.840	5.2
15.5	0.728	5.2
19.0	0.998	12.1
280.0	0.782	10.1

Source:: Adapted from Höfert M., and Ch. Raffnsøe. 1980. *Nuclear Instruments and Methods* 176:443–448.

all-terrain vehicle, called the Mobile Environmental Radiation Laboratory (MERL), and are powered by an on-board electrical generator. For mixed radiation fields, the MERL is also used for neutron measurements with a DePangher long counter and other instruments for measuring airborne radionuclide releases. A microwave telemetry system provides gating pulses and proton beam intensity information so that normalized beam-on and beam-off (background) measurements can be taken simultaneously. The scintillators were chosen to provide sufficient sensitivity to obtain statistical errors at the 20% level in remote locations receiving annual dose equivalents in the range of a few microsieverts (μSv) in a scan lasting an hour or two. In such a scan, the detectors are moved across a region of elevated muon flux density, stopping at several locations to acquire data and thus map out the spatial distribution of muons. In these detectors, a muon beam perpendicular to the detectors yields 1.72×10^4 counts minute^{-1} per μSv h^{-1} (or 1.03×10^6 counts μSv $^{-1}$). The normal singles background (i.e., the background of an individual scintillator not counted in coincidence with the other member of the pair) due to cosmic rays at Fermilab is approximately 400 counts per minute.

Smaller, more portable systems can be useful in conducting muon surveys. Fermilab has built such a system, called a *muon finder*, consisting of a pair of small plastic scintillators mounted in a compact package which is battery powered and can be carried by one person. It is read out by scalers and can record both singles and coincidence rates. The ratio of the two can be used to "find" unknown muon sources, hence the name of the detector. Also, the separation distance can be adjusted to enhance or reduce the directional sensitivity. This system has been described by Vaziri et al. (2004).

The parameters of this system are given in Table 9.9 where the dimensions, cone of sensitivity, and dose equivalent/effective dose calibration for muons at normal ($\perp$) incidence are given.

Of course, the use of such scintillators, especially in the "singles" mode, in mixed fields of muons and neutrons requires that one must be aware that the plastic scintillators have a nonzero detection efficiency for the neutrons. Vylet (1991) has used the values of total cross sections to calculate the neutron detection efficiency of the detectors previously described

TABLE 9.9

Parameters of the "Muon Finder" Used at Fermilab

Scintillator diameter	2.1 cm
Scintillator thickness	0.635 cm
Scintillator area	3.6 cm²
Scintillator spacing	0.5–8.9 cm
Half-angle cone of sensitive	0.9–0.2 radians (51–11.5°)
Dose calibration (muons ⊥ detectors)	90 muons μrem⁻¹
Dose rate calibration (muons ⊥ detectors)	25 muons s⁻¹ mrem⁻¹ h

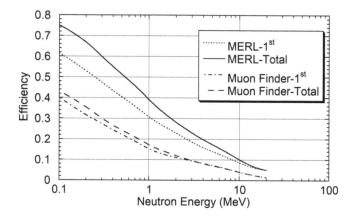

FIGURE 9.19

Calculated neutron efficiencies of scintillation counters used in the "singles" mode at Fermilab as a function of neutron energy as described in the text. (Adapted from Vylet, V. 1991. *Estimated sensitivity of the "muon gun" to neutrons.* Fermi National Accelerator Laboratory: Fermilab Radiation Physics Note. No. 92. Batavia, IL.)

for neutrons over a range of energies. The results are given in Figure 9.19. In this figure, effects due to the first and successive collisions (labeled "Total") as well as those due to just the first collisions (labeled "1st") with hydrogen atoms are given. The total efficiencies at the upper end of the energy region measured were an efficiency of 0.058 for the MERL scintillation counters and 0.0235 for the muon finders.

PROBLEMS

1. A cylindrical ion chamber is 5.0 cm in radius and 20 cm long. It is filled with methane (CH_4) at 1.0 atm absolute pressure. It is bombarded by a uniform flux density of high-energy (minimum-ionizing) muons incident perpendicularly to one of the ends. One can safely assume that the passage of the muons through the entire length of the chamber represents insignificant degradation of the muon energy or direction. The dose equivalent rate in the radiation field is 0.1 mrem hour⁻¹.

   a. Calculate the electric current that will be drawn from this chamber that represents the "signal" to be measured and correlated with the dose equivalent rate. One could use Table 1.2 to obtain values of $(dE/dx)_{min}$ and to obtain the density of CH_4.

b. If the charge liberated in the chamber is collected (i.e., integrated electronically) for 1.0 second and the chamber and circuit represent a capacitance of 10^{-10} Farads, calculate the size of the signal pulse in volts if one neglects any "pulse-shaping" of the readout electronics.

2. Consider the detector based on the 25.4 cm diameter moderating sphere for which the corresponding response curve is displayed in Figure 9.8.

   a. Calculate the approximate absolute intrinsic detection efficiency for neutrons. This is to be done for the $2.0 < E_n < 8.0$ MeV energy domain, and the sharp peaks in the detector response curve are to be ignored (i.e., averaged out). In this problem, 100% efficiency is defined to be 1.0 count generated for every neutron that strikes the sphere. Assume the incident neutrons to be aimed at the detector originating from a "point source", despite the fact that this is not quite true.

   b. Since the LiI detector only responds to thermal neutrons, calculate the efficiency with which the moderator transforms fast neutrons incident upon it into thermal neutrons present at the LiI. For this calculation, neglect any "dopants" in the LiI, assume that the Li is "natural" lithium with respect to isotopic abundance, and use the fact that the atomic weight of iodine is 127. The density of LiI is 3.5 g cm^{-3}. Assume that the detector is 100% efficient in detecting thermal neutron captures within its volume.

3. A BF$_3$ proportional chamber is used in a DePangher long counter. This detector, when placed in a certain neutron field that is known to be dominated by neutrons of approximately 5.0 MeV kinetic energy, has a response due to neutrons of 1.0 count min^{-1}. The detector sensitivity is that discussed in the text. The counter operates at 1.0 atm absolute pressure, the atomic weight of boron is 10.8 while the atomic weight of fluorine is 19. At standard temperature and pressure (STP) the density of BF$_3$ is 2.99 g liter^{-1}.

   a. What is the dose equivalent rate of this radiation field?

   b. If the radiation field persists full time, is this detector sufficiently sensitive to detect a dose rate of 10 mrem year^{-1}?

   c. In this radiation field, high-energy minimum ionizing muons pass through this detector, including the proportional counter. The largest muon signals in the proportional counter will obviously result when the muons pass lengthwise through the tube. If the tube is 40 cm long, what will be the size of the largest muon-induced signal relative to the neutron-induced signal? Is it likely that a simple discriminator circuit can be used to eliminate the muon-induced signals? It is quite permissible to estimate the value of $(dE/dx)_{min}$ by roughly interpolating among the values tabulated in Table 1.2.

4. One needs to understand the sensitivity of the technique of using the nuclear reaction $^{12}C(n,2n)^{11}C$ in a plastic scintillator to measure the dose equivalent rate external to thick concrete or earth shielding near a high-energy accelerator. The detector discussed in the text used by Moritz (1989) has a sensitive volume of approximately 100 cm^3 (a 5.0 cm diameter by 5.0 cm long cylinder). From Knoll (2010) the NE102A scintillator has a density of 1.032 g cm^{-3}. This detector is nearly 100% efficient at sensing the 0.511 MeV annihilation photons produced in the course of the ^{11}C decay.

a. This detector is irradiated in a particular radiation field external to such accelerator shielding. The irradiation, which is steady in time, is of sufficient length in time to result in saturation of the production of ^{11}C in the scintillator. After the beam is turned off, the detector records a rate of 10 counts per minute (including appropriate decay-correction to the instant of beam shutdown). Calculate the flux density of neutrons with $E_n > 20$ MeV during the irradiation and use the result along with Stevenson's conclusion (1984) concerning the conversion from the flux density of neutrons with $E_n > 20$ MeV to determine the dose equivalent rate.

b. Assuming this count rate is the smallest that can be reliably detected, how much smaller in volume can the detector be for it to barely be sensitive to a dose equivalent rate of 2.0 mrem hour^{-1}?

Appendix: Synopses of Common Monte Carlo Codes and Examples for High-Energy Proton-Initiated Cascades

A.1 Introduction

This appendix provides brief summary descriptions of some of the more prominent Monte Carlo codes used at modern particle accelerators. The reader should be cautioned that most of these codes are being constantly improved and updated. The wisest practice in using them is to consult with the authors of the codes directly to obtain detailed, current information. This appendix also provides some representative examples of Monte Carlo high-energy hadronic cascade calculations in the form of contour plots of equal star density as a function of longitudinal coordinate Z and radial coordinate R.

A.2 Synopses of Prominent Monte Carlo Codes

A.2.1 EGS

EGS, the "Electron Gamma Shower" code is a powerful code for calculating electromagnetic cascades. A recent version, current as of this writing, is EGS5. A complete description of this code system at an earlier version, EGS4, has been written by Nelson et al. (1985). Hirayama et al. (2016) has described the EGS5 version. This program provides a Monte Carlo analysis of electron and photon scattering including shower generation. In its standard usage, it does not calculate hadron or muon production directly. The lower limit of its validity is about 10 keV, while the upper limit of its validity is at least 1.0 TeV. Possible target materials span the periodic table. As the electron encounters target atoms, it is scattered randomly to mimic the known mechanisms of electron scattering. When secondary particles arise, they are loaded into a stack from which the program selects sequentially the lowest-energy particle and then traces out its further path until it leaves the target or until its energy falls below a selected cutoff value. The final kinematical and charge states of all the particles are noted and summed for all particles in the shower, concluding with a "history" of all of them. Improvements with the code are continuously being made. The code is a fundamental tool at many laboratories that have electron accelerators. It has been found to be extremely useful in applications in medicine and also in modeling the performance of high-energy physics apparatus. The EGS5 code system is available from the High Energy Research Organization of Japan (KEK). The official EGS5 website is http://rcwww.kek.jp/research/egs/egs5.html (accessed August 6, 2018).

A.2.2 CASIM

Van Ginneken and Awschalom (1975) developed this "Cascade Simulation" program. It was designed to simulate the average behavior of hadrons in the region 10–1000 GeV and was extended to 20 TeV (Van Ginneken et al. 1987). It used inclusive production distributions directly in order to obtain the particles to follow. The particle production algorithm was based on the Hagedorn-Ranft thermodynamic model. Only one or two high-energy particles were created in each simulated collision, and these carried a weight related to the probability of their production and the energy carried with them. Path length-stretching and particle-splitting were used. Electromagnetic showers resulting from π° production were calculated using the companion code AEGIS (Van Ginneken 1978). Magnetic fields could readily be included. The user generally wrote a FORTRAN subroutine to describe the geometry of interest with "logical" (i.e., "IF," etc.) statements used to deduce from the particle's spatial and directional coordinates the material type or magnetic field in which a particle being tracked was present at a given "time" in the calculation. A muon version called CASIMU, subsequently called MUSIM, was written (Van Ginneken et al. 1987). The accuracy of the hadron version was verified for energies up to 800 GeV (Awschalom et al. 1975, 1976; Cossairt et al. 1982, 1985a).

The muon version was verified for energies up to 800 GeV for production and transport of muons in real-life, complicated shields (Cossairt et al. 1989a,b). Normally, CASIM was not configured to follow particles with momenta less than 300 MeV/c, corresponding to a kinetic energy of 47 MeV for nucleons. All low-energy phenomena were then obtained by matching energy spectra and fluence at this energy with results of codes capable of tracking lower-energy particles (e.g., HETC, FLUKA, and MARS). Direct calculations of radioactivation were not available. These were severe limitations along with the tedious method for specifying the geometry. At Fermilab, CASIM was replaced by MARS as the code of choice. Results obtained using CASIM still serve as benchmarks.

A.2.3 FLUKA

FLUKA is an integrated, versatile multiparticle Monte Carlo program capable of handling a wide variety of radiation transport problems. Its energy range extends from 1.0 keV (for neutrons, thermal energies) to thousands of TeV. FLUKA can simulate with a similar level of accuracy the propagation of hadronic and electromagnetic cascades, cosmic muons, slowing-down neutrons, and synchrotron radiation in the keV region. An original treatment of multiple Coulomb scattering allows the code to handle accurately some challenging problems such as electron backscattering and energy deposition in thin layers. In a fully analog mode, FLUKA can be used in detector studies to predict fluctuations, coincidences, and anticoincidences. A rich supply of biasing options makes it well suited for studies of rare events, deep penetrations, and shielding in general. This code originated as a high-energy particle transport code developed by a CERN-Helsinki-Leipzig collaboration led by J. Ranft as discussed by Aarnio et al. (1986). More recently, it has been completely rewritten and extended to low energies as discussed by Fassò et al. (1993). It handles more than 30 different particles, including neutrons from thermal energies to about 20 TeV and photons from 1.0 keV to thousands of TeV. Several biasing techniques are available. Recoil protons and protons from A(n,p) reactions are transported explicitly. The official FLUKA website is http://www.fluka.org/fluka.php (accessed August 6, 2018).

A.2.4 HETC and LAHET

This code, developed over many years under the leadership of R. G. Alsmiller, Jr., and F. S. Alsmiller at the Oak Ridge National Laboratory, Tennessee, is considered by some to be the benchmark hadron shielding code of choice. It has been upgraded many times and can, in suitably augmented versions, follow particles from the 20 TeV region down to thermal neutron energies. It is an extremely flexible code but has the important disadvantage that the individual events are written to mass storage. It is the responsibility of the user to write a program to analyze the results. In terms of processor time, HETC is also relatively slow so that calculations to be done should be carefully selected. It is seen to be preferable to use selected HETC runs to calibrate other faster, but less accurate codes. It has been described by Armstrong (1980) and Gabriel (1985). It includes the same event generator used for FLUKA.

The Los Alamos Energy Transport Code (LAHET) variant of this code developed at the Los Alamos National Laboratory, New Mexico, has been described by Prael and Lichtenstein (1989). It is available from the Los Alamos National Laboratory. LAHET permits the transport of neutrons, photons, and light nuclei up to ^{4}He; employs the geometric capabilities of the MCNP code; and has been merged with more recent evolutions of the MCNP family of codes. The LAHET website is http://www.oecd-nea.org/tools/abstract/detail/ccc-0696/ (accessed August 6, 2018).

A.2.5 MCNP

MCNP is a general-purpose Monte Carlo N-particle code that can be used for neutron, photon, electron, or coupled neutron/photon/electron transport. This code has been developed at the Los Alamos National Laboratory and is well documented in LANL reports (e.g., Briesmeister 1997). The code treats an arbitrary three-dimensional configuration of materials in geometric cells bounded by first- and second-degree surfaces and fourth-degree elliptical tori. The neutron energy regime is from 10^{-11} to 150 MeV, and the photon and electron energy regimes are from 1.0 keV to 1000 MeV. For neutrons, all reactions given in a particular cross section evaluation exemplified by the ENDF/B-VI cross section database (NNDC 2018) are accounted for. Thermal neutrons are described by both the free gas and thermal particle scattering models. For photons, the code takes account of incoherent and coherent scattering, the possibility of fluorescent emission after photoelectric absorption, absorption in pair production with local emission of annihilation radiation, and bremsstrahlung. A continuous slowing-down model is used for electron transport that includes positrons, x-rays, and bremsstrahlung but does not include external or self-induced fields. Important standard features that make MCNP very versatile and easy to use include a powerful general source and a surface source; both geometry and output tally plotters; a rich collection of variance reduction techniques; a flexible tally structure (including a pulse-height tally); and an extensive collection of cross section data. The official MCNP website is http://mcnp.lanl.gov/ (accessed August 6, 2018).

A.2.6 MARS

The MARS Monte Carlo code system has been under continuous development over a number of years (Kalinovskii et al. 1989; Mokhov 1995; Kriovsheev and Mokhov 1998; Mokhov et al. 2004). Results have been compared by Mokhov and Cossairt (1986) with those obtained using then-current versions of CASIM and FLUKA with good agreement. The code allows exclusive and fast inclusive simulation of three-dimensional hadronic and electromagnetic cascades for shielding, accelerator, and detector components in the energy range from a fraction of an electron volt up to 100 TeV. The current version MARS15 uses the phenomenological model for inclusive hadron- and photon-nucleus interactions for $E > 5.0$ GeV and exclusive cascade-exciton model at 1.0 MeV $< E < 5.0$ GeV. By using the Los Alamos Quark-Gluon String Model (LAQGSM) event generator (Gudima et al. 2001), full theoretically consistent modeling of exclusive distributions of secondary particles, spallation, fission, and fragmentation products for hadron and heavy-ion beams of any energy can be done. MARS15 includes photo- and electro-production of hadrons and muons, advanced algorithms for the three-body decays, precise particle tracking in magnetic fields, synchrotron radiation by electrons and muons, extended histogram capabilities, and improved material description and computational performance. Along with direct energy deposition calculations, a set of dose conversion per fluence factors for all particles including neutrinos is incorporated. The code includes links to the MCNP4C code for neutron and photon transport below 20 MeV, to the ANSYS code for thermal and stress analyses, and to the MAD and STRUCT codes for lattice description for multiturn particle tracking in large synchrotrons and collider rings. MARS can also be linked to EGS5.

The geometry module allows the use of a set of the predefined shapes with the MARS extended geometry input data files or with the ROOT geometry package (http://root.cern.ch/drupal/content/users-guide, accessed August 6, 2018), arbitrary user-defined three-dimensional descriptions, or direct use of MCNP and FLUKA geometry input data files. The code is provided with a user-friendly graphical-use interface for geometry. A parallelized version of the code can run in a multiprocessor mode. The developments were induced by numerous challenging applications; Fermilab accelerator, detector, and shielding upgrades; Large Hadron Collider machine and detector studies; muon and electron-positron colliders, etc.; as well as by a continuous desire to increase code reliability, flexibility, and user friendliness. This code is continuously updated and improved and can be obtained from Fermilab. It is demonstrably the most advanced high-energy code available. The official MARS website is https://mars.fnal.gov/ (accessed August 6, 2018).

A.3 Sampling of Representative Star Density Plots

As discussed in Section 4.7.2, a convenient way to exhibit the "raw" output of Monte Carlo high-energy hadronic cascade calculations is in the form of contour plots of equal star density as a function of longitudinal coordinate Z and radial coordinate R. This section contains representative samples of such plots obtained using CASIM and MARS. This collection is illustrative in nature; it is clearly not intended to address all situations. The main text refers to more complete compilations of calculations. Individual calculations should be done for definitive results.

Figure A.1a through d presents results obtained using CASIM for protons incident along the axis of a solid CONCRETE cylinder perpendicular to one face of the cylinder. The concrete is of "standard" composition" and is taken to have a density of 2.4 g cm^{-3}.

Figure A.2a through d presents results obtained using CASIM for protons incident along the axis of a solid IRON cylinder perpendicular to one face of the cylinder. The iron is assumed to have a "textbook" density of 7.87 g cm^{-3}. As discussed in Chapter 6, most commercial sources of large pieces of iron have somewhat lower densities.

Figure A.3a through d presents results obtained using CASIM for 1.0 TeV protons incident on some examples of beamline components in enclosures or soil shields. For the components, the standard densities found in Table 1.2 were used. The density of concrete was taken to be 2.4 g cm^{-3}, and the density of soil was taken to be 2.24 g cm^{-3}. Beam pipes interiors were assumed to be evacuated. The captions describe the details of the beam loss scenarios used in the calculations.

Figure A.4a through c presents results of calculations obtained using MARS for 120 GeV protons incident on some examples of beamline components in enclosures or soil shields. For these calculations the concrete density was assumed to be 2.4 g cm^{-3}, while the soil density was taken to be 2.24 g cm^{-3}. The point losses were simulated by positioning the centroid of the proton beam one standard deviation (1.0 σ) from the inner surface of the chosen target aperture in a Gaussian distribution at an incident angle of zero. While a small nonzero incident angle would be more realistic, very small grazing angles on the order of $\sin \theta \leq 10^{-4}$ or less will produce only minimal increases (<20%) in the total dose. Larger grazing angles could potentially increase the dose by a factor of six or more.

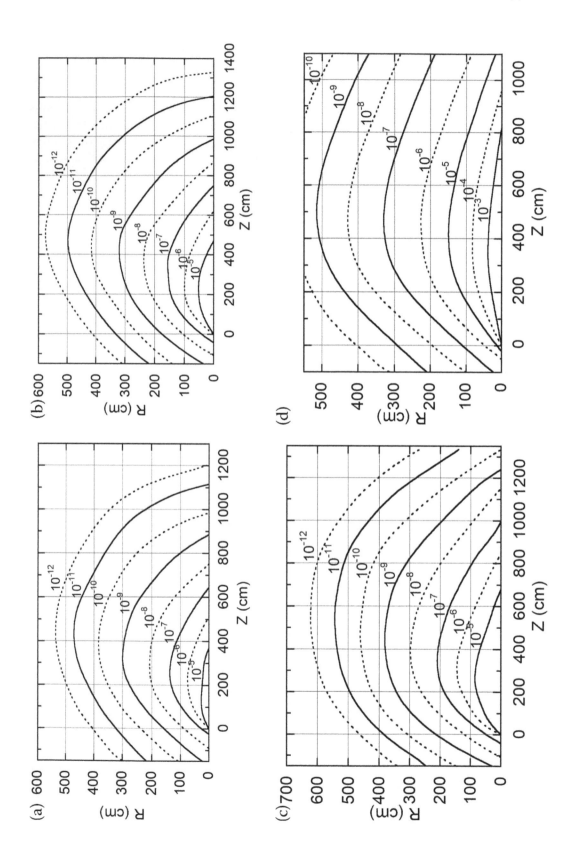

FIGURE A.1

(a) Monte Carlo results using CASIM for 30 GeV/c protons incident on a CONCRETE cylinder. Contours of equal star density (stars cm^{-3}) per incident proton are plotted. The beam of 0.3×0.3 cm^2 cross section is centered on the cylinder axis and starts to interact at zero depth ($Z = 0$). The star density includes only those due to hadrons above 0.3 GeV/c momentum. Contours of higher star density are not shown for clarity, while those of lower star density are not included due to statistical uncertainty. (b) Monte Carlo results using CASIM for 100 GeV/c protons incident on a CONCRETE cylinder. Contours of equal star density (stars cm^{-3}) per incident proton are plotted. The beam of 0.3×0.3 cm^2 cross section is centered on the cylinder axis and starts to interact at zero depth ($Z = 0$). The star density includes only those due to hadrons above 0.3 GeV/c momentum. Contours of higher star density are not shown for clarity, while those of lower star density are not included due to statistical uncertainty. (c) Monte Carlo results using CASIM for 1.0 TeV/c protons incident on a CONCRETE cylinder. Contours of equal star density (stars cm^{-3}) per incident proton are plotted. The beam of 0.3×0.3 cm^2 cross section is centered on the cylinder axis and starts to interact at zero depth ($Z = 0$). The star density includes only those due to hadrons above 0.3 GeV/c momentum. Contours of higher star density are not shown for clarity, while those of lower star density are not included due to statistical uncertainty. (Adapted from Van Ginneken, A., and M. Awschalom. 1975. *High energy particle interactions in large targets: Volume I, hadronic cascades, shielding, and energy deposition.* Fermi National Accelerator Laboratory. Batavia, IL.) (d) Monte Carlo results using CASIM for 10 TeV/c protons incident on a CONCRETE cylinder. Contours of equal star density (stars cm^{-3}) per incident proton are plotted. The beam of 0.3×0.3 cm^2 cross section is centered on the cylinder axis and starts to interact at zero depth ($Z = 0$). The star density includes only those due to hadrons above 0.3 GeV/c momentum. Contours of higher star density are not shown for clarity, while those of lower star density are not included due to statistical uncertainty. (Adapted from Van Ginneken et al. 1987. *Shielding calculations for multi-TeV hadron colliders.* Fermilab Report FN-447. Batavia, IL.)

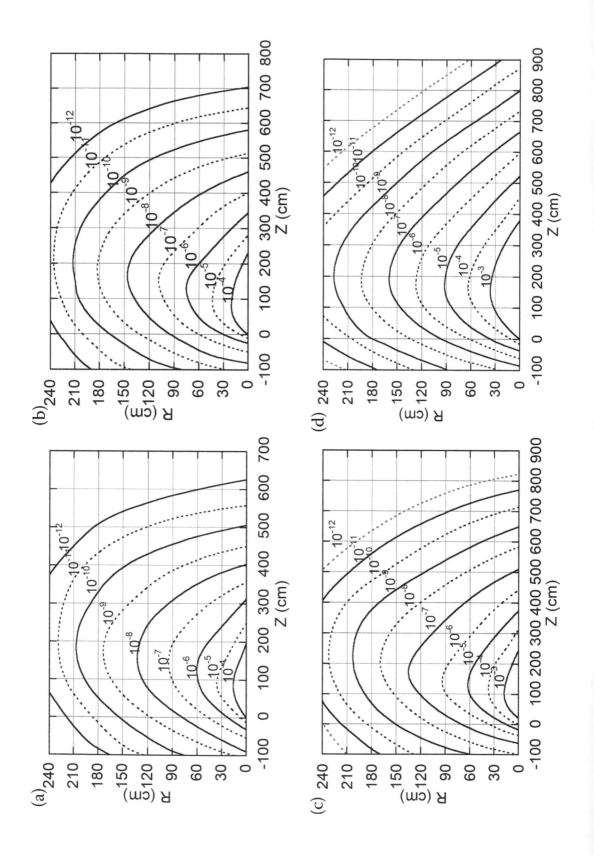

FIGURE A.2

(a) Monte Carlo results using CASIM for 30 GeV/c protons incident on an IRON cylinder. Contours of equal star density (stars cm^{-3}) per incident proton are plotted. The beam of 0.3×0.3 cm^2 cross section is centered on the cylinder axis and starts to interact at zero depth ($Z = 0$). The star density includes only those due to hadrons above 0.3 GeV/c momentum. Contours of higher star density are not shown for clarity, while those of lower star density are not included due to statistical uncertainty. (b) Monte Carlo results using CASIM for 100 GeV/c protons incident on an IRON cylinder. Contours of equal star density (stars cm^{-3}) per incident proton are plotted. The beam of 0.3×0.3 cm^2 cross section is centered on the cylinder axis and starts to interact at zero depth ($Z = 0$). The star density includes only those due to hadrons above 0.3 GeV/c momentum. Contours of higher star density are not shown for clarity, while those of lower star density are not included due to statistical uncertainty. (c) Monte Carlo results using CASIM for 1.0 TeV/c protons incident on an IRON cylinder. Contours of equal star density (stars cm^{-3}) per incident proton are plotted. The beam of 0.3×0.3 cm^2 cross section is centered on the cylinder axis and starts to interact at zero depth ($Z = 0$). The star density includes only those due to hadrons above 0.3 GeV/c momentum. Contours of higher star density are not shown for clarity, while those of lower star density are not included due to statistical uncertainty. (Adapted from Van Ginneken, A., and M. Awschalom. 1975. *High energy particle interactions in large targets: Volume I, hadronic cascades, shielding, and energy deposition*. Fermi National Accelerator Laboratory. Batavia, IL.) (d) Monte Carlo results using CASIM for 10 TeV/c protons incident on an IRON cylinder. Contours of equal star density (stars cm^{-3}) per incident proton are plotted. The beam of 0.3×0.3 cm^2 cross section is centered on the cylinder axis and starts to interact at zero depth ($Z = 0$). The star density includes only those due to hadrons above 0.3 GeV/c momentum. Contours of higher star density are not shown for clarity, while those of lower star density are not included due to statistical uncertainty. (Adapted from Van Ginneken et al. 1987. *Shielding calculations for multi-TeV hadron colliders*. Fermi National Accelerator Laboratory: Fermilab Report FN-447. Batavia, IL.)

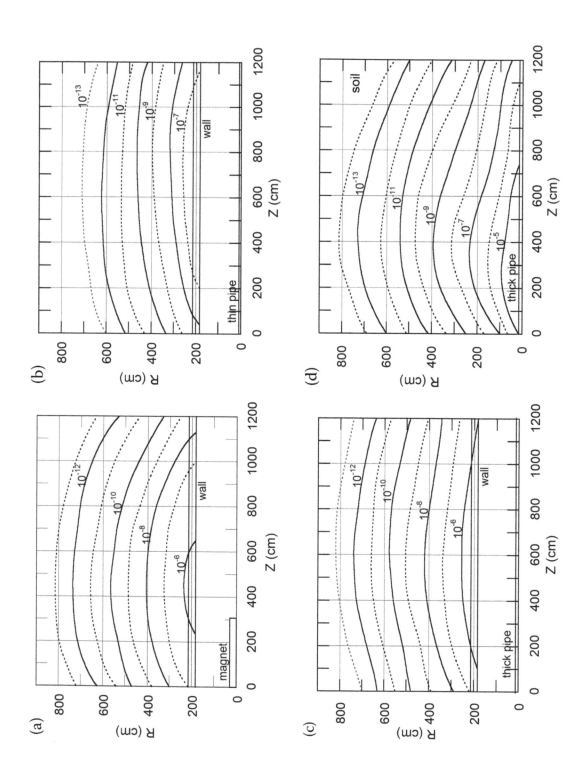

FIGURE A.3

(a) Contour plots of equal star density (stars cm^{-3}) per incident proton calculated using CASIM for a 1.0 TeV proton beam incident "head on" on the inner edge of one of the pole pieces one standard deviation of beam width beyond the inner aperture of the magnet at $Z = 0$. The cross section of the magnet was rectangular with an aperture of 3.8×12.7 cm^2 and outer dimensions of 31.8×40.6 cm^2. No magnetic fields were included in the model. The results were averaged over azimuth, and the magnet was centered in a cylindrical tunnel 182 cm in radius. The concrete wall was 30.48 cm thick and was surrounded by soil. (b) Contour plots of equal star density (stars cm^{-3}) per incident proton calculated using CASIM for a 1.0 TeV proton beam incident "head on" centered on a thin cylindrical aluminum pipe of 10.16 cm outside diameter with 0.318 cm thick walls at $Z = 0$. The results were averaged over azimuth, and the pipe was centered in a cylindrical tunnel 182 cm in radius. The concrete wall was 30.48 cm thick and was surrounded by soil. (c) Contour plots of equal star density (stars cm^{-3}) per incident proton calculated using CASIM for a 1.0 TeV proton beam incident "head on" centered on a thick cylindrical iron pipe of 30.48 cm outside diameter with 1.27 cm thick walls at $Z = 0$. The results were averaged over azimuth, and the pipe was centered in a cylindrical tunnel 182 cm in radius. The concrete wall was 30.48 cm thick and was surrounded by soil. (d) Contour plots of equal star density (stars cm^{-3}) per incident proton calculated using CASIM for a 1.0 TeV proton beam incident "head on" centered on a thick cylindrical iron pipe of 30.48 cm outside diameter with 1.27 cm thick walls at $Z = 0$. The pipe was surrounded by soil. (Adapted from Cossairt, J. D. 1982. *A collection of CASIM calculations.* Fermi National Accelerator Laboratory: Fermilab Report TM-1140. Batavia, IL.)

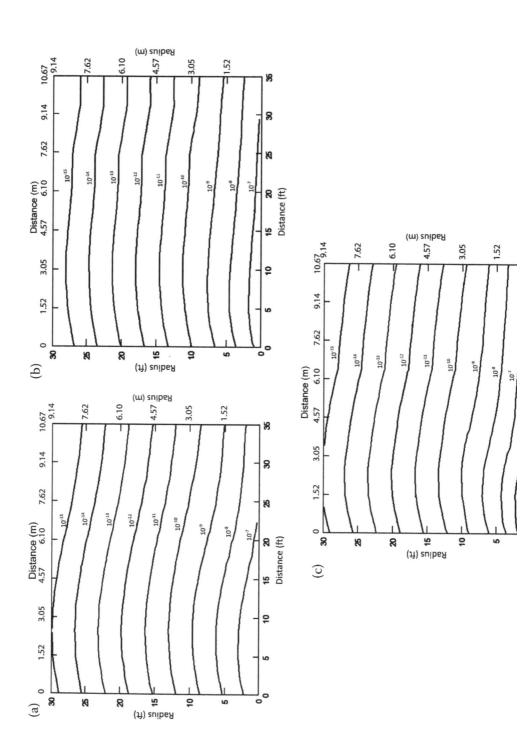

FIGURE A.4

(a) Contour plots of equal star density (stars cm⁻³) per incident proton calculated using MARS for a 120 GeV proton beam incident "head on" on the inner edge of one of the pole pieces of a magnet one standard deviation of beam width deep. The results are given as a function of distance along the beam direction and radius. The point loss was modeled to occur at a longitudinal distance of zero in the figure. The cross section of the magnet was rectangular with an aperture of 3.8×12.7 cm² and outer dimensions of 31.8×40.6 cm². No magnetic fields were included in the model. The results were averaged over azimuth, and the magnet was centered in a cylindrical tunnel 91.4 cm in radius. The concrete wall was 30.48 cm thick and was surrounded by soil. (b) Contour plots of equal star density (stars cm⁻³) per incident proton in soil calculated using MARS for a 120 GeV proton beam incident "head on" on a thin cylindrical aluminum pipe of 10.16 cm outside diameter with 0.318 cm thick walls. The results are given as a function of distance along the beam direction and radius. The point loss was modeled to occur at a longitudinal distance of zero in the figure. The results were averaged over azimuth, and the pipe was centered in a cylindrical tunnel 91.4 cm in radius. The concrete wall was 30.48 cm thick and was surrounded by soil. (c) Contour plots of equal star density (stars cm⁻³) per incident proton calculated using MARS for a 120 GeV proton beam incident "head on" on a thick cylindrical iron pipe of 30.48 cm outside diameter with 1.27 cm thick walls. The results are given as a function of distance along the beam direction and radius. The point loss was modeled to occur at a longitudinal distance of zero in the figure. The pipe was surrounded by soil. The lower star densities at very small radii result from buildup processes. These are not seen in the representative star density distributions for the other scenarios as the buildup processes take place in the concrete enclosure walls and are not included in these distributions in soil. (Adapted from Reitzner S. D. 2012. *Update to the generic shielding design criteria.* Fermi National Accelerator Laboratory: Fermilab Report TM-2550-ESH. Batavia, IL.)

References

Aarnio, P. A., A. Fassò, H. J. Moehring, J. Ranft, G. R. Stevenson. 1986. *FLUKA86 users guide*. European Organization for Nuclear Research: CERN Divisional Report TIS-RP/168. Geneva, Switzerland.

Aleinikov, V. E., A. P. Cherevatenko, F. B. Clapier, and V. I. Tsovbun. 1985. Neutron radiation field due to 6.6 MeV/amu ^{58}Ni ions bombarding a thick Cu target. *Radiation Protection Dosimetry* 11:245–248.

Alsmiller, Jr., R. G., and J. Barish. 1973. Shielding against the neutrons produced when 400 MeV electrons are incident on a thick copper target. *Particle Accelerators* 5:155–159.

Alsmiller, Jr., R. G., J. Barish, and R. L. Childs. 1981. Skyshine at neutron energies <400 MeV. *Particle Accelerators* 11:131–141.

Alsmiller, Jr., R. G., R. T. Santoro, and J. Barish. 1975. Shielding calculations for a 200 MeV proton accelerator and comparisons with experimental data. *Particle Accelerators* 7:1–7.

Anderson, M. P. 2007. Introducing groundwater physics. *Physics Today* 2007:42–47.

Andersson, I. O., and J. Braun. 1964. *A neutron rem counter*. U.S. Department of Energy Office of Scientific and Technical Information (OSTI). OSTI No. 4079749: Report AE-132. Washington, DC.

Apfel, R. E. 1979. The superheated drop detector. *Nuclear Instruments and Methods* 162:603–608.

Armstrong, T. W. 1980. The HETC hadronic cascade code. In *Computer Techniques in Radiation Transport and Dosimetry*, eds. W. R. Nelson and T. M. Jenkins, 373–385. New York, NY: Plenum Press.

Armstrong, T. W., and R. G. Alsmiller, Jr. 1969. Calculation of the residual photon dose rate around high energy proton accelerators. *Nuclear Science and Engineering* 38:53–62.

Armstrong, T. W., and J. Barish. 1969. Calculation of the residual photon dose rate due to the activation of concrete by neutrons from a 3-GeV proton beam in iron. *Nuclear Science and Engineering* 38:265–270.

Aroua, A., T. Buchillier, M. Grecescu, and M. Höfert. 1997. Neutron measurements around a high-energy lead ion beam at CERN. *Radiation Protection Dosimetry* 70:437–440.

Ashmore, A. G., Coccioni, A. N. Diddens, and A. M. Wetherell. 1960. Total cross sections of protons with momentum between 10 and 28 GeV/c. *Physical Review Letters* 5:576–578.

Atkinson, J. H., W. N. Hess, V. Perez-Mendez, and R. Wallace. 1961. 5-BeV neutron cross sections in hydrogen and other elements. *Physical Review* 123:1850–1859.

Awschalom, M., 1972. Bonner spheres and tissue equivalent chambers for extensive radiation area monitoring around a 1/2 TeV proton synchrotron. In *Proceedings of IAEA Symposium on Neutron Monitoring for Radiation Protection Purposes*. Volume 1, 297-313. Vienna, Austria: International Atomic Energy Agency (IAEA).

Awschalom, M., S. Baker, C. Moore, A. Van Ginneken, K. Goebel, and J. Ranft. 1976. Measurements and calculations of cascades produced by 300 GeV protons incident on a target inside a magnet. *Nuclear Instruments and Methods* 138:521–531.

Awschalom, M., T. Borak, and H. Howe. 1971. *A study of spherical, pseudospherical, and cylindrical moderators for a neutron dose equivalent rate meter*. Fermi National Accelerator Laboratory: Fermilab Report TM-291. Batavia, IL.

Awschalom, M., and L. Coulson. 1973. A new technique in environmental neutron spectroscopy. In *Proceedings of the Third International Conference of the International Radiation Protection Association*. CONF 730907-P2, 1464–1469. Oak Ridge, TN: U.S. Department of Energy Technical Information Center.

Awschalom, M., P. J. Gollon, C. Moore, and A. Van Ginneken. 1975. Energy deposition in thick targets by high energy protons: Measurement and calculation. *Nuclear Instruments and Methods* 131:235–241.

Awschalom, M., and R. S. Sanna. 1985. Applications of Bonner sphere detectors in neutron field dosimetry. *Radiation Protection and Dosimetry* 10:89–101.

Baarli, J., and A. H. Sullivan. 1965. Radiation dosimetry for protection purposes near high energy particle accelerators. *Health Physics* 11:353–361.

Baker, S. I., R. A. Allen, P. Yurista, V. Agoritsas, and J. B. Cumming. 1991. $Cu(p,x)^{24}Na$ cross section from 30 to 800 GeV. *Physical Review* C43:2862–2865.

Baker, S. I., C. R. Kerns, S. H. Pordes, J. B. Cumming, A. Soukas, V. Agoritsas, and G. R. Stevenson. 1984. Absolute cross section for the production of ^{24}Na in Cu by 400 GeV protons. *Nuclear Instruments and Methods* 222:467–473.

Ban, S., H. Hirayama, and S. Muiri. 1989. Estimation of absorbed dose due to gas bremsstrahlung from electron storage rings. *Health Physics* 57:407–412.

Barbier, M. 1969. *Induced radioactivity*. New York, NY: John Wiley and Sons.

Bathow G., E. Freytag, and K. Tesch. 1967. Measurements on 6.3 GeV electromagnetic cascades and cascade producing neutrons. *Nuclear Physics* B2:669–689.

Batu, V. 1998. *Aquifer hydraulics*. New York, NY: John Wiley and Sons.

Berger. M. J., and S. M. Seltzer. 1964. *Tables of energy losses and ranges of electrons and positrons*. National Aeronautics and Space Administration (NASA): Report NASA SP-3012. Washington, DC.

Berger, M. J., and S. M. Seltzer. 1966. *Additional stopping power and range tables for protons, mesons, and electrons*. National Aeronautics and Space Administration (NASA): Report NASA SP-3036. Washington, DC.

Berger, M. J., and S. M. Seltzer. 1970. Bremsstrahlung and photoneutrons from thick tungsten and tantalum targets. *Physical Review* C2:621–631.

Beringer, J. et al. (Particle Data Group). 2012. Review of particle properties. *Physical Review* D86:010001. This report is updated and periodically republished. Current tabulations are available at http://pdg.lbl.gov/ (accessed October 17, 2018).

Biersack J. P., and L. G. Haggmark. 1980. A Monte Carlo computer program for the transport of energetic ions in amorphous targets. *Nuclear Instruments and Methods* 174:257–269.

Borak, T. B., M. Awschalom, W. Fairman, F. Iwami, and J. Sedlet. 1972. The underground migration of radionuclides produced in soil near high energy proton accelerators. *Health Physics* 23:679–687.

Borak, T. B., and J. W. N. Tuyn. 1987. *Can uranium calorimeters become critical?* European Organization for Nuclear Research: CERN Report TIS-RP/194. Geneva, Switzerland.

Brackenbush, L. W., G. W. R. Endres, and L. G. Faust. 1979. Measuring neutron dose and quality factors with tissue-equivalent proportional counters. In *Advances in Radiation Monitoring. Proceedings of a Symposium Held by the International Atomic Energy Agency in Stockholm, Sweden*: Report IAEA-SM-229/52:231-240. Vienna, Austria: International Atomic Energy Agency.

Bramblett, R. L, R. I. Ewing, and T. W. Bonner. 1960. A new type of neutron dosimeter. *Nuclear Instruments and Methods* 9:1–12.

Briesmeister, J. F. ed. 1997. *MCNP: A general Monte Carlo N-particle transport code*. Los Alamos National Laboratory: LANL Report LA-12625-M, Version 4B. Los Alamos, NM.

Britvich, G. I., A. A. Chumakov, R. M. Ronningen, R. A. Blue, and L. H. Heilbronn. 1999. Measurements of thick target neutron yields and shielding studies using beams of ^4He, ^{12}C, and ^{16}O at 155 MeV/nucleon from the K1200 cyclotron at the National Superconducting Cyclotron Laboratory. *Review of Scientific Instruments* 70:2314–2324.

Britvich, G. I., A. A. Chumakov, R. M. Ronningen, R. A. Blue, and L. H. Heilbronn. 2001. Erratum: Measurements of thick target neutron yields and shielding studies using beams of ^4He, ^{12}C, and ^{16}O at 155 MeV/nucleon from the K1200 cyclotron at the National Superconducting Cyclotron Laboratory. *Review of Scientific Instruments* 72:1600.

Butala, S. W., S. I. Baker, and P. M. Yurista. 1989. Measurements of radioactive gaseous releases to air from target halls at a high-energy proton accelerator. *Health Physics* 57:909–916.

Carey, D. C. 1987. *The optics of charged particle beams*. New York, NY: Harwood Academic Publishers.

Cecil, R. A., B. D. Anderson, A. R. Baldwin, R. Madey, A. Galonsky, P. Miller, L. Young, and F. M. Waterman. 1980. Neutron angular and energy distributions from 710-MeV alphas stopping in water, carbon, steel, and lead, and 640-MeV alphas stopping in lead. *Physical Review* C21:2470–2484.

Cember H., and T. E. Johnson. 2009. *Introduction to health physics*, fourth edition. New York, NY: McGraw-Hill.

CFR. 1976. United States Code of Federal Regulations Title 40, Part 141.16. *National primary drinking water standard for β- and γ-emitting radionuclides*. Washington, DC: U.S. Government Printing Office.

CFR. 1989. United States Code of Federal Regulations Title 40, Part 61, Subpart H. *National emissions standard for hazardous air pollutants (NESHAP) for the emission of radionuclides other than radon from Department of Energy Facilities*. Washington, DC: U.S. Government Printing Office.

CFR. 1993. United States Code of Federal Regulations Title 10, Part 835. *Occupational radiation protection at Department of Energy facilities*. Washington, DC: U.S. Government Printing Office.

CFR. 2000. United States Code of Federal Regulations Title 40, Parts 9, 141, and 142. *National primary drinking water regulations*. Washington, DC: U.S. Government Printing Office.

CFR. 2007. United States Code of Federal Regulations Title 10, CFR Parts 820 and 835. *Procedural rules for DOE nuclear activities and occupational radiation protection; Final Rule*. Washington, DC: U.S. Government Printing Office.

CFR. 2017. United States Code of Federal Regulations Title 10, Part 835. *Occupational radiation protection at Department of Energy facilities*. Washington, DC: U.S. Government Printing Office.

Chao, A. W., K. H. Mess, M. Tigner, and F. Zimmermann, eds. 2013. *Handbook of accelerator physics and engineering*, second edition. Singapore: World Scientific.

Chen, F. G., C. P. Leavitt, and A. Shapiro. 1955. Attenuation cross sections for 860 MeV protons. *Physical Review* 99:857–871.

Chilton, A. B., J. K. Shultis, and R. E. Faw. 1984. *Principles of radiation shielding*. Englewood Cliffs, NJ: Prentice Hall.

Citron, A., L. Hoffmann, C. Passow, W. R. Nelson, and M. Whitehead. 1965. A study of the nuclear cascade in steel initiated by 19.2 GeV/c protons. *Nuclear Instruments and Methods* 32:48–52.

Clapier, F., and C. S. Zaidins. 1983. Neutron dose equivalent rates due to heavy ion beams. *Nuclear Instruments and Methods* 217:489–494.

Clark, F. H. 1971. *Shielding Data Appendix E. Protection against neutron radiation*. NCRP: NCRP Report No. 38. Washington, DC.

Cohen, B. L. 1978. Nuclear cross sections. In *Handbook of radiation measurement and protection, Section A: Volume I Physical science and engineering data*, ed. A. Brodsky, 91–212. West Palm Beach, FL: CRC Press/Taylor and Francis Group.

Cohen, M. O, W. Guber, E. S.Troubezkoy, H. Lichtenstein, and H. A. Steinberg. 1973. *SAM-CE, a three dimensional Monte-Carlo code for the solution of the forward neutron and forward and adjoint gamma ray transport equations*. Defense Nuclear Agency: Report DNA-2830-F, Rev. Washington, DC: B Defense Nuclear Agency.

Coor, T., D. A. Hill, W. F. Hornyak. L. W. Smith, and G. Snow. 1955. Neutron cross sections for 1.4-BeV neutrons. *Physical Review* 98:1369–1386.

Cossairt, J. D. 1982. *A collection of CASIM calculations*. Fermi National Accelerator Laboratory: Fermilab Report TM-1140. Batavia, IL.

Cossairt, J. D. 1983. Recent muon fluence measurements at Fermilab. *Health Physics* 45:651–658.

Cossairt, J. D. 1996. On residual dose rate within particle accelerator enclosures. *Health Physics*. 71:315–319.

Cossairt, J. D. 1998. *Rule of thumb for estimating groundwater activation from residual dose rates*. Fermi National Accelerator Laboratory: Fermilab Environmental Protection Note No. 15. Batavia, IL.

Cossairt, J. D. 2008. Accelerator and beam physics for health physicists. In *Topics in accelerator health physics*, eds. J. D. Cossairt, V. Vylet, and J. W. Edwards, 1–45. Madison, WI: Medical Physics. An updated version of this chapter is available as Fermi National Accelerator Laboratory: Fermilab Report FERMILAB-PUB-07-203-ESH-REV. 2015. Batavia, IL.

Cossairt, J. D. 2013a. *Approximate technique for evaluating the Moyer integral in the Moyer model of hadron shielding*. Fermi National Accelerator Laboratory: Fermilab Radiation Physics Note No. 117. Batavia, IL.

Cossairt, J. D. 2013b. *Approximate technique for estimating labyrinth attenuation of accelerator-produced neutrons*. Fermi National Accelerator Laboratory: Fermilab Radiation Physics Note No. 118. Batavia, IL.

Cossairt, J. D., S. W. Butala, and M. A. Gerardi. 1985a. Absorbed dose measurements at an 800 GeV proton accelerator; Comparison with Monte-Carlo calculations. *Nuclear Instruments and Measurements in Physics Research* A238:504–508.

Cossairt, J. D., J. G. Couch, A. J. Elwyn, and W. S. Freeman. 1985b. Radiation measurements in a labyrinth penetration at a high-energy proton accelerator. *Health Physics* 49:907–917.

Cossairt, J. D., and L. V. Coulson. 1985. Neutron skyshine measurements at Fermilab. *Health Physics* 48:171–181.

Cossairt, J. D., and A. J. Elwyn. 1987. Personal dosimetry in a mixed field of high-energy muons and neutrons. *Health Physics* 52:813–818.

Cossairt, J. D., A. J. Elwyn, and W. S. Freeman. 1989b. A study of the production and transport of high energy muons through a soil shield at the Tevatron. *Nuclear Instruments and Methods in Physics Research* A276:86–93.

Cossairt, J. D., A. J. Elwyn, W. S. Freeman, and S. W. Butala. 1989a. A study of the transport of high energy muons through a soil shield at the Tevatron. *Nuclear Instruments and Methods in Physics Research* A276:78–85.

Cossairt, J. D., A. J. Elwyn, W. S. Freeman, W. C. Salsbury, and P. M. Yurista. 1988. *Measurement of neutrons in enclosures and outside of shielding at the Tevatron.* Fermi National Accelerator Laboratory: Fermilab Report FERMILAB-CONF-88/106. Batavia, IL.

Cossairt, J. D., and M. A. Gerardi. 2009. *Measurements of radiation quality factors using a recombination chamber.* Fermi National Accelerator Laboratory: Fermilab Report TM-1248-REV. Batavia, IL.

Cossairt, J. D., N. L. Grossman, and E. T. Marshall. 1997. Assessment of dose equivalent due to neutrinos. *Health Physics* 73:894–898.

Cossairt, J. D., N. V. Mokhov, and C. T. Murphy. 1982. Absorbed dose measurements external to thick shielding at a high energy proton accelerator: Comparison with Monte-Carlo calculations. *Nuclear Instruments and Methods* 197:465–472.

Cossairt, J. D., and K. Vaziri. 2009. Neutron dose per fluence and weighting factors for use at high energy accelerators. *Health Physics* 96:617–628.

Cronin, J. W., R. L. Cool, A. Abashian. 1957. Cross sections of nuclei for high-energy pions. *Physical Review* 107:1121–1130.

DePangher J., and L. L. Nichols. 1966. *A precision long counter for measuring fast neutron flux density.* Battelle Memorial Institute Pacific Northwest Laboratory: Report BNWL-260. Richland, WA.

d'Hombres, M. M., C. Devillers, F. Gervaise, B. de Sereville, and P. Tardy-Joubert. 1971. *Propagation des neutrons dans les tunnes d'Accès à un accélérateur de haute energie à protons.* Centre d'études nucléaires de Saclay: Report CEA-R-3491. Saclay, France.

deSereville, B., and P. Tardy-Joubert. 1971. *Propagation des neutrons dans les tunnels d'access aux accelerateurs de haute energie a protons.* European Organization for Nuclear Research: CERN Report 71-16 Volume 2, 725–774. Geneva, Switzerland.

DeStaebler, H. 1965. *Similarity of shielding problems at electron and proton accelerators.* Stanford Linear Accelerator Center: Report SLAC-Pub-179. Menlo Park, CA. These results are also quoted by Fassò et al. (1990).

DeStaebler, H., T. M. Jenkins, and W. R. Nelson. 1968. Shielding and radiation. In *The Stanford two mile accelerator,* ed. R. B. Neal, 1029–1067. New York, NY: Benjamin.

Edwards, D. A., and M. J. Syphers. 1993. *An introduction to the physics of high energy accelerators.* New York, NY: John Wiley and Sons.

Elwyn, A. J., and J. D. Cossairt. 1986. A study of neutron leakage through an Fe shield at an accelerator. *Health Physics* 51:723–735.

Enge, H. A. 1966. *Introduction to nuclear physics.* Reading, MA: Addison-Wesley.

Fassò, A., A. Ferrari, J. Ranft, and P.R. Sala. 1993. FLUKA: Present status and future developments. In *Proceedings of the IVth International Conference on Calorimetry in High Energy Physics,* La Biodola, Italy, September 21–26, eds. A. Menzione and A. Scribano, 493–502. Singapore: World Scientific.

Fassò, A., K. Goebel, M. Höfert, J, Ranft, and G. Stevenson. 1990. *Landolt-Börnstein numerical data and functional relationships in science and technology new series; Group I: Nuclear and particle physics Volume II: Shielding against high energy radiation,* ed. H. Schopper. Berlin, Germany: Springer-Verlag.

Fassò, A., K. Goebel, M. Höfert, G. Rau, H. Schönbacher, G. R. Stevenson, A. H. Sullivan, W. P. Swanson, and J. W. N. Tuyn. 1984. *Radiation problems in the design of the large electron-positron collider (LEP)*. European Organization for Nuclear Research: CERN Report 84-02. Geneva, Switzerland.

Ferrari, A., M. Pelliccioni, and M. Pillon. 1996. Fluence to effective dose and effective dose equivalent conversion coefficients for photons from 50 keV to 10 GeV. *Radiation Protection Dosimetry* 67:245–251.

Ferrari, A., M. Pelliccioni, and M. Pillon. 1997a. Fluence to effective dose and effective dose equivalent conversion coefficients for electrons from 5 MeV to 10 GeV. *Radiation Protection Dosimetry* 69:97–104.

Ferrari, A., M. Pelliccioni, and M. Pillon. 1997b. Fluence to effective dose conversion coefficients for protons from 5 MeV to 10 TeV. *Radiation Protection Dosimetry* 71:85–91.

Ferrari, A., M. Pelliccioni, and M. Pillon. 1997c. Fluence to effective dose conversion coefficients for neutrons up to 10 TeV. *Radiation Protection Dosimetry* 71:165–173.

Fetter, C. W. 1988. *Applied hydrogeology*. Columbus, OH: Merrill.

Freeman W. S., and F. P. Krueger. 1984. *Neutron calibration tests of Fermilab radiation detectors*. Fermi National Accelerator Laboratory: Fermilab Radiation Physics Note No. 48. Batavia, IL.

Gabriel, T. A. 1985. *The high energy transport code HETC*. Oak Ridge National Laboratory: Report ORNL-TM-9727. Oak Ridge, TN.

Gabriel, T. A., D. E. Groom, P. K. Job, N. V. Mokhov, and G. R. Stevenson. 1994. Energy dependence of hadronic activity. *Nuclear Instruments and Methods in Physics Research* A338:336–347.

Gabriel T. A., and R. T. Santoro. 1973. Photon dose rates from the interactions of 200 GeV protons in iron and iron-lead beam stops. *Particle Accelerators* 4:169–186.

Gilbert, W. S. et al. 1968. *1966 CERN-LRL-RHEL shielding experiment at the CERN proton synchrotron*. University of California Radiation Laboratory: Report UCRL 17941. Berkeley, CA.

Goebel, K., G. R. Stevenson, J. T. Routti, and H. G. Vogt. 1975. *Evaluating dose rates due to neutron leakage through the access tunnels of the SPS*. European Organization for Nuclear Research: CERN Report LABII-RA/Note/75-10. Geneva, Switzerland.

Gollon, P. J. 1976. *Production of radioactivity by particle accelerators*. Fermi National Accelerator Laboratory: Fermilab Report TM-609. Batavia, IL.

Gollon, P. J. 1978. *Soil activation calculations for the anti-proton target area*. Fermi National Accelerator Laboratory: Fermilab Report TM-816. Batavia, IL.

Gollon P. J., and M. Awschalom. 1971. Design of penetrations in hadron shields. *IEEE Transactions in Nuclear Science* NS-18(3):741–745.

Greenhouse, N. A., T. M. De Castro, J. B. McCaslin, A. R. Smith, R. K. Sun, and D. E. Hankins. 1987. An evaluation of NTA film in an accelerator environment and comparisons with CR-39. *Radiation Protection Dosimetry* 20:143–147.

Griffith, R. V., and J. H. Thorngate. 1985. Neutron spectrometers for radiation monitoring at Lawrence Livermore National Laboratory. *Radiation Protection Dosimetry* 10:125–135.

Griffith, R. V., and L. Tommasino. 1990. Dosimetry for radiological protection at high energy particle accelerators. Chapter 1 in *The dosimetry of ionizing radiation*, Volume IV, eds. K. R. Kase, B. E. Bjärngard, and F. H. Attix. New York, NY: Academic Press.

Gudima, K. K., S. G. Mashnik, A. J. Sierk. 2001. *User manual for the code LAQGSM*. Los Alamos National Laboratory: Report LA-UR-01-6804. Los Alamos, NM.

Hagan, W. K., B. L. Colborn, T. W. Armstrong, and M. Allen. 1988. Radiation shielding calculations for a 70 to 250 MeV proton therapy facility. *Nuclear Science and Engineering* 98:272–278.

Hankins, D. E. 1962. *A neutron monitoring instrument having a response approximately proportional to the dose rate from thermal to 7.0 MeV*. Los Alamos Scientific Laboratory: Report LA-2717. Los Alamos, NM.

Hanson, A. O., and M. L. McKibben. 1947. A neutron detector having uniform sensitivity from 10 keV to 3 MeV. *Physical Review* 72:673–677.

Heilbronn, L. et al. 1999. Neutron yields from 155 MeV/nucleon carbon and helium stopping in aluminum. *Nuclear Science and Engineering* 132:1–15.

Hertel, N. E., and J. W. Davidson. 1985. The response of Bonner spheres to neutrons from thermal energies to 17.3 MeV. *Nuclear Instruments and Methods in Physics Research* A238:509–516.

Hikasa, K. et al. (Particle Data Group). 1992. Review of Particle Properties. *Physical Review* D45:S1–2.

Hirayama, H., Y. Namito, A. F. Bielajew, S. J. Wilderman, and W. R. Nelson. 2016. *The EGS5 code system.* SLAC National Accelerator Laboratory: Report SLAC-R-730. Menlo Park, CA. (Also available as High Energy Research Organization of Japan KEK Report No. 2005–89. Oho-machi, Tsukuba-gun, Ibaraki-ken:Japan.)

Höfert, M. 1969. Radiation hazard of induced activity in air as produced by high energy accelerators. In *Proceedings of the Second International Conference on Accelerator Radiation Dosimetry and Experience.* (Stanford, CA). National Technical Information Service: CONF-691101, 111–120. Washington, DC.

Höfert, M. 1983. *The NTA emulsion: An ill-reputed but misjudged neutron detector.* European Center for Nuclear Research: CERN Report TIS-RP/110/CF. Geneva, Switzerland.

Höfert M., and Ch. Raffnsøe. 1980. Measurement of absolute absorbed dose and dose equivalent response for instruments used around high-energy proton accelerators. *Nuclear Instruments and Methods* 176:443–448.

Höfert, M., and G. R. Stevenson. 1984. *The assessment of dose equivalent in stray radiation fields around high-energy accelerators.* European Organization for Nuclear Research: CERN Report TIS-RP/131/CF. Geneva, Switzerland.

Hubbard, E. L., R. M. Main, and R. V. Pyle. 1960. Neutron production by heavy-ion bombardments. *Physical Review* 118:507–514.

Huhtinen, M. 1998. *Method for estimating dose rates from induced radioactivity in complicated hadron accelerator geometries: Write-up of the FIASCO Code.* European Organization for Nuclear Research: CERN Report CERN/TIS-RP/IR/98-28. Geneva, Switzerland.

Huhtinen, M. 2003. Radioactivation at supercolliders. In *Proceedings of the 42nd Workshop on Innovative Detectors for Supercolliders.* September 28 to October 4. Erice, Italy.

Ing, H., and H. C. Birnboim. 1984. A bubble-damage polymer detector for neutrons. *Nuclear Tracks Radiation Measurement* 8:285–288.

Ipe, N. E., and A. Fassò. 1994. Gas bremsstrahlung considerations in the shielding design of the Advanced Photon Source synchrotron radiation beam lines. *Nuclear Instruments and Methods* A351:534–544.

International Commission on Radiation Units and Measurements (ICRU). 1971. *Radiation protection instrumentation and its application.* ICRU Publication 20. Bethesda, MD.

International Commission on Radiation Units and Measurements (ICRU). 1978. *Basic aspects of high-energy particle interactions and radiation dosimetry.* ICRU Report 28. Washington, DC.

International Commission on Radiation Units and Measurements (ICRU). 1979. *Average energy required to produce an ion pair.* ICRU Report 31. Bethesda, MD.

International Commission on Radiological Protection (ICRP). 1959. *Report of Committee II on permissible dose for international radiation.* ICRP Publication No. 2, Washington, DC: Pergamon Press.

International Commission on Radiological Protection (ICRP). 1973. *Data for protection against ionizing radiation from external sources: Supplement to ICRP Publication 15. ICRP Publication 21: Data for protection against ionizing radiation from external sources.* Washington, DC: Pergamon Press.

International Commission on Radiological Protection (ICRP). 1991. *1990 Recommendations of the International Commission on Radiological Protection.* ICRP Publication 60. New York, NY: Pergamon Press.

International Commission on Radiological Protection (ICRP). 1996. *Conversion coefficients for use in radiological protection against external radiation.* ICRP Publication 74. Oxford, UK: Pergamon Press.

International Commission on Radiological Protection (ICRP). 2007. *The 2007 recommendations of the International Commission on Radiological Protection.* ICRP Publication 103. Oxford, UK: Elsevier.

Jackson, J. D. 1998. *Classical electrodynamics,* third edition. New York, NY: John Wiley and Sons.

Jackson, J. D. ed., M. G. D. Gilchriese, D. E. Groom, J. R. Sanford, G. R. Stevenson, W. S. Freeman, T. E. Toohig, K. O'Brien, R. H. Thomas. 1987. *SSC environmental radiation shielding.* Superconducting Super Collider Central Design Group: Report SSC-SR-1026. Lawrence Berkeley National Laboratory, Berkeley, CA.

Jaeger, R. G. ed.-in-chief, E. P. Blizard, A. B. Chilton, M. Grotenhuis, A. Hönig, T. A. Jaeger, H. H. Eisenlohr, coordinating ed. 1968. *Engineering compendium on radiation shielding*, Volume 1 Shielding fundamentals and methods. New York, NY: Springer-Verlag.

Kalinovskii, A. N., N. V. Mokhov, and Yu. P. Nikitin. 1989. *Passage of high-energy particles through matter*. New York, NY: American Institute of Physics.

Kaye and Laby. 2018. *Tables of physical constants; section 4.5.4, radiation quantities and units*. UK: Institute of Physics, National Physical Laboratory . Available at http://www.kayelaby.npl.co.uk/toc/ (accessed October 17, 2018).

Knoll, G. F. 2010. *Radiation detection and measurement*, fourth edition. New York, NY: John Wiley and Sons.

Konobeyev, A. Yu., and Yu. A. Korovin. 1993. Tritium production in materials from C to Bi irradiated with nucleons of intermediate and high energies. *Nuclear Instruments and Methods in Physics Research* B82:103–115.

Konopinski, E. J. 1981. *Electromagnetic fields and relativistic particles*. New York, NY: McGraw-Hill.

Kosako, T., T. Nakamura, and S. Iwai. 1985. Estimation of response functions of moderating type neutron detectors by the time-of-flight method combined with a large lead pile. *Nuclear Instruments and Methods in Physics Research* A235:103–122.

Krivosheev, O. E., and N. V. Mokhov. 1998. *A new MARS and its applications*. Fermi National Accelerator Laboratory: FERMILAB-CONF-98/043 (1998). Batavia, IL.

Krueger, F., and J. Larson. 2002. Chipmunk IV: Development of and experience with a new generation of radiation area monitors for accelerator applications. *Nuclear Instruments and Methods in Physics Research* A495:20–28.

Kurosawa, T., N. Nakao, T. Nakamura, H. Iwase, H. Sato, Y. Uwamino, and A. Fukumura. 2000. Neutron yields from thick C, Al, Cu, and Pb targets bombarded by 400 MeV/nucleon Ar, Fe, Xe, and 800 MeV/nucleon Si ions. *Physical Review* C62:044615-1–044615-11.

Kurosawa, T., N. Nakao, T. Nakamura, Y. Uwamino, T. Shibata, N. Nakanishi, A. Fukumura, and K. Murakami. 1999. Measurements of secondary neutrons produced from thick targets bombarded by high-energy helium and carbon ions. *Nuclear Science and Engineering* 132:30–57.

Leake, J. W. 1968. An improved spherical dose equivalent neutron detector. *Nuclear Instruments and Methods* 63:329–332.

Lee, S.-Y. 2012. *Accelerator physics*, third edition. Singapore: World Scientific.

Levine, G. S., D. M. Squier, G. B. Stapleton, G. R. Stevenson, K. Goebel, and J. Ranft. 1972. The angular dependence of dose and hadron yields from targets in 8 and 24 GeV/c extracted proton beams. *Particle Accelerators* 3:91–101.

Lide, D. R. ed. 2000. *CRC handbook of chemistry and physics*. Boca Raton, FL: CRC Press/Taylor and Francis Group.

Lindenbaum, S. J. 1961. Shielding of high energy accelerators. *Annual Review of Nuclear Science* 11:213–258.

Liu, J. C., A. Fassò, H. Khater, A. Prinz, and S. H. Rokni. 2005. *Radiation safety design for SPEAR3 Ring and synchrotron radiation beamlines*. Stanford Linear Accelerator Center: SLAC Report RP-05-33. Menlo Park, CA.

Liu, J. C., W. R. Nelson, and K. R. Kase. 1995. Gas bremsstrahlung and associated photon-neutron shielding calculations for electron storage rings. *Health Physics* 68:205–213.

Liu, J. C., and V. Vylet. 2001. Radiation protection at synchrotron radiation facilities. *Radiation Protection Dosimetry* 96:345–357.

Lowry, K. A., and T. L. Johnson. 1984. *Modifications to interactive recursion unfolding algorithms and computer codes to find more appropriate neutron spectra*. U.S. Naval Research Laboratory: Report NRL-5430. Washington, DC.

Maerker, R. E., and V. R. Cain. 1967. *AMC, a Monte-Carlo code utilizing the albedo approach for calculating neutron and capture gamma-ray distributions in rectangular concrete ducts*. Oak Ridge National Laboratory: Report ORNL-3964. Oak Ridge, TN.

Malensek, A. J., A. A. Wehmann, A. J. Elwyn, K. J. Moss, and P. M. Kesich. 1993. *Groundwater migration of radionuclides at Fermilab*. Fermi National Accelerator Laboratory: Fermilab Report TM-1851. Batavia, IL.

Margaritondo, G. 1988. *Introduction to synchrotron radiation*. New York, NY: Oxford University Press.

Marion, J. B., and F. C. Young. 1968. *Nuclear reaction analysis, graphs and tables*. Amsterdam, Netherlands: North Holland.

McCaslin, J. B. 1960. A high-energy neutron-flux detector. *Health Physics* 2:399–407.

McCaslin, J. B., P. R. LaPlant, A. R. Smith, W. P. Swanson, and R. H. Thomas. 1985. Neutron production by Ne and Si ions on a thick Cu target at 670 MeV/A with application to radiation protection. *IEEE Transactions on Nuclear Science* 32(5):3104–3106.

McCaslin, J. B., H. W. Patterson, A. R. Smith, and L. D. Stephens. 1968. Some recent developments in technique for monitoring high-energy accelerator radiation. In *Proceedings of the First International Congress of Radiation Protection. International Radiation Protection Association, Part 2*, 1131–1137. New York, NY: Pergamon Press.

McCaslin, J. B., R.-K. Sun, W. P. Swanson, J. D. Cossairt, A. J. Elwyn, W. S. Freeman, H. Jöstlein, C. D. Moore, P. M. Yurista, and D. E. Groom. 1987a. *Radiation environment in the tunnel of a high-energy proton accelerator at energies near 1.0 TeV*. Lawrence Berkeley Laboratory: Report LBL-24640. Berkeley, CA.

McCaslin, J. B., W. P. Swanson, and R. H. Thomas. 1987b. Moyer model approximations for point and extended beam losses. *Nuclear Instruments and Methods in Physics Research* A256:418–426.

Mokhov, N. V. 1995. *The MARS code system user's guide, version 13*. Fermi National Accelerator Laboratory: Fermilab Report FN-628. Batavia, IL.

Mokhov, N. V., and J. D. Cossairt. 1986. A short review of Monte Carlo hadronic cascade calculations in the multi-TeV energy region. *Nuclear Instruments and Methods in Physics Research* A244:349–355.

Mokhov, N. V., K. K. Guidima, C. C. James, M. A. Kostin, S. G. Mashnik, E. Ng, J.-F. Ostiguy, I. L. Rakhno, A. J. Sierk, and S. I. Striganov. 2004. *Recent enhancements to the MARS15 Code*. Fermi National Accelerator Laboratory: Fermilab Report FERMILAB-CONF-04/053-AD. Batavia, IL.

Mokhov, N. V., E. I. Rakhno, and I. L. Rakhno. 2006. *Residual activation of thin accelerator components*. Fermi National Accelerator Laboratory: Fermilab Report FERMILAB FN-0788-AD. Batavia, IL.

Mokhov, N.V., and A. Van Ginneken. 1999. *Neutrino-induced radiation at muon colliders*. Fermi National Accelerator Laboratory: Report FERMILAB-CONF-99-067. Batavia, IL.

Moritz, L. E. 1989. Measurement of neutron leakage spectra at a 500 MeV proton accelerator. *Health Physics* 56:287–296.

Nagel, H. H. 1965. Electron-photon-kaskaden in Blei: Monte-Carlo rechnungen für primärelektronenergien zwischen 100 and 100 MeV. *Zeitschrift für Physik* 186:319–346.

Nakamura, T. 1985. Neutron energy spectra produced from thick targets by light-mass heavy ions. *Nuclear Instruments and Methods in Physics Research* A240:207–215.

Nakamura, T., and L. Heilbronn. 2005. *Handbook on secondary particle production and transport by high-energy heavy ions*. Singapore: World Scientific.

Nakamura T., and T. Kosako. 1981. A systematic study on neutron skyshine from nuclear facilities, Part I: Monte-Carlo analysis of neutron propagation in air-over-ground environment from a monoenergetic source. *Nuclear Science and Engineering* 77:168–181.

Nakamura, T., M. Yoshida, and T. Shin. 1978. Spectra measurements of neutrons and photons from thick targets of C, Fe, Cu, and Pb by 52 MeV protons. *Nuclear Instruments and Methods* 151:493–503.

National Council on Radiation Protection and Measurements (NCRP). 1959. *Maximum permissible body burdens and maximum permissible concentrations of radionuclides in air and in water for occupational exposure*. NCRP: Report No. 22. Also listed as U.S. National Bureau of Standards Handbook No. 69. 1959. Bethesda, MD. Essentially the same information was published by ICRP Committee II. 1960. *Health Physics* 3:1–233.

National Council on Radiation Protection and Measurements (NCRP). 1971. *Protection against neutron radiation*. NCRP Report No. 38. Bethesda, MD.

National Council on Radiation Protection and Measurements (NCRP). 2003. *Radiation protection for particle accelerator facilities*. NCRP Report No. 144. Bethesda, MD.

National Institute of Standards and Technology (NIST). 2018a. Stopping power and range tables for protons. Available at https://physics.nist.gov/PhysRefData/Star/Text/PSTAR.html (accessed October 17, 2018).

National Institute of Standards and Technology (NIST). 2018b. X-ray and gamma-ray data. Available at https://www.nist.gov/pml/x-ray-and-gamma-ray-data (accessed October 17, 2018).

National Nuclear Data Center (NNDC). 2018. National Nuclear Data Center at Brookhaven National Laboratory. Available at http://www.nndc.bnl.gov/ (accessed October 17, 2018).

Nelson, W. R. 1968. The shielding of muons around high energy electron accelerators: Theory and measurement. *Nuclear Instruments and Methods* 66:293–303.

Nelson, W. R., H. Hirayama, and D. W. O. Rogers. 1985. *The EGS4 code system.* Stanford Linear Accelerator Center: Report SLAC-265 UC-32. Menlo Park, CA.

Nelson, W. R., and K. R. Kase. 1974. Muon shielding around high-energy electron accelerators. Part I, Theory. *Nuclear Instruments and Methods* 120: 401–411.

Nelson, W. R., K. R. Kase, and G. K. Svensson. 1974. Muon shielding around high-energy electron accelerators. Part II, Experimental investigation. *Nuclear Instruments and Methods* 120:413–429.

Northcliffe, L. C., and R. F. Schilling. 1970. Range and stopping-power tables for heavy ions. *Nuclear Data Tables* A7:233–463.

O'Brien, K. 1971. *Neutron spectra in the side-shielding of a large particle accelerator.* U.S. Atomic Energy Commission: Health and Safety Laboratory: Report HASL-240. New York, NY.

O'Brien, K. 1980. The physics of radiation transport. In *Computer techniques in radiation transport and dosimetry,* eds. W. R. Nelson and T. M. Jenkins, 17–56. New York, NY: Plenum Press.

O'Brien, K., and R. Sanna. 1981. Neutron spectral unfolding using the Monte-Carlo method. *Nuclear Instruments and Methods* 185:277–286.

O'Dell, A. A., Jr., C. W. Sandifer, R. B. Knowlen, and W. D. George. 1968. Measurements of absolute thick-target bremsstrahlung spectra. *Nuclear Instruments and Methods* 6:340–346.

Ohnesorge, W. F., H. M. Butler, C. B. Fulmer, and S. W. Mosko. 1980. Heavy ion target area fast neutron dose equivalent rates. *Health Physics* 39:633–636.

Patterson, H. W., and R. H. Thomas. 1973. *Accelerator health physics.* New York, NY: Academic Press.

Patterson, H. W., and R. Wallace. 1958. *A method of calibrating slow neutron detectors.* University of California Radiation Laboratory: Report UCRL-8359. Berkeley, CA.

Pelliccioni, M. 2000. Overview of fluence-to-effective dose and fluence-to-ambient dose equivalent conversion coefficients for high energy radiation calculated using the FLUKA Code. *Radiation Protection Dosimetry* 88:279–297.

Prael, R. E., and H. Lichtenstein. 1989. *User guide to LSC: The LAHET code system.* Los Alamos National Laboratory: Report No. LA-UR-89–3014. Los Alamos, NM.

Quinn, M., J. D. Cossairt, M. Schoell, N. Chelidze, K. Graden, S. McGimpsey, D. Reitzner, W. Schmitt, and K. Vaziri,. 2018. *Review of control of occupational exposure to airborne radioactivity at Fermilab.* Fermi National Accelerator Laboratory: Fermilab Radiation Physics Note No. 158 Revision 2. Batavia, IL.

Rakhno, I., N. Mokhov, A. Elwyn, N. Grossman, M. Huhtinen, and L. Nicolas. 2001. *Benchmarking residual dose rates in a NuMI-like environment.* Fermi National Accelerator Laboratory: Fermilab Report FERMILAB-Conf-01/304-E. Batavia, IL.

Ranft, J. 1967. Improved Monte-Carlo calculation of the nucleon-meson cascade in shielding material: I. Description of the method of calculation. *Nuclear Instruments and Methods* 48:133–140.

Ranft, J., and J. T. Routti. 1972. Hadronic cascade calculations of angular distributions of integrated secondary particle fluxes from external targets and new empirical formulae describing particle production in proton-nucleus collisions. *Particle Accelerators* 4:101–110.

Reitzner, S. D. 2012. *Update to the generic shielding design criteria.* Fermi National Accelerator Laboratory: Fermilab Report TM-2550-ESH. Batavia, IL.

Rindi, A. 1982. Gas bremsstrahlung from electron storage rings. *Health Physics* 42:187–193.

Rindi, A., and R. H. Thomas. 1975. Skyshine—A paper tiger? *Particle Accelerators* 7:23–39.

Rogers, D. W. O., and A. F. Bielajew 1990. Monte Carlo techniques of electron and photon transport for radiation dosimetry. *The dosimetry of ionizing radiation,* eds. K. R. Kase, B. E. Bjärngard, and F. H. Attix, Volume III, Chapter 5. New York, NY: Academic Press.

Rohrig, N. 1983. Plotting neutron fluence spectra. *Health Physics* 45:817–818.

Rossi, B. 1952. *High energy particles.* Englewood Cliffs, NJ: Prentice Hall.

Rossi, B., and K. Greisen. 1941. Cosmic-ray theory. *Reviews of Modern Physics* 13:240–315.

Rossi, H. H. 1968. Microscopic energy distribution in irradiated matter. In *Radiation dosimetry*, Volume 1, eds. F. H. Attix and W. C. Roesch. New York, NY: Academic Press.

Rossi, H. H., and W. Rosenzweig. 1955. A device for the measurement of dose as a function of specific ionization. *Radiology* 64:404–411.

Routti, J. T., and J. V. Sandbert. 1980. General purpose unfolding LOUHI78 with linear and nonlinear regularization. *Computer Physics Communications* 21:119–144.

Routti, J. T., and R. H. Thomas. 1969. Moyer integrals for estimating shielding of high-energy accelerators. *Nuclear Instruments and Methods* 76:157–163.

Ruffin, P., and C. Moore. 1976. *A study of neutron attenuation in the E-99 labyrinth.* Fermi National Accelerator Laboratory: Fermilab Radiation Physics Note No. 9. Batavia, IL.

Shen G. et al. 1979. Measurement of multiple scattering at 50 to 200 GeV/c. *Physical Review* D20:1584–1588.

Shin, K., K. Miyahara, E. Tanabe, and Y. Uwamino. 1995. Thick-target neutron yield for charged particles. *Nuclear Science and Engineering* 120:40–54.

Silari, M., S. Agosteo, J-C. Gaborit, and L. Ulrici. 1999. Radiation produced by the LEP superconducting RF cavities. *Nuclear Instruments and Methods in Physics Research* A432:1–13.

Slade, D. A. ed. 1955. *Meteorology and atomic energy.* U.S. Atomic Energy Commission. Washington, DC.

Slade, D. A. ed. 1968. *Meteorology and atomic energy.* U.S. Atomic Energy Commission, Office of Information Services TID-24190. Washington, DC.

Slaughter, D. R., and D. W. Rueppel. 1977. Calibration of a DePangher long counter from 2 keV to 19 MeV. *Nuclear Instruments and Methods* 145:315–320.

Stevenson, G. R. 1983. *Dose and dose equivalent from muons.* European Organization for Nuclear Research: CERN Report TIS-RP/099. Geneva, Switzerland.

Stevenson, G. R. 1984. The estimation of dose equivalent from the activation of plastic scintillators. *Health Physics* 47:837–847.

Stevenson, G. R. 1986. *Dose equivalent per star in hadron cascade calculations.* European Organization for Nuclear Research: CERN Report TIS-RP/173. Geneva, Switzerland.

Stevenson, G. R. 1987. *Empirical parametrization of the shielding of end-stops at high-energy proton accelerators.* European Organization for Nuclear Research: CERN Report TIS-RP/183/CF.

Stevenson, G. R. 2001. Induced activity in accelerator structures, air and water. *Radiation Protection Dosimetry* 96:373–380.

Stevenson, G. R., P. A. Aarnio, A. Fassò, J. Ranft, J. V. Sandberg, and P. Sievers. 1986. Comparison of measurements of angular hadron energy spectra, induced activity, and dose with FLUKA82 calculations. *Nuclear Instruments and Methods in Physics Research* A245:323–327.

Stevenson, G. R., L. Kuei-Lin, and R. H. Thomas. 1982. Determination of transverse shielding for proton accelerators using the Moyer model. *Health Physics* 43:13–29.

Stevenson, G. R., and D. M. Squier. 1973. An experimental study of the attenuation of radiation in tunnels penetrating the shield of an extracted beam of the 7 GeV proton synchrotron NIMROD. *Health Physics* 24:87–93.

Stevenson, G. R., and R. H. Thomas. 1984. A simple procedure for the estimation of neutron skyshine from proton accelerators. *Health Physics* 46:115–122.

Sudicky, E. A., T. D. Wadsworth, J. B. Kool, and P. S. Huyakorn. 1988. *PATCH 3D; Three dimensional analytical solution for transport in a finite thickness aquifer with first-type rectangular patch source.* University of Waterloo, Ontario, Canada.

Sullivan, A. H. 1989. The intensity distribution of secondary particles produced in high energy proton interactions. *Radiation Protection Dosimetry* 27:189–192.

Sullivan, A. H. 1992. *A guide to radiation and radioactivity levels near high energy particle accelerators.* Ashford, Kent, UK: Nuclear Technology.

Sullivan, A. H., and T. R. Overton. 1965. Time variation of the dose-rate from radioactivity induced in high-energy particle accelerators. *Health Physics* 11:1101–1105.

Sullivan, A. H. 1984. *The use of detectors based on ionisation recombination in radiation protection.* European Organization for Nuclear Research: CERN Report TIS-RP/130/CF

Swanson, W. P. 1979a. *Radiological safety aspects of the operation of electron linear accelerators.* International Atomic Energy Agency: IAEA Technical Report No. 188. Vienna, Austria.

Swanson, W. P. 1979b. Improved calculation of photoneutron yields released by incident electrons. *Health Physics* 37:347–358.

Swanson, W. P., and R. H. Thomas. 1990. Dosimetry for radiological protection at high energy particle accelerators. In *The dosimetry of ionizing radiation*, eds. K. R. Kase, B. E. Bjärngard, and F. H. Attix, Volume III, Chapter 1, 1–161. New York, NY: Academic Press.

Tesch, K. 1982. The attenuation of the neutron dose equivalent in a labyrinth through an accelerator shield. *Particle Accelerators* 12:169–175.

Tesch, K. 1983. Comments on the transverse shielding of proton accelerators. *Health Physics* 44:79–82.

Tesch, K. 1985. A simple estimation of the lateral shielding for proton accelerators in the energy range from 50 to 1000 MeV. *Radiation Protection and Dosimetry* 11:165–172.

Tesch, T. 1988. Shielding against high energy neutrons from electron accelerators-a review. *Radiation Protection Dosimetry* 22:27–32.

Thomas, R. H., and G. R. Stevenson. 1985. Radiation protection around high energy accelerators. *Radiation Protection Dosimetry* 10:283–301.

Thomas, R. H., and G. R. Stevenson. 1988. *Radiological safety aspects of the operation of proton accelerators.* International Atomic Energy Agency: IAEA Technical Report No. 283. Vienna, Austria.

Thomas, R. H., and S. V. Thomas. 1984. Variation and regression analysis of Moyer model parameter data—A sequel. *Health Physics* 46:954–957.

Torres, M. M. C. 1996. Neutron radiation fields outside shielding at the Fermilab Tevatron. Ph.D. Thesis. University of Michigan, Ann Arbor, MI.

Tromba G., and A. Rindi. 1990. Gas bremsstrahlung from electron storage rings: A Monte Carlo evaluation and some useful formulae. *Nuclear Instruments and Methods* A292:700–705.

Tsai, Y-S. 1974. Pair production and bremsstrahlung of charged leptons. *Reviews of Modern Physics* 46:815–851.

Tuyn, J. W. N., R. Deltenre, C. Lamberet, and G. Roubaud. 1984. *Some radiation aspects of heavy ion acceleration.* European Organization for Nuclear Research: CERN Report TIS-RP/125/CF. Geneva, Switzerland.

U.S. Department of Energy (DOE). 1990. *U.S. Department of Energy Order 5400.5: Radiation protection of the public and the environment.* Washington, DC. (For reference only. This order was superseded by DOE 2011a.)

U.S. Department of Energy (DOE). 2011a. *U.S. Department of Energy Order 458.1: Radiation protection of the public and the environment.* Washington, DC.

U.S. Department of Energy (DOE). 2011b. *U.S. Department of Energy Standard DOE-STD-1196-2011: Derived concentration technical standard.* Washington, DC.

Uwamino, Y., T. Nakamura, and K. Shin. 1982. Penetration through shielding materials of secondary neutrons and photons generated by 52 MeV protons. *Nuclear Science and Engineering* 80:360–369.

Van Ginneken, A. 1971. ^{22}Na production cross section in soil. Fermi National Accelerator Laboratory: Fermilab Report TM-283. Batavia, IL.

Van Ginneken, A. 1978. *AEGIS, a program to calculate the average behavior of electromagnetic showers.* Fermi National Accelerator Laboratory: Fermilab Report FN-309. Batavia, IL.

Van Ginneken, A., and M. Awschalom. 1975. *High energy particle interactions in large targets: Volume I, hadronic cascades, shielding, and energy deposition.* Fermi National Accelerator Laboratory. Batavia, IL.

Van Ginneken, A., P. Yurista, and C. Yamaguchi. 1987. *Shielding calculations for multi-TeV hadron colliders.* Fermi National Accelerator Laboratory: Fermilab Report FN-447. Batavia, IL.

Vasilik, D. G., G. H. Erkkila and D. A. Waechter. 1985. A portable neutron spectrometer/dosemeter. *Radiation Protection Dosimetry* 10:121–124.

Vaziri, K., V. R. Cupps, D. J. Boehnlein, J. D. Cossairt, and A. J. Elwyn. 1996. *A detailed calibration of a stack monitor used in the measurement of airborne radionuclides at a high energy proton accelerator.* Fermi National Accelerator Laboratory: Fermilab Report FERMILAB-Pub-96/037. Batavia, IL.

Vaziri, K., V. Cupps, D. Boehnlein, D. Cossairt, A. Elwyn, and T. Leveling. 1993. *AP0 stack monitor calibration*. Fermi National Accelerator Laboratory: Fermilab Radiation Physics Note No. 106. Batavia, IL.

Vaziri, K., B. Hartman, S. Hawke, F. Krueger, and J. Larson. 2004. *Muon scope: characteristics and operation*. Fermi National Accelerator Laboratory: Fermilab Radiation Physics Note No. 149. Batavia, IL.

Völkel, U. 1965. *Electron-photon-kaskaden in Blei für Primärteilchen der Energie 6 GeV*. Deutches Elektronen-Synchrotron Report No. DESY 65/6. Hamburg, Germany.

Vylet, V. 1991. *Estimated sensitivity of the "muon gun" to neutrons*. Fermi National Accelerator Laboratory: Fermilab Radiation Physics Note No. 92. Batavia, IL.

Wiedemann, H. 2003. *Synchrotron radiation*. New York, NY: Springer-Verlag.

Wiedemann, H. 2013. Synchrotron radiation. In *Handbook of accelerator physics and engineering*, second edition, eds. A. W. Chao, K. H. Mess, M. Tigner, and F. Zimmermann, 215–227. Singapore: World Scientific.

Wilson, R. R. 1952. Monte Carlo study of shower production. *Physical Review* 86:261–269.

Yurista, P. M., and J. D. Cossairt. 1983. *Concrete shielding exterior to iron*. Fermi National Accelerator Laboratory: Fermilab Report TM-1204. Batavia, IL.

Zazula, J. M. 1987. Derivation of accelerator shielding parameters from adjoint high energy neutron transport calculations. In *Theory and Practices in Radiation Protection and Shielding, Proceedings of the Radiation Protection and Shielding Division*. Volume 2, 434–442. American Nuclear Society, Knoxville, TN.

Zel'chinskij, M., and K. Zharnovetskij. 1967. *Differential recombination chamber*. Badan Jadrowych Nuclear Research institute: Report INR No. 739/XIX/D. Sverk, Poland. Translated at the European Organization for Nuclear Research (CERN), Geneva, Switzerland.

Zerby C. D., and H. S. Moran. 1963. Studies of the longitudinal development of high-energy electron-photon cascade showers in copper. *Journal of Applied Physics* 34:2445–2457.

Ziegler, J. F., J. P. Biersack, and U. Littmark. 1996. *The stopping and range of ions in solids*. New York, NY: Pergamon Press. The SRIM code is available at http://www.srim.org (accessed October 17, 2018).

Zielczynski, M. 1963. Recombination method for determination of linear energy transfer of mixed radiation. In *Proceedings of Symposium Neutron Dosimetry*. Vienna, Austria: International Atomic Energy Agency.

Index

Milton Keynes UK
Ingram Content Group UK Ltd.
UKHW051947071024
449327UK00026B/2210